Building Services Design for Energy-Efficient Buildings

The role and influence of building services engineers are undergoing rapid change and are pivotal to achieving low-carbon buildings. However, textbooks in the field have tended to remain fairly traditional with a detailed focus on the technicalities of heating, ventilation and air conditioning (HVAC) systems, often with little wider context. This book addresses that need by embracing a contemporary understanding of the urgent challenge to address climate change, together with practical approaches to energy efficiency and carbon mitigation for mechanical and electrical systems, in a concise manner.

The essential conceptual design issues for planning the principal building services systems that influence energy efficiency are examined in detail. These are HVAC and electrical systems. In addition, the following issues are addressed:

- background issues on climate change, whole-life performance and design collaboration
- generic strategies for energy-efficient, low-carbon design
- health and wellbeing and post occupancy evaluation
- building ventilation
- air conditioning and HVAC system selection
- thermal energy generation and distribution systems
- low-energy approaches for thermal control
- electrical systems, data collection, controls and monitoring
- building thermal load assessment
- building electric power load assessment
- space planning and design integration with other disciplines.

In order to deliver buildings that help mitigate climate change impacts, a new perspective is required for building services engineers, from the initial conceptual design and throughout the design collaboration with other disciplines. This book provides a contemporary introduction and guide to this new approach, for students and practitioners alike.

Paul Tymkow was at Hoare Lea for 30 years where he was Director of Learning and Knowledge from 2007 to 2018. He has led design groups, project teams and projects in a wide range of sectors, in building services and related areas, for Hoare Lea and other organisations. He was a visiting academic at Brunel University London from 2005 to 2019, where he lectured on the postgraduate programme, and was a Royal Academy of Engineering visiting professor from 2016 to 2019.

Savvas Tassou is Professor of Energy Engineering and Director of the Institute of Energy Futures at Brunel University London. He has over 30 years' academic and research experience in the areas of building services engineering, energy conservation technologies for the built environment and process industries, the design optimisation of these technologies and their optimum integration and control to minimise energy consumption and greenhouse gas emissions.

Maria Kolokotroni is Professor of Energy in the Built Environment in the Department of Mechanical and Aerospace Engineering and Institute for Energy Futures, Brunel University London. She teaches ventilation, low-energy technologies and building performance at a postgraduate level and carries out research on UK- and European-funded research projects in these fields. She works very closely with industry to transfer research results to application and has contributed to International Energy Agency research projects.

Hussam Jouhara is Professor of Thermal Engineering at Brunel University London. He is an established academic author and an internationally recognised expert in heat exchangers, fluid dynamics and two-phase heat transfer processes. Over a 16-year career, he has developed innovative solutions to heat exchange/heat-pipe problems, resulting in many filed patents, and has attracted substantial research funding from various UK/EU-based research councils and industrial partners.

Building Services Design for Energy-Efficient Buildings

Second Edition

**Paul Tymkow, Savvas Tassou,
Maria Kolokotroni and Hussam Jouhara**

LONDON AND NEW YORK

Second edition published 2021
by Routledge
2 Park Square, Milton Park, Abingdon, Oxon, OX14 4RN

and by Routledge
52 Vanderbilt Avenue, New York, NY 10017

Routledge is an imprint of the Taylor & Francis Group, an informa business

First edition published by Routledge 2013

British Library Cataloguing-in-Publication Data
A catalogue record for this book is available from the British Library

Library of Congress Cataloging-in-Publication Data
Names: Tymkow, Paul, author. | Tassou, Savvas, author. | Kolokotroni, Maria, author. | Jouhara, Hussam, author.
Title: Building services design for energy-efficient buildings / Paul Tymkow, Savvas Tassou, Maria Kolokotroni and Hussam Jouhara.
Description: Second edition. | New York : Routledge, 2020. | Includes bibliographical references and index.
Identifiers: LCCN 2020008356 (print) | LCCN 2020008357 (ebook) | ISBN 9780815365600 (hardback) | ISBN 9780815365617 (paperback) | ISBN 9781351261166 (ebook) | ISBN 9781351261159 (adobe pdf) | ISBN 9781351261135 (mobi) | ISBN 9781351261142 (epub)
Subjects: LCSH: Sustainable buildings—Design and construction. | Architecture and energy conservation.
Classification: LCC TH880 .T96 2020 (print) | LCC TH880 (ebook) | DDC 720/.47—dc23
LC record available at https://lccn.loc.gov/2020008356
LC ebook record available at https://lccn.loc.gov/2020008357

ISBN: 978-0-8153-6560-0 (hbk)
ISBN: 978-0-8153-6561-7 (pbk)
ISBN: 978-1-351-26116-6 (ebk)

Typeset in Baskerville
by Apex CoVantage, LLC

Contents

Acknowledgements

For this second edition, the authors wish to express their thanks to Mark Ryder, Steve Wisby, John Pietrzyba and Keith Horsley of Hoare Lea and Les Norman for reviewing selected chapters and to Dominic Meyrick and Simon Russett of Hoare Lea for reviewing specialist parts of chapters. Their valuable insight, comments and guidance are much appreciated.

For the first edition, the authors wish to again express their thanks to Mark Ryder, Ashley Bateson, John Pietrzyba and Keith Horsley of Hoare Lea for reviewing selected chapters and to Steve Wisby, Dominic Meyrick and Simon Russett of Hoare Lea, as well as John O'Leary of Trend Control Systems Ltd, for reviewing specialist parts of chapters. Their valuable comments and guidance are much appreciated. They would also like to thank Louise Gillane of Hoare Lea for her assistance with word processing and illustrations. The authors would like to acknowledge contributions made in various ways by Mark Robinson, Dr Mohammed B. Ullah and Dr I. Nyoman Suamir, Dr Ian Pegg, Dr David Warwick and the partners of the European projects Vent DisCourse and Building AdVent.

Acknowledgements

Introduction

In recent decades there has been growing awareness of the environmental impact of man's activities and concerted efforts to identify and address the key factors that give rise to the most damaging impacts. It is clear that buildings are one of the principal sources of environmental degradation. This is primarily due to the carbon impact of the energy used in both the operation and construction of buildings, which is a major contributor to climate change. There is an essential challenge for building services designers to influence the design of buildings and to plan their active engineering systems so that carbon emissions are minimised. This requires an awareness of all the factors that give rise to carbon emissions throughout the life cycle of a building.

The design of buildings is a complicated process of synthesis and iteration involving a range of disciplines. Building services engineering is one of the principal design disciplines alongside architecture and civil/structural engineering. Building services engineering is itself made up of a range of sub-disciplines, the principal disciplines for energy performance being mechanical and electrical engineering. Many of those entering the building services engineering profession are from traditional mechanical and electrical engineering undergraduate courses and have usually had only limited exposure to the wider issues involved in building design more generally, and energy-efficient buildings in particular. Many undertake master's degrees in building services engineering, or similar courses, which often provide an element of conversion into their new field. For buildings to be successful in meeting the climate challenge alongside other aspirations such as the health and wellbeing of occupants, it is essential for all design professionals to work in a collaborative way. This requires mutual respect and a clear understanding of the wider context and common objectives. An essential feature of courses in building services engineering should therefore be to impart sufficient awareness of the environmental challenges and the nature of collaborative design, together with the range of interdisciplinary influences that need to be resolved and developed into a satisfactory design resolution. The emphasis throughout is on design approaches to achieve good whole-life performance.

This book has largely arisen from a module in Brunel University London's full-time and distance-learning MSc programmes in Building Services Engineering and Building Services Engineering with Sustainable Energy that cover the general issues related to building services design. In common with many undergraduate and postgraduate courses in this subject area, individual mechanical and electrical building services systems are taught as separate modules. So, for this reason, the building services design module was developed, which brings together the necessary background on sustainability with an understanding of the interdisciplinary collaboration required to achieve integrated design for low-carbon buildings and engineering systems.

The book focuses on energy efficiency and carbon mitigation as a central and fundamental strand of wider sustainable objectives for the built environment. It can be considered a general introduction that is primarily aimed at students who have not previously studied built environment subjects. It does not cover the many other aspects of sustainability that are necessary considerations in building design (such as materials management, water management, transport policy, biodiversity and self-sufficiency) or the more general environmental and societal aspects. It only covers sustainability issues related to designing for low-carbon performance throughout the life of the building. It assumes a basic understanding of the main mechanical and electrical engineering systems in buildings.

Building Services Design for Energy-Efficient Buildings is primarily concerned with the general strategies and design approaches for designing low-carbon solutions for buildings. It only refers to those building services systems that are energy consuming, so it does not cover other non-energy systems that form part of the wide spectrum of building services engineering. Lighting and lifts are only covered in relation to energy and load assessment aspects. It only partially covers the background principles or detailed design processes for engineering systems sufficient to convey the key energy efficiency aspects. Numerous academic textbooks and institutions' guides provide comprehensive coverage of detailed design aspects for individual systems (or groups of related systems), including heating, ventilation, air conditioning, hot water systems, electric power, lighting, lift engineering and automatic control systems. Such books cover, in considerable depth, aspects such as comfort criteria, indoor climate systems, system concepts and analysis, plant and equipment sizing, health and safety aspects, ductwork sizing, pipework sizing, cable sizing, lighting design and so on. This book does not seek to cover these detailed design aspects, but is instead primarily concerned with the wider concepts of systems and the design approach that should be adopted. This is necessary so that a building and its systems can be developed as an integrated whole that minimises environmental impact and is adaptable to the predicted climate changes.

Chapter 1 provides a brief introduction to the background issues on the environmental impact of human activities, particularly climate change and the urgent need to minimise impacts, together with the need to maintain security of the energy supply. The context is provided for energy and materials usage in the built environment, including the urgent need for designs to adapt to address climate change impacts. This chapter outlines the challenges of designing energy-efficient buildings as a key strand of creating a sustainable built environment and introduces key design concepts for sustainability.

Chapter 2 provides a brief outline of the principal disciplines in a design team and how they collaborate to create integrated solutions. Development of the brief is a key activity and is described in some detail. Key objectives of the building services designer are described, together with specific design considerations that can help in achieving low-carbon performance. A brief outline is provided for legislation and codes for designers, including health and safety management and quality management.

Chapter 3 describes generic strategies for achieving energy-efficient and low-carbon buildings. This includes an initial focus on elements of passive design. A range of generic measures are described for active engineering systems and renewable technologies, which are explored in more detail in Chapters 6–10, together with management regimes. Selected aspects of the Building Regulations in England and Wales Approved Document Part L are briefly introduced as an example of national legislation for conservation of energy.

A key requirement for undertaking design is an understanding of how existing buildings perform in practice, so that relevant operational feedback can inform the design for future buildings. Chapter 4 describes how a formal post occupancy evaluation (POE) can play a key part in optimising the energy and environmental performance of buildings. This includes case studies, methodologies for environmental assessment and European directives on building energy performance.

Chapter 5 addresses health and wellbeing for the building occupants. It describes the four elements of the indoor environment as perceived by occupants (indoor air quality, thermal, visual and acoustics) and currently used indoor environmental quality assessment methods and tools.

Chapters 6–10 describe design strategies for the main 'active' energy-using engineering systems. Chapter 6 explains energy-efficient methods of ventilation. This includes requirements and strategies, together with natural ventilation, ventilation for cooling and traditional methods of ventilation. Chapter 7 gives details on air conditioning systems, including system classifications and types for different applications. A method is presented for system selection and evaluation. Chapter 8 describes the principal components of HVAC equipment, including the plant for generating heating and cooling, heat pumps and solar thermal systems. Chapter 9 covers the distribution of thermal energy, including hydraulic systems, ductwork systems, variable-volume circuits, and low-energy heating and cooling systems for optimal energy and environmental performance. Chapter 10 describes energy efficiency considerations for designing electrical systems, together with data collection, controls and monitoring and for all the energy-using systems. This includes power distribution, lighting, lifts, building management systems and renewable technologies for electricity generation.

An essential part of engineering systems design, and a key component in the decision-making process for designing and selecting appropriate engineering systems, is load assessment. Chapters 11 and 12 cover the key aspects of load assessment for thermal systems (heating and cooling) and electric power systems, respectively.

The active engineering systems must be integrated with the architectural and structural designs in such a way that promotes good energy and carbon performance over the life of the building. Chapter 13 outlines the key principles of space planning and design integration for services. This includes planning plant spaces, together with vertical and horizontal distribution.

With the wider recognition of the 'climate emergency' and the urgent need to drastically limit impacts, there will be fast-moving changes internationally, with significant regulatory and infrastructural changes in the UK and elsewhere. It is difficult to predict the specific outcomes. There will inevitably be a transition period in which some traditional approaches remain and need to be designed in the most energy-efficient way, so the focus here is more on key concepts, principles and anticipated trends. It is clearly not possible to cover design approaches that would be applicable to all types of buildings (domestic and non-domestic, new and refurbished/retrofitted/re-purposed), in different locations and jurisdictions, during a period of likely fundamental change. Instead, the focus here is primarily on new commercial and public (i.e. non-domestic buildings). A variety of approaches and systems are described, some of which might be more or less appropriate for different applications, but this will largely depend on location and the prevailing regulatory frameworks and carbon scenarios for electricity grids. Where examples are given (such as for regulations), they are mainly for the UK to provide context only.

It is hoped that *Building Services Design for Energy-Efficient Buildings* will be of use to those studying at a master's level in building services engineering and related built environment, architectural engineering, sustainability and energy subjects. It should also be of use to those studying in the final year of BSc, BEng and MEng courses in these subject areas. It is hoped that the subject matter will also be of more general use to practitioners in the field, together with architects and other building design professionals, as a useful text that brings together a broad coverage of building services design and energy efficiency matters in a single volume.

1 Background for an energy-efficient and low-carbon built environment

1.1 Introduction

This chapter provides a brief introduction and background on the urgent need for energy-efficient, low-carbon buildings to mitigate climate change impacts arising from the use of fossil fuels. It starts by looking at the principal threats to the global environment and the need to undertake development in a sustainable way. The background issues of global warming and climate change are summarised briefly, together with an outline of the likely impacts, which indicate the immense scale and urgency of the challenge. The recent and emerging trends in the energy supply scenario for buildings are outlined, together with the implications for designers as the energy supply infrastructure goes through a transition to reduce carbon impacts. In order to show the wider relevance of these issues for building services design, some of the other environmental impacts of the built environment are also described.

A separate impact of fossil fuel usage is examined through looking at the adequacy of infrastructures to meet the anticipated energy requirements in the near future. A further aspect of energy strategy relates to materials used in building construction and operation and to identifying ways in which their impact can be reduced.

The final part of the chapter looks at general holistic principles that apply to the design of sustainable products and services. These are put into the context of energy performance of buildings, with an emphasis on the need for an integrated and interdisciplinary whole-life approach, with a primary focus on demand reduction.

It should be noted that this chapter focuses primarily on the energy and carbon aspects of built environment sustainability, rather than the wider factors.

1.2 Principal threats to the global environment

To understand the exceptional challenge for sustainable development in the context of the built environment, it is necessary to start from a perspective of the wider nature and range of environmental factors that threaten man's continued habitation on Earth and then identify those causes that specifically arise from the built environment. By identifying the priorities for attention, it is possible to focus on the key aspects related to design for buildings, while accepting that a wider range of environmental factors will require attention to contribute to the broader objective of sustainability in practice.

A major study was undertaken by a group of scientists in 2009 to identify the principal environmental processes that could cause significant disruption to human life on Earth. The study, which was under the auspices of the United Nations, also sought to calculate boundaries for these processes which, if exceeded, could limit the planet's ability to sustain human life. Summaries of this study were reported by Foley (2010) and Pearce (2010) and identified nine key environmental processes:

- climate change
- ocean acidification
- stratospheric ozone depletion
- nitrogen and phosphorous cycles
- fresh water use
- biodiversity loss
- land use
- aerosol loading
- chemical pollution.

The study identified target values for all the processes (except for the last two, because it was felt that there was insufficient understanding to do this). Of the seven processes where targets were set, it was found that three processes have already passed their safe limit: climate change, biodiversity loss and nitrogen pollution. It was also found that the others were moving closer to their safe boundary level. For two of the processes – climate change and the increasing acidification of the oceans – the principal cause is increased levels of CO_2 in the atmosphere arising from mankind's use of fossil fuels (Foley 2010; Pearce 2010). As the movement of any of these key processes towards their threshold levels could result in significant environmental damage, the imperative for society is to ensure that each process be maintained as safely within the boundary figure as is practically possible. While the processes and limits were presented separately, it was acknowledged that they were interconnected in many ways. For example, the increased acidification of the oceans could have a severe impact on their ecosystems, with implications for biodiversity and, as a consequence, threaten the food chain. A further consequence is that the ocean's ability to absorb CO_2 would reduce with acidification, with implications for the rate of climate change as a positive feedback relationship (Foley 2010; Pearce 2010). Climate change is likely to have severe implications for fresh water, biodiversity and land use.

The key message that emerges is that mankind now has a clearer indication of the limitations of the Earth's resources and their rate of usage. There is also better understanding about the ability of the Earth to absorb the waste and emissions arising from their use. There is, therefore, an urgent need for society to ensure that its activities are maintained within the limiting operational boundaries of the Earth's environmental systems.

The issues outlined previously relate to the key environmental processes that needed to be addressed on an urgent basis. It has, however, been recognised that development in all senses needs to adopt principles that will allow continued and sustainable habitation on Earth. Sustainable development has been defined in the 1987 Brundtland Report as 'development that meets the needs of the present without compromising the ability of future generations to meet their own needs' (UN 1987). This introduced the concept of sustainable development being a satisfactory balance between environmental protection, social equity and economic development, sometimes known as the 'triple bottom line', as shown in Figure 1.1. While a wide range of challenges was involved in achieving such a balance, there is now a compelling scientific view that the most significant threat to a sustainable future is climate change arising from 'anthropogenic' (i.e. caused by mankind's activities) global warming. This is mainly due to the presence of greenhouse gases (GHGs) in the troposphere, as described in the following section.

To stimulate international efforts to address the broad range of sustainability challenges, in 2015, the UN created an agenda with a series of objectives and targets. Their General

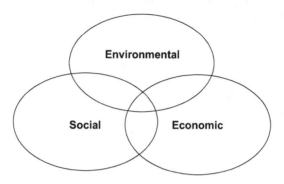

Figure 1.1 Sustainable development: the 'triple bottom line'

Source: UN (1987)

Assembly adopted the 2030 Agenda for Sustainable Development, which included 17 Sustainable Development Goals (SDGs) (UN 2015). These objectives include several goals that are directly relevant to the built environment and construction industry. These include good health and wellbeing; clean water and sanitation; affordable and clean energy; industry, innovation and infrastructure; sustainable cities and communities; and climate action. Two other objectives are more general and are indirectly relevant: responsible consumption and production; and partnerships to achieve the goal (UN 2015). In September 2019 the UN held its first Sustainable Development Goals Summit since adoption to discuss the 2019 Global Sustainable Development Report, which included an assessment of progress towards the goals. The report noted that progress made in the previous two decades was in danger of being reversed, in part due to potentially irreversible declines in the natural environment, and that the present development model had brought the global climate system and biodiversity loss close to tipping points. The report highlighted the need for developed countries to change their patterns of production and consumption, which needs to include limiting the use of fossils fuels (UN 2019a).

1.3   The greenhouse effect, global warming and climate change

A variety of natural factors have altered the climate of the Earth in the past, and there is a natural 'greenhouse effect' that has warmed the Earth's surface for millions of years. However, the scientific consensus in the past two decades has increasingly indicated that these natural factors alone cannot account for the extent of warming that has been observed. The evidence from observation and modelling indicates that this warming has largely been due to increased emission and accumulation of GHGs caused by mankind's activities. This has created an 'enhanced greenhouse effect', alongside which are feedback processes that cause amplification of the warming (Romm 2018).

The physics and mechanisms associated with the interaction between the Earth's atmosphere and GHG emissions are, obviously, extremely complicated and well beyond the scope of this book. For the purpose of understanding the basic concepts and terminology related to climate change, a highly simplified approach is sufficient, as shown in Figure 1.2.

The sun's rays are the sole energy input, and the radiation is mostly in the visible part of the spectrum (Coley 2008). This short-wave, high-frequency radiation reaches the Earth's

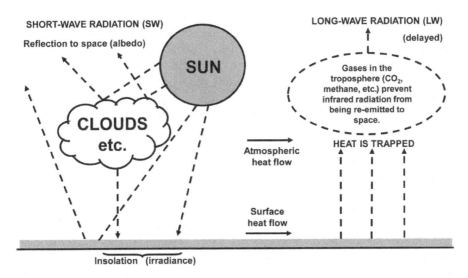

Figure 1.2 The greenhouse effect relationships (greatly simplified)

Source: Derived from Christopherson (1997: figure 4.1)

atmosphere, where it is absorbed, reflected or scattered. A proportion (about 31%) is reflected back into space by the atmosphere, clouds and the lighter-coloured areas of the Earth's surface. This fraction is termed the 'albedo' (Coley 2008). Therefore, the whiteness of cloud cover and the relative lightness of the different parts of the Earth's surface play a significant factor in the proportions reflected or absorbed. The residual radiation heats the atmosphere and the Earth's surface. The relatively low temperature of the Earth's surface gives rise to long-wave, low-frequency (infrared) radiation back to space.

The radiation from the Earth is thus a mix of radiation in the visible range that is directly reflected and radiation in the infrared range that is re-radiated. Over time, the Earth re-radiates an average 69% of incoming energy to space. The presence of certain GHGs in the troposphere traps the heat and delays it in the atmosphere. Some of the energy is re-radiated back towards the Earth. This is known as 'back radiation' and warms the surface and the lower atmosphere. This warming phenomenon is known as the 'greenhouse effect', as the GHGs create an effect similar to that created by the glass in a greenhouse, letting heat in but preventing it from escaping. These climate mechanisms, outlined in very simple terms earlier, are described in detail in various texts (Boyle 2004; Christopherson 1997; Coley 2008; Goudie 2000; Romm 2018). To provide a perspective in terms of heat alone, it has been estimated that the total heat arising from all human activities is only about 0.01% of the solar energy absorbed at the surface (Goudie 2000).

The overall increase in heat that results is called 'global warming'. It is estimated that the Earth is warmer than it has been at any time during the past 2,000 years (Coley 2008). The atmosphere close to the surface of the Earth has risen in temperature by about 1 degree C since the Industrial Revolution, and much of this temperature rise is estimated to have occurred in recent decades. After the year 2000, emissions growth began to accelerate (Romm 2018). There has been other evidence linked to global warming in recent times, including rises in sea level and reduced levels of sea ice in the Arctic, which have been widely reported in the scientific literature.

'Global warming' is a term that is generally used to refer to observed warming of the Earth due to GHG emissions caused by humans. Global warming has already affected, and will continue to affect, the world's climate in numerous complicated ways, giving rise to climate change. 'Climate change' is a more useful term than 'global warming', because it refers more generally to a variety of long-term changes in climate, including rises in sea level, extreme aspects of weather and acidification of the oceans, and is considered to be more accurate, as other non-temperature weather impacts may be bigger than those just caused by increases in temperatures (Romm 2018). It is also necessary to distinguish between weather and climate. Weather relates to the short-term variability in a particular location, whereas climate relates to the long-term pattern of statistics for conditions in a particular region (Coley 2008).

The principal international body on this subject is the United Nations Intergovernmental Panel on Climate Change (IPCC). This panel is made up of respected scientists from around the world and has been publishing comprehensive reports every few years on the scientific viewpoint. The IPCC's 4th Report (IPCC 2007) (also known as Assessment Report 4 or AR4) of 2007 was regarded as pivotal because it provided a consensus of certainty on the influence of human activities on climate change. The report concluded that most of the global warming that had been seen since the mid-20th century was very likely to have been due to the increase which has been observed in the concentration of GHGs caused by human activity creating an 'enhanced greenhouse effect' (IPCC 2007). Subsequent IPCC reports have provided increased levels of certainty and raised growing concerns about the urgency and scale of the challenge to prevent the likely impacts on human societies and the natural world. A particular concern is that it is not only the absolute level of CO_2 that is unprecedented but also the rate of change, meaning that all life on Earth would have to adapt faster, which would be extremely difficult, so climate change has been described as an 'existential issue for humanity' (Romm 2018).

The most important GHGs are carbon dioxide, methane, chlorofluorocarbons, nitrous oxide and ozone. The relative contribution of each gas to the enhanced greenhouse effect depends on its global warming potential and the level of concentration in the atmosphere. The relative contributions have been noted by Coley (2008):

CO_2	65%
Methane	20%
CFCs, HCFCs	10%
N_2O	5%

Because so much impact arises from CO_2 emissions, a key question is: What is the upper threshold of CO_2 that can be tolerated in the long term (in relation to anticipated impacts) and that can therefore be a target? The generally assumed figure for pre-industrial CO_2 is about 280 parts per million (ppm). Scientists had previously indicated that a level of about 350 ppm by volume might be a target that would avoid the worst impacts, typically related to a temperature rise of 2 degrees C above the pre-industrial level. Recent data on CO_2 levels and temperatures have indicated the scale of impacts already experienced. In November 2019, the World Meteorological Organisation reported (WMO 2019) that the concentration of CO_2 in the atmosphere, averaged across the planet, reached 407.8 ppm in 2018. The increase since 2017 was higher than the average rise for the previous decade. The increases in recorded levels for methane and nitrous oxide in one year were also higher than the increases in the previous decade. In early 2020, the UK Meteorological Office (Met Office 2020) confirmed that, based on annual global temperature figures, the 'cardinal decade' 2010–2019 was the warmest on record and each of the decades from the 1980s has, in turn, been warmer than all the previous decades. They also confirmed that in Australia, the year 2019 was the warmest and driest year on record, and it was also one of the warmest years on record in Europe.

1.4   Likely impacts of climate change and the challenge of mitigation

There has been continuing debate in the scientific community during the past 20 years or more about the likely impacts of global warming and the associated climate change, and international agreements have come into place to help avoid the most damaging impacts. The United Nations Framework Convention on Climate Change (UNFCCC) was established in 1994 as an environmental treaty to stabilise GHG concentrations such that they would protect the climate system from manmade interference. The intention was that the timescale for achievement would support natural adaptation for ecosystems and safeguard food production. The IPCC is an international body of respected scientist that reviews climate change science and provides evidence to the UNFCCC.

The UNFCCC has been meeting annually since 1995 – the meetings are termed the 'Conference of Parties' (COP) – with scientific evidence provided by the IPCC. The series of climate conferences led to Paris in 2016 (COP21). This was considered to be a landmark, as the outcome, known as the 'Paris Agreement', was the first legallybinding global climate deal, achieving a consensus of nearly 200 countries on the urgent and far-reaching need to make drastic reduction in their GHG emissions. The conference recognised that a central feature of their objectives would be to prevent levels of climate change that would be at a dangerous level and could be irreversible. The consensus among the leading scientists advising the UN was that this would relate to warming of about 2 degrees C above the prevailing level in pre-industrial times. The agreement was to restrict global temperature rises by 2100 to well below the dangerous level of 2 degrees C and to seek to restrict rises in temperature to 1.5 degrees C. Further objectives included a long-term target of reducing the rise in GHG emissions so that they reach a peak as soon as practical. Alongside this, there was an intention to reduce GHG emissions attributable to human activities to net zero, i.e. to balance the emissions with natural absorption (such as from vegetation, soils and oceans) by the second half of the 21st century. The conference recognised that countries would need to markedly increase their mitigation activities to achieve the objectives and the level of their mitigation ambitions would need to be increased over time.

Following the Paris Agreement, the IPCC was asked to produce a special report with guidance to governments on ways in which they could develop their plans to meet the objectives. Their report (IPCC 2018) – which reviewed the practicalities, costs and impacts of restricting rises in temperature to 1.5 degrees C – highlighted concerns that the carbon reduction pledges that had been made by countries would result in warming by more than 3 degrees C by 2100, so actions to reduce carbon would need to be intensified. There was a consensus that there were only 12 years available (i.e. until 2030) to restrict warming to 1.5 degrees C. They reported that if warming increased by a further 0.5 degrees C, many millions of people would have an increased risk of extreme levels of heat, drought and poverty. In relation to sea level rises, many low-lying countries would disappear.

The consensus view was that achieving the 1.5 degrees C objective would relate to the ambitious end of the Paris Agreement, but was considered to be affordable and achievable. However, it would require fundamental changes – by governments and individuals – that were both urgent and unprecedented, across a range of political, physical and social activities. These include lifestyles, energy generation, food, reforestation, land use, cities and industry. The IPCC advised that the solutions will require a departure from using fossil fuels, with carbon emissions needing to be reduced by about half by 2030 and reduced to zero by 2050. Notwithstanding this stark assessment, the IPCC report, due to its need to achieve consensus across so many countries, was considered by many to be highly conservative. For example, it did not consider potential extreme impacts from climate tipping points or increases in refugees from climate-affected areas.

In the UK, the Met Office reported concerns about the increased incidence of heatwaves, noting that heatwaves have recently been the deadliest global weather hazard and that it is expected that most regions will experience more intense heatwaves. In the UK the ten warmest years on record have occurred since 2002, and they predicted that heatwaves similar to that in 2018 were now 30 times more likely to happen (Met Office undated).

In the United States, NASA's Jet Propulsion Laboratory reported in December 2019 (NASA 2019) that the Greenland ice sheet was melting rapidly, which they predicted would cause an increase in flooding impacts compared with previous assumptions. The study, based on 26 independent satellite datasets, found that annual rates of ice mass loss in Greenland had accelerated from 25 billion tonnes in the 1990s to 234 billion tonnes. It stated that if present trends were to continue, then by 2100 flooding from rising sea levels could affect 400 million people each year (NASA 2019).

The UN Environment Programme (UNEP) Emissions Gap Report 2019 (UN 2019b) stated that even if all the unconditional commitments made in response to the Paris Agreement were implemented, temperatures would be expected to rise by 3.2 degrees C. This would result in climate impacts that were more wide-ranging and destructive. The report warned that global GHG emissions would need to fall by 7.6% annually during the 2020s to be on track to limit temperature rise to 1.5 degrees C. It was also stated that CO_2 emissions in 2018 were a new high of 55.3 gigatonnes CO2e (including from land-use changes, such as deforestation) (UN 2019b).

In response to the growing concerns, in 2019 the UK amended the Climate Change Act of 2008 with a new target to achieve 'net zero' GHGs by 2050 (the previous target was to reduce by 80% compared to the levels in 1990);and the European Parliament declared a 'climate emergency'.

The recent global and UK performance for energy and carbon has shown that a transition is under way from fossil fuels and towards low-carbon alternatives. There have been positive steps in the growth of renewable energy contributions to global energy supply systems, which averaged over 5% annually during the decade to 2019, alongside which the price of renewable electricity dropped, from 2009 to 2019, by 77 % for photovoltaics and by 38% for onshore wind energy (UN 2019c). The provisional estimate of the UK's total GHG emissions for 2018 was 448.5 million tonnes of carbon dioxide equivalent, which is a reduction of 44% since 1990. In 2018, 19% of the UK's primary energy came from low-carbon sources, increasing from 12% in 2011 and 9.4% in 2000. The largest contributions were from nuclear power (about 39%), bioenergy (about 37%) and wind power (about 14%). In 2018, 11% of final energy consumption was from renewable sources, up from 9.9% in 2017 (BEIS 2019a).

The COP26 meeting was due to be held in Glasgow in late December 2020, but has been postponed until 2021. The expectation is that individual countries will outline their detailed longer-term climate plans and targets, to significantly enhance the commitments in the Paris Agreement to avoid catastrophic impacts. It is clear that the predicted impacts would be extreme across the three main focuses for sustainability – environmental, social and economic. It seems inevitable that some of the widest-ranging and most severe impacts will be in developing countries, although the cause to date is largely due to carbon emissions from the developed countries. People in poorer and more remote countries, as well as low-lying land areas and islands, can be highly exposed, and the nature of their environments and lifestyles can make it difficult to change and harder to adapt.

1.5   **Energy use and carbon emissions from buildings**

In order to understand its relevance to building services engineering design, the wider background outlined earlier needs to be set in the context of the built environment. A good way to start is to see energy in the built environment in the context of all other usage. The breakdown of UK-delivered energy consumption in 2018 by sector (BEIS 2019a) was:

Transport	40%
Industry	16%
Domestic Buildings	29%
Services (mainly non-domestic buildings)	15%

The services sector figure includes agriculture, but this has traditionally been only about 1%, so buildings are responsible for about 43% of energy consumed. Additional environmental impacts arise from energy used in construction materials and the construction process and from waste, so the total energy usage and carbon impact is on the order of 50%. An important factor to appreciate is that the turnover of the building stock in the UK is relatively low (traditionally it has been at about 3% per annum). Therefore, while there is inevitably a significant focus on sustainable construction for new buildings, the greatest potential for reduction in carbon emissions is actually from refurbishment of existing buildings, particularly in the domestic sector. It has been estimated that about 80% of the UK's residential 2050 stock already exists and 85% of households currently use fossil fuel–based natural gas, so deep retrofitting for existing housing stock is likely to be essential if the required carbon savings are to be achieved (IET undated).

In the UK there have been government-led initiatives to improve the energy efficiency of homes by installing cavity wall and loft insulation. These are considered to be cost-effective where cavity walls and lofts, respectively, exist and where there is a practical way of installing these measures. Government figures (BEIS 2019a) show that by December 2018, the progress for homes with cavities was 14.1 million of 20.3 million were insulated, and for homes with lofts 16.5 million of 25.0 million were insulated. This situation had improved within the previous ten years; however, to give an indication of the scale of improvements still required, 6.2 million homes with cavities were not insulated and 8.5 million homes with lofts were not insulated (BEIS 2019a). This means that by early 2019 more than 14 million actions were still needed to achieve this basic level of thermal insulation, and until addressed, this will represent a continuing energy and carbon burden.

Based on UK government definitions, in England in 2017, the proportion of households in fuel poverty was estimated to be 10.9% in 2017 (which equates to about 2.53 million households). As a measure of the domestic impact of fuel costs, in 2017/18 households spent an average of 3.9% of their total expenditure on fuel. However, for households in the lowest income group, expenditure was considerably higher, at 8%, than for households in the highest income group, where expenditure was only 2.6% (BEIS 2019a). As the cost of fuel is a higher proportion of expenditure for lower-income groups, they would be the main beneficiaries of energy efficiency measures, so there would be a clear societal benefit from pursuing this energy efficiency initiative.

To address the energy and carbon issues related to buildings, it is necessary to understand the processes through which energy is delivered for consumption in buildings. The inefficiencies of energy supply do not relate to generation processes alone, but to the whole delivery process from source to consumer. This is a rapidly changing scenario, which is improving considerably. However, it is worthwhile appreciating the inherent inefficiencies of conventional electricity delivery from fossil fuel power stations to grid infrastructure. Energy conversion (i.e. from one form to another) and distribution inevitably incur losses, and it is important to understand the true operational efficiency of an energy system. This is because energy usage in buildings usually involves a diversity of fuel inputs and systems of utilisation and a highly dynamic load pattern, both daily and annually. Energy consumption information is only really meaningful if it relates to primary energy on an annual basis, so that it represents the operational performance, taking account of diurnal and seasonal patterns and influences. It should not be confused with load assessment, which usually relates to peak or worst-case load. An important concept to understand in relation to the efficiency of energy systems in buildings is the difference between 'primary energy' (the energy of the fuel input at the source) and 'delivered energy' or 'final energy' (the energy delivered to a consumer) and to refer to these correctly when evaluating or comparing energy systems or using data about energy and carbon performance. Carbon emissions relate to the primary energy.

The carbon impact of electricity is a consequence of the way in which it is generated and distributed. In many developed countries electricity generation has traditionally been dominated by thermal power stations using fossil fuels, although there have been major changes in recent years in many parts of the world. In the UK in 2010, 28% of the total electricity was supplied from coal and 46% was supplied from gas, with only about 7% supplied from renewable sources (BEIS 2019a). Since then, there has been a major beneficial change in the fuel mix for electricity generation, with a sharp decline in coal usage and a significant growth in renewables. In 2018, the breakdown of the UK's total electricity generated by fuel type (BEIS 2019a) was:

Gas	39.5%	(fossil fuel)
Wind and solar	21.0%	(renewable)
Nuclear	19.5%	(low-carbon)
Other renewable	10.4%	(renewable)
Coal	5.1%	(fossil fuel)
Oil and other fuels	2.9%	(fossil fuel)
Hydro	1.6%	(renewable)

Renewable sources provided 33% of the electricity generated, which was a record, but it means that two-thirds were still from non-renewable sources. More specifically for carbon content, nearly half (47.5%) was generated from fossil fuel sources emitting CO_2, with consequent impacts.

Most conventional thermal power stations usually have conversion efficiencies of about 30%, on average, at best (higher for combined-cycle gas turbines), so about 70% of the primary fuel input is wasted as heat to the atmosphere, sea or rivers. The overall efficiency of the grid will depend on the mix of generation sources. This will be higher than the efficiency of thermal power stations alone due to the contribution from zero and low-carbon sources: renewables and nuclear. There has been continuing evolution of the fuel mix for electricity generation in many countries in recent decades, and this is undergoing further rapid change in the UK and elsewhere. The specific mix that will arise, year on year, is not predictable; this will have a considerable impact on efforts to relate environmental impacts and decision-making to energy usage metrics for buildings.

The useful electricity generated from the widely dispersed sources is then passed through an extensive system – the transmission grid and local distribution networks – before it reaches consumers. This gives rise to further distribution losses, typically in the order of 10–12%. There is also some energy industry usage. The final delivered energy to all consumers (as measured at their intakes) is therefore only a modest proportion of the total primary energy input. The specific efficiency balance between primary and delivered (final) energy will vary on a diurnal and seasonal basis, as the generation fuel mix varies year by year, as the demand changes and as the infrastructure and operating arrangements change. It is clear that there are inherent inefficiencies on the energy supply side in general, although this is most pronounced for electricity. In 2018, nearly 26% of all UK inland energy consumption (i.e. electricity and other forms) was related to conversion and distribution losses and energy industry use, so only about three-quarters of primary energy were delivered to meet energy needs, but this is a big improvement from 2000, when total losses were nearly 32% (BEIS 2019a).

The CO_2 content of electricity will vary with the mix of primary fuels used in generation. In the UK, the effective CO_2 content of electricity is considered to have fallen considerably in recent decades, largely due to the major reduction in coal usage and the rapid growth in renewables. The primary energy carbon factor used in Building Regulations Approved Document Part L 2013 for gas is 0.216 kg CO_2/kWh, compared with the figure for consumed electricity (from the grid) of 0.519 kg CO_2/kWh. On this basis, electricity is considered to have about 2.4 times the carbon impact of gas. There is continuing debate about the appropriateness of the carbon emission factor for grid-derived electricity and the extent to which it properly represents the decarbonisation of the grid that has already happened and the planned further decarbonisation in coming years. This is also related to the need to address the required upgrading of the grid infrastructure, as outlined in Section 1.6.

All fossil fuels have an environmental impact related to the resources they take from the Earth and the emissions they give out to deliver energy to consumers. These are set out in a simplified sense for conventional electricity generation in Figure 1.3, which illustrates the wide range of impacts. It is also necessary to consider the extensive range of materials used throughout the energy generation and distribution process, covering industrial facilities such as mines, rigs, pipelines, power stations, transmission lines, refineries and their related roads, railways, shipping, transport, etc. (Coley 2008). This would include materials such as steel,

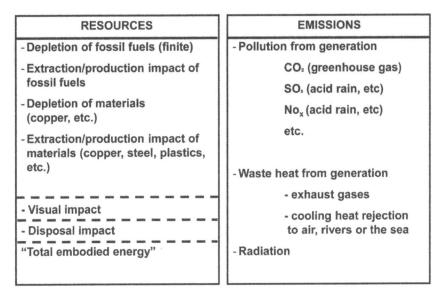

Figure 1.3 The environmental issues in fossil fuel electricity generation and transmission

Source:Author's elaboration[1]

concrete, copper and plastics. Each of these materials is produced by an industrial process that has an environmental impact in a similar way to the extraction of the fuels. A useful indicator of the effectiveness of an energy generation technology is the 'energy payback ratio', which relates whole-life energy expended in the facility's materials compared with the useful energy produced during the plant's lifetime (Coley 2008). It has been estimated that the energy payback ratio for fossil fuel power stations is low, at about 5–7 for coal, and 5 for natural gas (Coley 2008). This provides a good illustration of the need to take account of all relevant factors when considering the environmental impact of any particular system or activity.

To illustrate the wasteful energy pattern for conventional fossil fuel usage, Figure 1.4 is a notional Sankey diagram for the annual energy consumption of a building with a traditional energy provision, using grid-derived electricity primarily from thermal power stations, and gas-fired heating. This represents a situation of a type prevailing in the UK earlier this century. It is intended as an example only of a typical annual energy flow format. In this example, the annual delivered energy to the building is shown for simplicity as 100 units, split 42:58 between electricity and gas. There would be some losses within the building due to distribution and heat generation from a boiler system, but these have been omitted for simplicity. The actual useful energy usage would therefore be a little lower than 100. The diagram shows notional proportionate losses in generation and transmission that have been typical in many countries for conventional electrical energy grid provision. In this case, 125 units of primary electrical energy are required to give 42 units of delivered electrical energy, primarily due to the poor overall conversion efficiency of the mix of generation technologies, resulting in heat wasted to the environment. It can be seen that the total primary energy input to the building is 188 units. So in this notional example, nearly twice as much primary energy is required to provide the total delivered energy. Moreover, to provide 100 units of useful energy to the building, the carbon emissions would be proportionate to the primary energy input of 188 units.

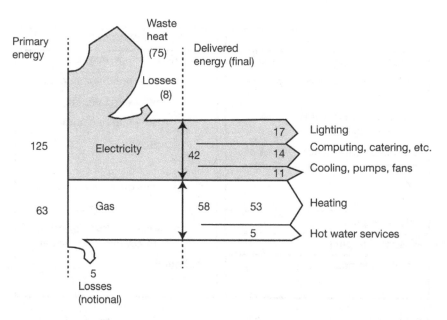

Figure 1.4 Notional Sankey diagram for (now outdated) conventional energy delivery to a building (not to scale)

For an energy supply picture like that shown notionally in Figure 1.4, a rational perspective would question the logic of throwing away waste heat from electricity generation that is greater than the whole thermal energy requirements for the building. It is obvious that if the waste heat could be made available in the building, it could offset the separate energy demand from gas for heating and hot water services, significantly reducing fossil fuel consumption, CO_2 emissions and fuel costs. This illustrates the highly wasteful overall energy performance of conventional thermal power stations. For this type of outdated energy supply scenario, a co-generation approach using a combined heat and power (CHP) plant can reduce overall energy consumption and carbon by recovering a large proportion of the heat that is normally wasted in electricity generation, which is then available (in suitable applications) for heating and/or process purposes. However, due to the progress in reducing the proportion of fossil fuel electricity generation in recent years in the UK and elsewhere, the relative carbon impacts of electricity and gas are narrowing and are set to reduce further. This decarbonisation of electricity means that a CHP approach becomes less viable. As decarbonisation of electricity continues, the major challenge is to 'decarbonise heating', and this has started to happen with the use of heat pumps for space heating and water heating. However, this 'electrification of heat' is not without its challenges, as it will require a significant expansion of the electricity system, especially the peak capacity (IET undated), and this will be required when the grid will be facing many other challenges, as outlined in Section 1.6.

1.6 Energy grids: adequacy of infrastructure and security of supply

The delivery of energy to buildings (and other consumers) is not just related to energy generation, but is also dependent upon the sufficiency of fuel reserves (where appropriate) and the adequacy of the energy delivery infrastructure, both of which will affect security of supply. While the use of fossil fuels must be rapidly reduced and then stopped to mitigate climate impacts, during the transition there will be a need to maintain a certain level of usage until sufficient capacities of more benign sources are operational to meet needs. Wind and solar sources are intermittent, so it is not clear how quickly it will be possible to eliminate fossil fuel generation, recognising the need to satisfy demands and their dynamic load requirements.

Fossil fuels are a finite resource. Whether we consider coal, oil or gas, these fuels have been created in complex organic and geological processes lasting millions of years. Their depletion through usage as a fuel has been taking place at a significant rate since the Industrial Revolution. Exploration has continually identified new reserves, but these are increasingly in locations where 'winning' the fuel is ever more difficult (Quaschning 2005) and where there is often a risk of creating environmental damage. As these locations are often in areas of considerable remoteness, natural beauty and ecological importance, extraction of fuels is likely to have increased environmental impact. Differing views prevail about the extent of unknown reserves, but there is a view that availability of fossil fuels can only be extended by decades at best (Quaschning 2005). So, notwithstanding any of the other environmental issues related to fossil fuel usage, there is the fundamental issue that reserves will simply become unable to meet demand, so security of supply is threatened. Moreover, in the near term there are complicated geo-political issues related to the locations of the major sources of oil and gas, which could have an impact on reliability of supply in some parts of the world and further threaten security of the supply.

Alongside the limited reserves of fossil fuels and the need to rapidly reduce fossil fuel usage, there are serious concerns about the electricity grids in many countries and their suitability to maintain the necessary flows of energy from generation locations to consumer locations in the short term. In the UK this includes particular concerns over the infrastructure of electricity generation plant and transmission systems, which is outdated for the contemporary and emerging patterns of supply and demand. The electricity infrastructure concerns include its lack of suitability for incorporating the expansion of large-scale renewable generation, including offshore wind generation, and the need to replace the supply capacity to replace the ageing nuclear capacity and other generation plant that is being shut down (coal power stations) or nearing the end of its planned life. The Institution of Engineering and Technology(IET) has reported that the necessary transition to a low-carbon electricity grid will require flexibility and interactive control for local distribution networks to allow continuous balancing with large amounts of renewable generation and load increases for electric vehicles and heat pumps, as well as dynamic interactions between energy provider and consumer (IET 2011).

An indication of the availability of the energy supply infrastructure to meet demand is the operating margin between the maximum supply capacity and the maximum demand. For electricity, during the period from 2007–2008 up to 2014–2015, there was mainly a year-on-year increase in the electricity capacity margin. This was caused by a reduction in peak demand and a rise in capacity. However, the situation changed between 2013–2014 and 2016–2017, when the capacity of major power producers reduced at a faster rate than the fall in peak demand (as a result of closures and conversions of plants). As a consequence, the capacity margin fell from a high point of 44% in 2013–2014 to 29% in 2016–2017. This gap was the lowest since 2009–2010 (BEIS 2019a). There was an increase in embedded renewables capacity over that period. However, due to the intermittent nature of solar and wind generation, the increase in renewable capacity was not sufficient to cover the reduction due to the fall in capacity of the major power producers. The margin increased to 42% in 2018–2019 due to a reduction in peak demand and a rise in renewable capacity. The UK gas capacity margin has mainly been widening since about the mid-2000s, and the margin in recent years has remained at about 60–80% of usage of maximum capacity (BEIS 2019a).

Because such a high proportion of UK electricity generation is currently from gas (about 40% in 2018), its reliability is strongly linked to gas security. UK continental shelf gas production peaked in 2000; however, since 2000, there have been major reductions in the levels of gas production, of around 5% annually (BEIS 2019a). In 2018 production was only about one-third of the record levels seen in 2000. As a result, the UK is increasingly reliant on the import of gas in order to meet demand. The UK is now a net importer of all main fuel types, and all EU countries are now net importers of energy (BEIS 2019a)

The Royal Society (with others) reported in 2017 (Royal Society 2017) that the magnitude of the task of achieving the Paris Agreement goals, while also maintaining security of energy supplies, was not fully understood. It reported that new developments will be required for a decarbonised energy system, including storage, to maintain a balanced load and ensure security of energy supply, and that the UK will almost certainly require substantial carbon capture and storage (CCS). The report noted that decarbonising heat was a major challenge and that in the UK it would be undesirable to build an infrastructure that had a long life expectancy and would thus commit the country to emissions well into the future (Royal Society 2017). The Royal Society also reported that while the costs of electricity generated from solar photovoltaics and offshore wind generation had fallen unexpectedly fast, and with much greater increases in capacities of these renewables expected, their intermittent nature would create difficulties in managing very high levels of dependency (which might provide a limitation to this expansion). They noted that the expansion of these renewable sources, when considered alongside the growth in electric transport, will dramatically change energy movements on networks and could potentially impact on the stability of the grid (Royal Society 2017).

Concerns about reserves, infrastructure and hence security of supply are compelling reasons in themselves to reduce energy consumption in every sense, but particularly fossil fuel consumption (Quaschning 2005). Taken together with the paramount and enormous challenge to mitigate climate change, they provide further impetus to drastically and quickly reduce energy consumption and reliance on fossil fuels.

1.7   Other environmental impacts of the built environment

While climate change is the paramount issue arising from the emission of CO_2 into the atmosphere, it is not the only environmental impact of using fossil fuels. Burning fossil fuels also emits sulphur dioxide and nitrous oxides and particulates, with considerable impact, including 'acid rain' and air quality (Coley 2008). Conventional energy generation using fossil fuels is also a major contributor to air pollution, and there is increasing evidence of the impact on health. In 2018, Public Health England stated that poor air quality was the largest environmental risk to public health in the UK, because when people have long-term exposure to air pollution, it can cause a range of conditions which lead to reduced life expectancy (Public Health England 2018). Two major components of urban air pollution are particulate matter (PM) and nitrogen dioxide (NO_2). It was noted that there is, at present, no clear evidence of a safe exposure level with no risk of adverse health effects. The main sources of manmade PM include the combustion of fuels. NO_2 is a gas that is produced, along with nitric oxide (NO), by combustion processes.

There are many other direct consequences arising from energy use, such as resource depletion, despoliation of the landscape, heat, radiation, noise, etc., together with indirect consequences (Coley 2008). Other environmental issues related to energy consumption are the manufacture, use, disposal and recycling of materials, which are discussed in Section 1.8. A separate environmental impact arises from the use of chemicals in the refrigerants of air conditioning systems. These have identifiable ozone depletion potential, global warming potential (GWP) and total equivalent warming impact. There is more about the GWP of refrigerants and the selection of refrigerants for air conditioning and heat pump systems in Section 8.2.2.

It is also useful to see the sustainability challenges in the context of the nature and growth of the urban environment in a global sense. The UN estimated that by 2008, half of the world's population lived in towns and cities (Barley 2010) and that this was a trend that was likely to continue. The UN reported in 2019 that the proportion of the global population living in cities is projected to increase to two-thirds by 2050, and to achieve the UN's SDGs, cities will need to be more compact and efficient and be better served by infrastructure, including public transport (UN 2019c). Studies of the relative impacts of urban and rural areas have shown that urban areas are efficient in land use, as they only take up about 3% of the land surface of the Earth (Barley 2010). Moreover, because cities have high population densities, they provide an opportunity for efficient energy, water and sanitation infrastructure, together with viable public transport networks, all of which reduce the relative CO_2 impact (Barley 2010). Studies of CO_2 emissions in the United States have indicated that citizens of New York produce, on average, only about 30% of the average emissions for the country (Barley 2010).

A further impact of the growth of urban areas is the difference in thermal response when compared with rural areas. The surfaces of a city absorb more solar radiation during the day and release it at night. This is due to the high thermal capacity and arrangement of high walls and dark-coloured roofs in city centres, as outlined by Goudie (2000). The densely built-up centres of cities provide the highest temperature anomalies, generally called the 'urban heat island effect' (Christopherson 1997; CIBSE 2007; Goudie 2000); thus, urban expansion can have localised impacts on climate.

To limit environmental impacts, the most direct focus for building services engineers will be on how all these issues can be addressed for individual buildings, or groups of buildings, as part of a site development proposal. In terms of wider solutions, however, there will be a need to make our towns and cities sustainable in a more general sense, through design and planning that address interrelated aspects in an integrated way. It is therefore useful to see the future scenario as requiring sustainable urban planning for communities whose arrangement and facilities encourage sustainable lifestyles and to think conceptually about a built environment, rather than groups of buildings. At its most simplistic level this would involve the relationships between buildings and the spaces between buildings, and how they perform as a unified whole. In a wider sense, the built environment can be considered as the complex interrelationships of all the elements and influences from the built forms mankind has created and how it relates to the natural environment. While there has been high societal awareness of the environmental impact of transport, the awareness of built environment impact was generally low at the end of the last century (Smith *et al.* 1998), although this has been changing in the past two decades. Smith *et al.* (1998) have proposed that the built environment requires more holistic models of development management and planning that recognise the complex interrelationships. This still seems a suitable aspiration.

1.8   Materials usage and embodied energy/carbon

The preceding sections have mainly focused on the environmental impact of operational energy consumption in buildings that gives rise to CO_2, and hence contributes to climate change and threatens security of the supply. Another important consideration, however, is the impact of the usage of materials within the built environment. Materials themselves require energy in every aspect of their life cycle, from extraction, transportation, production and delivery, to construction, maintenance and, where appropriate, eventually demolition, disposal and recycling. This energy is known collectively as 'embodied energy' or 'embodied carbon'. Embodied energy has been defined as the total primary energy consumed by a product or service from both direct and indirect processes, while embodied carbon is the amount of CO_2 emitted by a building, product or system during the entire life cycle (BSRIA 2012). As with operational energy, embodied energy involves inputs and emissions. The Inventory of Carbon & Energy database (Hammond and Jones 2011) includes these examples of typical levels of embodied energy for materials widely used in the industry:

- primary glass: 15.00 MJ/kg
- steel: 20.10 MJ/kg (general UK (EU))
- copper: 42.00 MJ/kg (EU tube and sheet)
- PVC: 77.20 MJ/kg (general)
- aluminium: 155 MJ/kg (general)

For buildings and their engineering services, the embodied energy in materials was previously estimated to typically represent about 10% of the total energy usage during the life of a building, the rest being operational energy. However, as operational energy usage has been and will continue to reduce due to the adoption of carbon mitigation measures, the relative proportion of whole-life energy (or carbon emissions) due to embodied energy will increase. There is, therefore, an increasing focus on materials selection and usage to further reduce carbon impact.

The relevance of material usage as one measure of the effectiveness of a conventional energy generation technology was outlined in Section 1.6. Another important aspect requiring consideration of embodied energy is the viability of renewable energy systems. To take account of the embodied energy in the renewable technology equipment, it is necessary to assess the 'energy payback' in addition to the financial payback. This will show how long it will take for the renewable technology to deliver sufficient 'free' energy to offset the energy expenditure that has been incurred by the materials throughout their life cycle. There is, quite obviously, no benefit in incorporating renewable technologies if they take much of their expected life to recover the energy expenditure that they have already incurred in their usage. In such cases, it would be misleading to view their contribution as a low-carbon feature.

As an indication of the wider concerns about materials usage, UN projections show that the use of all materials globally is likely to almost double in the period from 2017 to 2060, with corresponding increases in GHG emissions (UN 2019c).

The wider consideration of materials that is relevant to the construction industry is to re-think the whole approach to materials usage during the construction and installation processes, so that materials efficiency is designed into the system from the outset (CIBSE 2007). This is in addition to the desirability of using materials that are environmentally benign and derived from local, sustainable sources. In industrial nations, material usage in construction has traditionally been inefficient, and at the beginning of the 21st century, it was estimated that waste from building accounted for 50% of packaging and 44% of landfill (Birkeland 2002), but methodologies have improved considerably in recent years. There is a new focus on planning the processes during a building's life cycle so that waste is minimised and recycling is maximised. This is part of a wider sustainable aspiration to adopt a 'circular economy'. In simple terms, a circular economy minimises impact on nature and its resources by using fewer resources in the first place; maximising usage during life cycles; and, at the end of the life cycle, fully exploiting materials through reclaiming, re-using or recycling (BSRIA undated). Modern buildings typically have intended life cycles of 60–70 years, which compares with less than 9 years for many manufactured products. Nevertheless, material usage in the construction industry has an enormous environmental impact. For example, it has been reported that construction in Europe is responsible for about half of materials extracted, one-third of waste produced and one-third of water usage, alongside its energy impacts (BSRIA undated).

The sustainable approach is for all materials to be seen as a useful resource and utilised accordingly to maximise their potential, so that there is no longer a concept of 'waste'. This will include reducing obsolescence of products and increasing their longevity – doing more with less – and using local materials and skills (Smith *et al.* 1998), or maximising performance with minimal means (Moe 2008). To promote a sustainable approach, construction projects should have a comprehensive materials management plan as part of their whole-life methodology. Modern methods of construction (MMC) are becoming more widespread and can significantly reduce materials usage, construction time and cost, as well as potentially contributing to health and safety management.

1.9   The need to plan for adaptation

It should be recognised that the full effects of GHGs take a considerable time to create their impact, with CO_2 remaining in the atmosphere for about 100 years (Goudie 2000). It is therefore inevitable that climate change will arise from CO_2 that has already been emitted over previous decades, which will have a considerable impact on the built environment. It means that, as well as a focus on mitigating carbon emissions to minimise future impacts, building services engineers will also be involved in adaptation strategies so that buildings can cope with the anticipated climate change impacts over the coming decades. So, for designers in the built environment, climate change has often been described as presenting a dual challenge of 'mitigation and adaptation'.

It is clear that there is an urgent need to accelerate adaptation policies and actions so that resilience to climate change can be embedded within communities and societies. There is no certainty on the exact climate change impacts in different locations, but predictions for the UK include a general increase in temperatures, increased intensity of rainfall and wetter winters, drier summers and increased daily mean wind speeds in winter (CIBSE 2007). It is also likely that there will be enhanced effects of 'urban heat islands' (CIBSE 2007), as outlined in Section 1.5, resulting in even higher temperatures in city centres. While there will be a need for specific design considerations for each predicted impact, more generally there will be a need for design to allow adaptability in the future. This is likely to include contingency allowances, such as additional space in plant rooms (CIBSE 2004a) and adopting a more flexible and modular approach to systems design to aid durability. This could allow plant arrangements to be matched to the prevailing needs in an incremental way as climate change impacts are manifested.

Recognising the anticipated increase in incidence of flooding in the UK and many other parts of the world, this will significantly influence decisions on the locations of mechanical and electrical equipment (as a considerable amount has traditionally been located in ground floors and basements) and will influence building design more generally.

1.10   General concepts for whole-life holistic design

1.10.1   Developing a holistic approach

To develop a suitable approach to the design of building services for minimising energy usage and carbon emissions, it is necessary to adopt some general concepts that could apply equally to the design of any product or facility. The principles are much the same for sustainability in the wider sense as for the more focused attention here on energy and carbon reduction. Concepts that can help in the understanding and development of a more sustainable approach are:

- *Contextual Awareness*. Engineers require an awareness of the background issues related to the built environment, climate change impacts, carbon mitigation and adaptation. This is so they can contribute to astute interdisciplinary decision-making and see energy and carbon in the wider context of sustainability and building performance.
- *Interrelationships*. The physical interrelationships between energy and materials usage in buildings and their carbon and environmental impacts.
- *Interconnectedness*. There should be a clear understanding that people (and their health and wellbeing), engineering systems, buildings and the environment are interconnected. As such, it is useful to adopt a wider systems-type perspective to inform design principles and decision-making, so that the implications across multiple systems can be considered concurrently. The best starting point is usually people and their behaviours and the ways through which they interact with the engineering and building systems.
- *Holistic whole-life approach*. Buildings should be seen as complex, multi-faceted entities the success of which must be judged on several levels, including aesthetic, environmental

impact, functional performance, health and comfort of the occupants, economy and longevity. For buildings to be truly successful and be of benefit to society as well as stakeholders, they must continue to perform at an optimum level throughout their life. 'Whole-life performance' is a key objective and requires a different mind-set from considerations of performance at completion and handover only. Adopting a broad outlook makes it possible to recognise the wider whole-life potential for mutual benefit and synergy between different aspects. This can help to avoid the narrow decision-making and outcomes that have often arisen from specialisation and entrenched viewpoints.

- *Commitment.* While regulatory frameworks are in place to set targets for minimising energy consumption and carbon emissions, the creation of a successful low-carbon built environment that is suitably adapted for climate change will ultimately depend on individual and collective commitment to make it happen.
- *Collaboration and communication.* All aspects of the building life cycle involve close teamwork across different organisations and disciplines. It is essential that a collaborative approach is adopted from the outset. Clear and professional communication is a key element to ensure that concepts and proposals are understood, as practitioners from different cultural backgrounds can differ widely in their understanding and interpretation of the issues.

A useful conceptual starting point for reducing resources and impacts more generally is the well-known mantra of 'reduce, re-use, recycle'. This is important for establishing the priority order for attention and relates closely to the 'circular economy' described in Section 1.8. It is always preferable to reduce the demand, or the need, for something at the outset through critically questioning the root cause of the demand. Only when the demand has been reduced rigorously to the practical minimum level to satisfy needs or objectives should the focus switch to re-use of the product or service when it has expired. It is important to understand that re-use means using something again in its existing form, for its original purpose, and therefore involves some further usage of energy and materials. Recycling usually involves considerable materials, energy and cost debits to return something that would otherwise be discarded into a usable commodity. It should be seen as the third priority and is much less attractive than reducing or re-using. For buildings, 'reduce' relates to all aspects, including initially questioning the need and scale of a new development to see if a development of lesser environmental impact could meet objectives, reducing the amount of materials and embodied carbon in construction and installation, reducing all the factors that bring about operational energy usage and carbon emissions and reducing water usage at all stages. 'Re-use' relates to re-using buildings (through refurbishment, retrofitting or re-purposing, where this can meet objectives and reduce impacts) and infrastructure (where this is practical); re-using materials, both construction materials and building services equipment; and re-using heat (through heat recovery), waste water and air (where this might be appropriate). 'Recycling' relates to provision of recycling facilities and encouraging their use, as well as potentially recycling existing construction materials and engineering equipment.

There are various understandings and definitions in use for sustainable, environmentally benign and 'green buildings' in different contexts, some of which are quite loose. Key factors that have featured include energy, water, materials, ecology, pollution and management (for efficient operation) (Bleicher 2019). But using 'green' and similar terms on their own has limited value in communication unless they can be related to specific objectives and metrics and relate to performance over a building's entire life.

A useful metric for planning and managing reductions in carbon in general is a 'carbon footprint'. This is measured in tonnes of carbon dioxide equivalent (tCO2e). The figure for tCO2e is obtained from the multiples of the emissions for each of six GHGs by its 100-year GWP (Carbon Trust 2018). A 'product footprint' relates to a product (such as a piece of equipment) or service and covers the accumulated emissions over the whole life. This would encompass the initial extraction of raw materials;the manufacturing or construction phase;the use in operation; and then eventual re-use, recycling and, finally, disposal (Carbon Trust 2018).

For buildings numerous environmental assessment methodologies are used in different parts of the world, with examples described in Chapter 4, and compliance criteria for national building regulations, with an example described in Chapter 3. With the urgent need to make huge improvements in line with the Paris Agreement, a new framework definition for 'Net Zero Carbon Buildings' was published in the UK in 2019 (UKGBC 2019). This has been issued as guidance initially, with an aim to provide a mechanism to achieve net zero across the whole life of buildings. It has separate definitions for in-use operational energy and for embodied emissions from construction. The intention is that the framework will develop to include tighter standards and targets. This framework will have a fundamental influence on design approaches.

To achieve the scale and pace of improvements required for the present challenges, while making wise decisions recognising the inherent complexities, it is likely that design concepts will benefit from a 'systems thinking' approach (Stasinopoulos *et al.* 2009). A 'whole-system approach' has been described as one in which active consideration is given to the interconnections between sub-systems and systems and where solutions seek to address multiple problems concurrently (Stasinopoulos *et al.* 2009). Under this type of approach, designers have to think not just beyond the scope of their discipline, or even the integrated disciplines in building design, but to the much wider systems' relationships between occupants, processes, buildings, infrastructure and environment. The goal of such an approach is to consider the whole system, in its environment, throughout its whole life cycle (Stasinopoulos

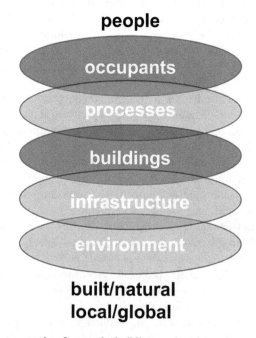

Figure 1.5 Simplified systems realms for people, buildings and environment

et al. 2009). Birkeland (2002) notes that 'we need to redesign not only the built environment, but the nature of development itself'. The breadth of such an all-embracing approach is well beyond the scope of this book, but the underlying principles inform many of the proposals.

As a simplistic example of how a systems-type approach might work, we could consider the engineering systems and energy needs for a printing machine that is operating almost continuously within an office environment. A narrow approach would see this as a fixed client design requirement, develop a suitable ventilation system to deal with the particulate emissions and heat generated and accept the environmental impact of operational and embodied energy and the less-than-satisfactory air quality. Systems thinking would involve analysing relationships and looking for better questions to seek prevention (Birkeland 2002). A systems approach would explore the business process and behaviours to see whether extensive printing was really required or whether it could be located elsewhere and potentially create a solution requiring less usage of carbon for energy and materials (including paper) and with better air quality within the office area.

The complicated systems interrelationships for buildings through their whole life can be expressed in a greatly simplified form as a set of 'realms', as shown in Figure 1.5. People – when they use buildings – become occupants and adopt, individually and collectively, certain behaviours that will influence the internal environment and operation. They have certain needs, including health and wellbeing, which should be satisfied by the building. Buildings usually also host processes related to the function and purpose of the building. The processes will influence behaviours and the facilities, arrangement and operation of the building. Buildings host people and processes and are integrally linked to the local and national infrastructure that supports the building and occupants – such as energy, water, sewerage, communications and transport. The infrastructure, in turn, is influenced by the demands of the buildings and has to integrate with the buildings. The buildings and infrastructure together constitute the built environment, which interacts with, and should co-exist harmoniously with, the natural environment. These environments can be considered together at local and global levels. So, while previous sections have identified the impacts of the built environment on the global environment, there is a significant influence from the occupants and processes in buildings. A systems-type approach would look at these wider interrelationships to inform design concepts.

As building services engineers seek to address energy matters at the briefing stage, it is necessary to understand the fundamental impact of occupancy behaviour on the eventual environmental impact. A traditional approach largely accepts the building's function as being pre-ordained, and the designers seek to accommodate the client's defined requirements with the minimum impact. A more participatory and user-centred approach would include consulting widely with and involving communities. This could lead to a systems-type approach, seeking ways to address the root cause by encouraging behaviours that will minimise the need for active engineering systems, operational and embodied carbon. So, the designer's attention focuses initially on the human activity behaviour, ergonomics and any associated processes.

Related to the whole-systems approach is a deeper ecological awareness that requires changing the way people think and seeing the way we live, and our environment, as part of an ecosystem (Pearson 2005). Wines (2000) and Pearson (2005) describe buildings that have been designed to be harmonious with their environment by adopting these organic concepts.

Some of the concepts outlined earlier may appear, at first sight, to be too abstract and aspirational for such a pragmatic field as engineering. However, as solutions are sought for our future buildings – and the built environment more generally – it will be necessary to have a mind-set that questions and challenges prevailing technological solutions in a creative but pragmatic way.

In a much wider sense, as an example of systems thinking applied to a national energy strategy for the complex challenges ahead, the IET Energy Principles 2013 (IET 2013) are based on seeing the 'big picture' of all the interrelationships influencing the whole energy system. This breadth includes energy consumers, suppliers, people, technology and resources. These principles emphasise the need to think across not just the whole physical system but also the policy environment when formulating solutions. The principles propose an energy hierarchy to reduce impacts and waste within the energy system, with priorities for sustainability in this order: energy conservation (including influencing occupancy behaviours); energy efficiency; renewable, sustainable energy sources; and conventional energy sources (using low-/no-carbon technologies) (IET 2013). The principles note that there is likely to be a need for collaboration between suppliers and consumers to manage peaks in electricity demand.

1.10.2 *Harnessing renewable energy*

Many locations are likely to have some potential for renewable energy, although this will vary considerably from location to location. For an appreciation of the potential, it is worth starting by looking at solar power, which is the single energy input driving the Earth–atmosphere system (including winds) and the most significant source of renewable energy (Christopherson 1997; Quaschning 2005). Figure 1.6 provides a greatly simplified conceptual illustration of global energy potential and usage with the area of the square approximating to quantity. The total annual solar energy received is about 9,000 to 10,000 times greater than the annual primary energy demand (about 409,000 million GJ) (Coley 2008), even allowing for the present highly wasteful usage of energy. Yet to meet the demand, developed societies have mainly used finite, polluting and costly energy reserves – with their associated impacts – rather than the free, clean and perpetual energy from the sun. The sun provides radiant power of about

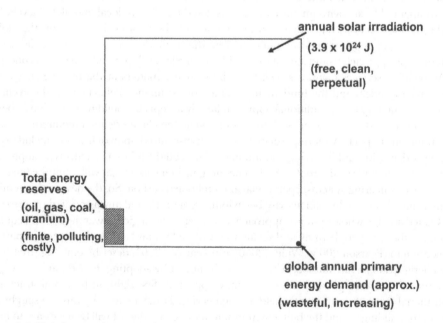

Figure 1.6 Global energy in perspective

Source: Derived from Quaschning (2005: Figure 1.9)

1kW/m² when directly overhead, with a clear sky. The amount of radiation received will vary depending on location – and hence the solar path variation throughout the year – as well as the extent to which clear or overcast conditions prevail. Examples of average annual solar radiation on a horizontal surface at ground level (Christopherson 1997) are:

- UK 90–125W/m²
- Spain 160–220W/m²
- North Africa/Arabia 240–280W/m²

The average for the whole planet is about 240W/m² (Coley 2008).

There are several interesting statements (derived from simple, average pro rata calculations) that can be useful if understood in context. For example, in one hour the Earth receives sufficient energy from the sun to power mankind's activities for about 12 months.

It is, of course, not always practical to utilise available renewable energy, but a key part of the design strategy is to assess its potential usage in the context of the other options and design imperatives to reduce usage of fossil fuels.

1.10.3 Adopting a whole-life approach

An important concept for sustainability is to address issues, and make decisions, with respect to the entire life cycle of the building and its services. This whole-life approach is equally applicable for any manufactured product. Figure 1.7 shows a greatly simplified building life cycle; the inception to handover period is, of course, a much smaller portion than that shown. The stages from inception to handover include the brief development stage to determine and

Figure 1.7 Simplified building life cycle (not to scale)

clarify the client requirements. They also include the various design stages, procurement, construction, testing and commissioning. Typical design stages are described in Chapter 2. The operational part of the life cycle is where the building is occupied and the active engineering systems are operated and maintained. As the typical economic life expectancy of a building is typically two to four times the life expectancy of the active components of engineering systems, items of plant and equipment will undergo replacement at different times, either due to life expiry or as part of refurbishments or upgrades. The whole building will, eventually, be considered for part or complete renewal. The proposals developed at the inception stage, and the arrangements at handover stage, should address the needs during the subsequent continual operate/refurbish/operate cycle and the eventual renewal strategy. Adaptation for inevitable climate change should form part of the proposals. To properly address whole-life aspects, designers should, ideally, consider the ways in which similar buildings and systems have performed following occupation, as outlined inChapter 4. Traditionally, only limited data were available on the energy performance of UK buildings across the whole building stock. More recently, the National Energy Efficiency Data-Framework (NEED) reports have provided useful information about energy usage and efficiency for domestic and non-domestic buildings in Great Britain (National Statistics 2019).

In order to deliver buildings that will perform in accordance with the design intentions, it is necessary to have a particular focus on the period from handover to initial occupancy. To achieve this, there should be a planned engagement that ensures the testing and commissioning results are validated and that the performance during the settling-in period is monitored, so that adjustments can be made to optimise the building performance. This extended involvement is outlined in the BSRIA Soft Landings Framework (BSRIA 2010). Chapter 4 describes the benefits that can arise from undertaking post occupancy evaluation as part of this extended approach.

1.10.4   *Focusing on demand reduction as the priority*

From the awareness outlined earlier, a simple concept model for addressing sustainability in relation to energy usage is shown in Figure 1.8. In this diagram, the demand (i.e. final energy consumed by the building loads) can be considered the output of a range of energy processes whose input is fuel and materials and whose unwanted side effect is the production of emissions: pollution, waste heat, etc., that have previously been described for conventional supply from fossil fuels. For the purposes of this model, the processes can be considered to be all the processes involved lumped together, covering both the supply side (energy generation, transmission and distribution) and the demand side (usage in the building's systems). So in this model, all energy wasted through inefficiencies can be considered as emissions. The total environmental impact can, for convenience, be considered to be that made by the combination of the input and the emissions. The emissions have an impact that includes the paramount threat of climate change, but also other impacts from other pollutants, waste heat and so on. While climate change is without question the principal environmental threat, the resources have an impact arising from the 'winning' of fuel and materials extraction for the whole energy systems infrastructure. The continuing resources usage also brings about a reduction in fuel reserves, and hence the threat to energy security.

It should be emphasised that the three arrows have different measures, as this is a concept diagram and not a conventional flow diagram. However, it should be apparent that the most appropriate starting point for reducing the environmental impact is to reduce the demand; the input and emissions are only there to meet the demand (and – for the same processes – would be roughly proportional to the demand). The next priority is to make all the energy

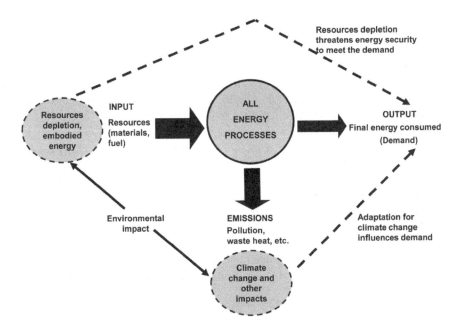

Figure 1.8 Concept model

processes as efficient as possible and use sources that have low or zero carbon impact, so that the proportions of energy input and carbon and other emissions per unit of useful energy consumed are reduced. An aspect that will influence the demand arises from the inevitable climate change impacts and the need to design for adaptation to suit. Solutions will therefore need to be energy-efficient for adaptation scenarios during the life cycle.

Energy efficiency – meaning the mix of demand reduction and improving the performance of systems – has been called the most important climate solution, because it is the biggest resource, the cheapest resource, the fastest to deploy and its potential never runs out (Romm 2018). There are, of course, certain limitations in the extent to which building services designers can influence energy processes on the supply side, which are primarily determined by international and national political scenarios, conventional energy suppliers and regulatory frameworks, and on the demand side, which relate in large part to the clients' functional needs for the building. Strategies for building services designers will largely focus on those aspects of occupancy behaviours, functional needs, building construction characteristics, energy processes and systems where they can exert an influence.

1.11   Summary

In this chapter we have reviewed some of the key background issues related to the climate change aspects of energy usage and the likely impacts.

The most significant threats to the Earth's life support systems have been identified in recent scientific studies. Anthropogenic climate change is the paramount threat. Human-induced climate change is a global and unprecedented challenge that could have catastrophic impacts, the primary cause being the emission of carbon dioxide from the use of fossil fuels. The built environment has a carbon impact due to both operational energy in use and embodied energy related to materials usage. The scale and urgency of the 'climate emergency' and the need to drastically reduce carbon emissions require major changes in the design of buildings and their energy-consuming systems. Because of the delayed effect of CO_2 in the upper atmosphere, climate change is an inevitable outcome from carbon emissions in previous decades. There are, therefore, dual imperatives for building design related to climate change: rapid carbon mitigation to prevent additional future climate change and adaptation to the climate change that is inevitable.

There has been a growth of renewable electricity generation in many countries, but there are concerns that infrastructures for generation and transmission of electricity may not be adequate for the necessary further expansion of renewable and low-carbon sources and changes in electricity usage. Rapid reduction in energy demand is therefore essential not only to mitigate climate change impacts but also to maintain security of supply. To understand the potential to reduce carbon impact from buildings, it is necessary to understand the patterns of operational energy usage and material usage, which will inform proposals for energy efficiency and carbon reduction. Design approaches should adopt key principles for holistic and sustainable design, including a whole-life approach to decision-making. The starting point for strategies for energy-efficient and low-carbon buildings should be demand reduction.

Note

1   Unless stated otherwise, all further figures are attributable to the authors.

2 Interdisciplinary design collaboration for energy-efficient buildings

2.1 Introduction

This chapter provides an outline of the various collaborative activities through which the design of building services takes place as an integral part of the wider interdisciplinary development of the building design, alongside wider involvement in influencing the energy performance. Particular emphasis is given to the key aspects and decision-making that influence the building services design strategy, and hence the energy and functional performance of buildings. As the design of buildings is an interdisciplinary activity, the chapter starts with a brief description of the traditional design team structures and the roles of the principal design disciplines.

The development of the brief, from the client's initial brief, is an essential early-stage activity. Key elements are outlined, together with their implications for the engineering systems and energy performance. The essential objectives for this development require an early involvement to influence design concepts. The building services engineer's typical design involvement at each stage is described as a series of interconnected activities that can form part of the collaborative development by all disciplines. Typical areas of involvement are loosely outlined against each design stage, based on the intended outcome only, rather than the software tools, digital methodology and information exchange that might be used.

Designers have to work within relevant codes, standards, regulations and legislation. Key aspects are described, with particular emphasis on the designers' duties for health and safety (H&S) management. Design is a highly professional undertaking, with considerable potential liabilities, so all design work has to be methodical and meet specific quality criteria. The nature of quality management processes for designers is outlined, with an emphasis on the need for auditability of design decisions through all stages.

It is emphasised that this chapter is about the generality of what needs to be done to achieve the energy performance objectives, rather than the specifics of how it is done – the sequence, methods and tools that are used. As such, the chapter does not cover the design process or building information modelling (BIM).

There is widespread use of BIM software as a design tool and throughout the whole project process to construction. BIM uses a variety of software tools to create a single information source for use by the project team, with all designers contributing to the 3D model. It creates an intelligent model in which each component has a set of pre-defined properties, so it is like building in the computer, allowing an accurate representation of the design (Makstutis 2018). This allows a high level of integrated and coordinated design, which allows the processes for design, and for construction, to be better managed with a greater degree of accuracy and consistency. This facilitates closer collaborative working across all parties involved in the project to make the whole process more effective and hence more cost-effective. For all disciplines, this allows improved flow and accuracy of information. It would be expected that the majority of project teams would be collaborating using BIM to achieve the interdisciplinary design outcomes described in this chapter, but other project teams might still be using more traditional methods.

2.2   Design team structures and roles

2.2.1   Design teams

The design team – or professional team – on a construction project has traditionally provided a consulting service that links the client – as the procurer of the building – with the contractor, who constructs the building. This relationship has become more blurred and complicated in recent times with the expansion of 'design and build' procurement methods, where the contractor undertakes the design as well as the construction, often through 'novation' of the consultants who prepared the initial design proposals. In a similar way, the construction for certain types of buildings, particularly commercial office buildings, has become more complicated in the UK with a separation of the 'base-build' or 'shell and core' construction stage for the developer from the fit-out stage(s) for the tenant(s). Sometimes it is even more complicated, with a landlord's 'Category A' fit-out stage and then a tenant's fit-out stage. However, for the purposes of simplicity, to understand the typical design team responsibilities and to convey the underlying principles, it is still useful to think of them in terms of their traditional roles.

The design team is, collectively, the designers for the building, but it is usually the case that they also act, to some extent, as agents and advisers to the client. The traditional construction contract is between the client and the contractor. The design team also acts, in effect, as intermediaries and protectors of the client's interests in their relationship with the contractor, as shown in Figure 2.1.

The design team is multi-disciplinary and varies in size depending on the type, scale and complexity of the building project. It can vary in size and composition and will change with the stage of the design process. As with any multi-disciplinary team, it will only succeed if all members understand their own role, and that of the other members, and the disciplines work together with mutual respect. Figure 2.2 shows a traditional design team structure of consultancy organisations.

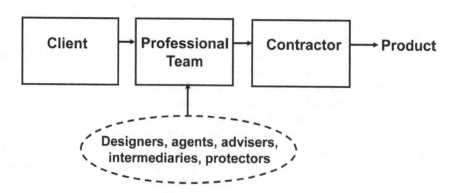

Figure 2.1 Traditional design team relationships

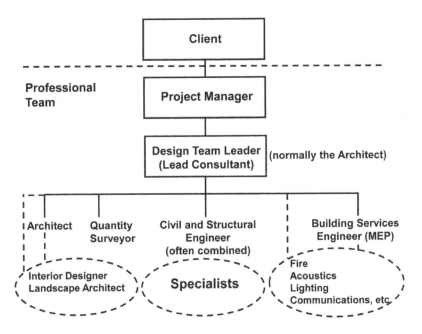

Figure 2.2 A traditional design team structure

The principal design disciplines are architecture, civil and structural engineering, and building services engineering. The cost consultant (which has largely superseded the term 'quantity surveyor') also forms part of the team, although cost management is not a design discipline. The architect is usually appointed as the lead consultant, and hence the design team leader (although other disciplines can hold this role, usually for projects of a non-standard nature where the architectural content of the work is much less than the engineering content). In many cases there are also a variety of specialist designers. These could be appointed directly by the client or could be sub-consultants to one of the main consultants.

It is essential for the appointments to be integrated so that they collectively provide coverage for all the required duties and so that each consultant has a clear understanding of their own duties and how these relate to the duties of the other designers. It is also essential for the delivery team and the client to work together in a collaborative and coordinated manner.

In order to understand the role of the building services engineer in the context of the team, it is first necessary to understand, in simple terms, the roles of the other key design team disciplines.

2.2.2   The role of the architect

Architects have a wide-ranging role that usually includes responsibility for the appearance, shape, form, space usage, finishes and planning for the building, and may include other responsibilities. There is a need for a spatial solution of some form, and architects design such solutions (Makstutis 2018). They have traditionally been the lead designers and professionals on building projects, and their appointment is usually as the lead consultant. Much of the documentation and procedures in the industry reflect this role. As design team leaders they are therefore concerned with all design disciplines. The role of the architect includes strategic organisation of a wide range of knowledge and processes (Moe 2008), so there is a coordinating role alongside the design role (Makstutis 2018). Architects are the primary coordinators, having a high level of involvement in all aspects of the building design process.

The architect will lead with the initial concepts and will, in effect, often act as the overall arbiter on design decisions (although project managers might have this responsibility). From an aesthetic viewpoint, the architect will aspire to a consistent statement or style and seek to maintain the integrity of the design intent (Makstutis 2010). The architect will seek to create balance, harmony and unity of appearance, and therefore try to minimise any visual intrusion of structure and services that might detract from the aesthetic aspirations. In some cases, for certain types of buildings, an architect might seek to express the structure and services elements to achieve a particular aesthetic. Architects have specific responsibility for obtaining the approvals for planning and building regulations. They usually undertake the contract administration role through the construction period, although other disciplines can also act as the contract administrator.

At the commencement of a project, the architect has an underlying need to understand the nature of the client's requirement at the most fundamental level to provide a starting point for developing the initial concepts. This can include observing existing operational patterns and use of spaces and the nature and type of functional activities that will inform the briefing requirements.

From the outset, the architect will seek to create an overall vision or concept to define the character of the proposals. This means the idea behind the form that is eventually developed into a coherent appearance (Makstutis 2010). The architect will usually undertake initial research from existing building designs and produce sketches to assess options for the basic massing and form. The architect will often create models (physical as well as computer) and be involved in discussions with interested parties and stakeholders. These will lead to the first, fairly sketchy, outline design representations as a proposed response to the site and the brief. This will include consideration of the location on-site of the building, or buildings; how they integrate with the landscape and circulation routes within and outside the site; and their massing, shape and texture. From an architectural perspective, each project tends to be considered a unique design challenge. The development of the design evolves through a continuous process of iteration for all disciplines toward a resolution (Makstutis 2010).

As the design develops, the architect will progress from loose sketches that represent the project's basic idea, through numerous stages, to more detailed drawings or a BIM model. This will start to identify and quantify the sub-divisions, circulation routes and adjacencies between spaces to meet the functional relationships. The model or drawings will also illustrate key aspects of the proportions, volume and relationship of the building(s) to the site and the surroundings. All design team members add their design information to the model, or drawings, to create their designs.

It is an architect's responsibility to specify the range and style of construction materials. The objective will be to choose a suite of materials and textures that create a coherent visual resolution in relation to the overall aesthetic adopted (Makstutis 2010). As the design team takes the project through each successive stage, the design will evolve until it represents a more definite proposal. It will be the architect's duty, when acting as design team leader, to take overall responsibility for directing the interdisciplinary design process, accepting that each discipline will have a team or project leader in their own right. This will normally involve chairing design meetings on a regular basis to handle the flow of design information to the other consultants. It will also include managing the resolution of client comments, brief development and other brief design changes (Makstutis 2010).

Architects have a unique role among design professionals by virtue of their involvement in the planning process and often act as the master planners for large site developments (Makstutis 2010). 'Planning' is the process by which a proposed building or development is given consent or approval by the local authority. All other design professionals will normally be involved in contributing to the planning proposals and the planning drawings, but the architect provides the leadership and coordination. This is described in more detail in Section 2.7.1.

2.2.3   The role of the civil and structural engineer

It is normally the case that a single consultancy will be appointed for both the civil and structural engineering design duties for buildings, and this is often under the title of 'structural engineer'. In its simplest sense, the civil engineering aspects relate to construction work within the ground below and around buildings. In the wider construction field, this scope covers everything from roads, railways, runways and tunnels to drainage, sewerage, sea walls, docks and harbours. In relation to buildings, civil engineering normally relates to all below-ground or 'sub-structure' work, including:

- site excavation, re-distribution of spoil and levelling
- foundations for the building, including piling, to suit the particular ground conditions, or 'geotechnics'
- basements and below-ground enclosures
- below-ground drainage and sewerage and site drainage (although this depends on the appointment and is sometimes the responsibility of the building services engineer)
- watertightness of the sub-structure in relation to the groundwater conditions to prevent water ingress into the building
- hard surfaces, such as roads and car parks
- concrete ducts and other construction within the ground to house building services plant, equipment or distribution elements.

Civil/Structural engineers normally take responsibility for the external design coordination of all disciplines, covering the work they design along with landscape architecture features and external work designed by the building services engineer. This is similar to the role of the architect who deals with internal design coordination.

Civil/Structural engineers have an objective to provide stable foundations to support the load of the building. There can be a lot of building services equipment located below ground, together with incoming services and site distribution, as well as the associated penetrations of the sub-structure. There is therefore a considerable amount of design integration involved.

The civil engineering work is the first activity as part of the construction process. It can often be lengthy and represent a significant proportion of the overall construction period. It is therefore essential for the design of civil engineering work to be completed as early as possible. This is so that construction can commence on-site, even if the design of other above-ground elements is still being resolved. The building services engineer will therefore be required to provide key information for sub-structure elements at an early stage.

The structural engineering aspects often refer to the structure of a building above ground level. This is generally known as the 'super-structure'. This normally comprises a structural frame built up from, and supported by, the sub-structure foundations. In a multi-storey building, the structural frame is usually made of one or more cores, steel or reinforced concrete columns (or a mixture of the two), reinforced concrete floor slabs and steel or reinforced concrete beams.

Civil/Structural engineers are usually also responsible for miscellaneous structures within the overall structural frame of a building, such as primary supports, bases and platforms for plant and equipment. They need to have an awareness of all load disposition and load movement within the building, whether designed by themselves or not. This will include architectural elements, building services equipment (and often the vibration and thermal expansion associated with it), people and external forces such as wind. The structural engineer will seek to guarantee the structural integrity of the building in an economic way, while also seeking to realise the other design aspirations. This would include trying to create clear open spaces to maximise flexibility for usage of the space by minimising the number of columns within occupied areas. To achieve this in multi-storey buildings, there is usually a trade-off between the number of columns and the depth of the floor slab and/ or downstand beams, and hence the floor-to-floor height.

From a building services engineer's perspective, in addition to the design integration for plant rooms, a key aspect of integration is the incorporation of vertical and horizontal spaces for services distribution. This requires vertical risers, usually incorporated within structural cores, and pre-formed openings in structural elements such as reinforced concrete and steel. All such spaces need to be designed in, and allowing for these clear spaces and holes is a major design consideration for the structural engineer. This is discussed in Chapter 13.

2.2.4 *The role of the building services engineer*

From the brief descriptions of the roles of the other principal design professionals outlined earlier, it can be seen that the architectural and civil/structural engineering designs alone would create, in effect, enclosed spaces, aesthetic treatment and internal finishes, together with the associated sub- and super-structures. However, without active engineering systems, such spaces would, for the most part, be inert. In its simplest sense, the role of the building services engineer can be considered to be primarily about designing features to make the internal spaces habitable and safe. This could perhaps be better considered as facilitating the desired functional performance within spaces through integrated features that form part of an overall coordinated building design. But it is much more than that; it is the work of the building services engineer to allow spaces to function in their intended manner throughout their period of occupancy, with a high level of performance, while minimising environmental and resources impacts, so sustainability is a common theme in their scope. It is also about promoting good health and wellbeing through a user-centred approach to design. An essential aspect of the role is the interdisciplinary involvement so that the resulting outcome satisfies the objectives of the different disciplines.

Building services engineers are involved in building physics and influencing the design of the passive features of the building that help to create building performance without the use of active engineering systems and energy. They therefore have a crucial involvement at an early stage to contribute to and influence the development of those aspects of the architecture and structural engineering that are integral to building performance, such as facades and the building envelope more generally and thermal mass. Beyond this, the design role largely relates to the mechanical, electrical and public health (MEP) systems that create the required internal environment, or climate, to satisfy comfort criteria (thermal, visual and aural) and indoor air quality; provide electrical services for power-consuming equipment, as well as water and sanitary services; and provide life-safety facilities to protect life in an emergency. This is shown in a simple diagrammatic form in Figure 2.3 as the traditional part of the spectrum of building services engineering. In addition, there has always been a need to provide facilities for

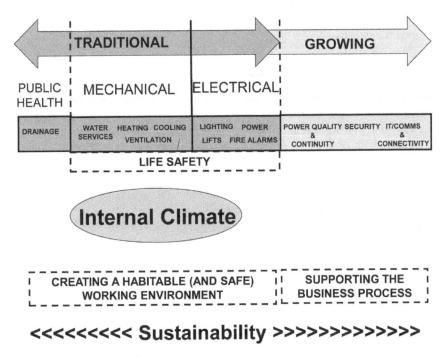

Figure 2.3 The spectrum of building services engineering

the processes undertaken in the spaces (although the process equipment itself does not normally form part of the fixed building services). More recently, there has been a growth in the spectrum of systems, with a particular focus on the quality and continuity of power supplies and a range of security and communications systems to realise the concept of an intelligent and responsive building. This often arises from a need to support modern, and often complex, business processes alongside maintaining a suitable internal environment. The imperative to achieve sustainability requires an involvement to consciously design proposals to address energy efficiency, carbon mitigation, energy security and adaptation, as outlined in Chapter 1, and wider aspects of sustainability, such as those relating to water and waste. Sustainability is therefore an overarching consideration across the whole spectrum.

The engineer will be seeking to satisfy the client's functional objectives in an economic way while supporting the overall architectural aspirations and needs of the users for health and wellbeing. For the outcome to be successful, the role should not be limited to design, construction and handover; it is equally about creating buildings that will achieve predictable performance for the likely climate scenarios and that will be amenable to effective operation and maintenance. Thus, building services engineers need to consider the whole-life issues, including aspects such as testing, commissioning and optimising initial occupation, and the strategies for operation, maintenance, upgrading and renewal throughout the building's life.

It is usually the case that engineers leading projects have either mechanical or electrical engineering as their primary discipline. Mechanical and electrical engineers tend to act as generalists, but with an expectation that they should have sufficient awareness of the whole building services discipline in order to make overall decisions covering the whole spectrum of systems. Alongside the generalists, many specialist sub-disciplines have emerged, such as public health (plumbing), acousticians, lighting designers, vertical transportation and communications engineers.

2.2.5 The role of the cost consultant

The cost consultant (previously more often described as the quantity surveyor [QS]) is responsible for all aspects of the cost management of construction projects. Most projects will have a formal cost plan that sets out the plan for expenditure on each element of the construction works, together with costs for associated aspects, including fees, applications and consents. The scope of the role covers a wide range of financial and contractual issues, including:

- preliminary advice on economic and investment aspects of a construction project
- initial estimating and creating the cost plan
- updating cost estimates and the cost plan through each design stage
- contract selection and creation, including creating the contract preliminaries
- cost monitoring, control and reporting throughout the design and construction stages
- staged evaluations of progress during the construction stage to allow staged payments to contractors in accordance with the contract
- creating the final account at the conclusion of a contract
- general project financial advice to the client on matters such as taxation.

The cost consultant will maintain the cost plan as the primary reference document for cost control throughout each stage of the project's design iteration.

2.2.6 The role of the project manager

Within the construction field, the traditional role of a professional team's project manager usually relates to an organisation or individual whose specific role and responsibility to the client is to manage the overall process for delivering a complete project. It is separate from the internal management of a project within one of the design disciplines (which tends to be known as 'project leadership').

The project manager's role has grown considerably in prominence in recent years. The project manager is usually appointed at an early stage to work with the client to understand their intentions and help them formulate a suitable plan for delivering the objectives of the complete project. In this context, 'complete project' relates to the overall business objective of the client. This could, for example, cover the movement of a department within an organisation and not just the creation of a new building or buildings for them, although this might be the principal element of the project.

A project manager's role within the client's professional team usually covers a wide range of management matters, including:

- direct and close liaison with the client to ensure that the project is delivered in line with the client's requirements, programme and budget
- assembling the design team and arranging their appointments
- determining the client's initial objectives and requirements and briefing the design team
- directing the design team leader as the client's objectives and requirements evolve
- determining the procurement method for a contractor, in liaison with the cost consultant and other design team members

- determining the programme, together with the project procedures for the design team and others, which would usually include meetings, communication methods, information exchange and management of change
- overall control of the programme of design, procurement, demolition and enabling works (if required), construction, removals, occupation, phasing, etc.

2.2.7   Other specialist design disciplines

Depending on the type, scale and nature of the project, there might be a need for other specialist designers and consultants.

In the UK, all projects above a certain size will have a construction design and management (CDM) coordinator to effect the coordination of H&S management for the project. This role is described in Section 2.7.3.

Where an environmental assessment methodology is being used, there will also be an assessor for the chosen method of environmental assessment. This role is described in Chapter 4.

Interior designers are often involved in the spaces within buildings that require particular attention to the design ambience, aesthetic, style and finishes. This could include most areas in high-quality hotels and foyers and reception areas in office buildings. They work closely with the architect to achieve an overall coherent interior design solution. They can play a major role in prestige hotel projects and the more high-quality residential projects.

Landscape architects are involved in external areas of the site covering aspects such as planting, landscape form, paving and water features. They could be involved in selecting materials for soft and hard landscape surfaces.

On some projects there might be catering consultants, audio-visual consultants or various other specialists.

2.2.8   Facilities managers and operational engineers

All organisations that own or lease property have to operate and maintain their facilities once they have been constructed and 'handed over' by the contracting team. This function is sometimes called 'aftercare' and, in its broadest sense, is known professionally as facilities management (FM). The specific aftercare management role for the MEP engineering systems is sometimes known as 'operational engineering'. Operational engineers need to have a thorough understanding of their property and MEP systems and maintain up-to-date record information. They usually have or employ specialist personnel to:

- operate the plant
- maintain the plant
- undertake an appropriate maintenance regimen, such as planned preventative maintenance, to make effective use of the facilities and provide a reliable service, which would normally include planning to minimise the incidence of failures and the need for breakdown maintenance
- undertake minor works, adaptations and modifications.

The record information would include drawings of the installation 'as-built', together with logbooks and operational and maintenance manuals, which, taken together, should provide all the necessary technical information required to run the systems as intended by the designers.

Although they are not part of the design team, it can be highly beneficial for the design team to engage with the client's operational engineering team at an early stage in the design process, so that operational and maintenance issues can be properly addressed in the design. See Chapter 3.

2.3 Design appointments and work stages

The designers appointed by clients to undertake design services for building projects often have a standard general contractual agreement for their professional services. In the UK, standard appointments for consulting engineers in building services engineering are often based on the suite of agreements of the Association of Consultancy and Engineering (ACE). The appointments for the whole design team should be arranged so that the full range of duties is covered across all the disciplines and so that there is an unambiguous allocation and demarcation of responsibilities. In the majority of cases it is likely that the architect will be appointed as the lead consultant under the appropriate Royal Institute of British Architects (RIBA) agreement, with the building services engineer appointed as a non-lead consultant for mechanical and electrical engineering services design.

The standard range of engineering systems to be covered would typically comprise:

- mechanical systems
- electrical systems
- public health systems
- fire protection systems
- incoming utility supplies
- lifts and escalators
- acoustic control of noise arising from engineering systems.

It is often the case that other responsibilities will be included as a schedule of 'additional services'.

In order to achieve a sustainable outcome, building services engineers should seek to be involved in projects as early as possible so that relevant sustainability issues can be raised and so that opportunities are not lost to influence the disposition of buildings on-site, as well as the form and envelope of buildings, as outlined in Chapter 3 (CIBSE 2007). All design team members should work in a collaborative way to deliver a design that is integrated and energy-efficient (CIBSE 2004a).

Design is usually undertaken in a series of defined stages, or work stages, and in the UK this is usually based on the RIBA Plan of Work 2013 (RIBA 2013). The plan has eight stages, 0–7, of which five are related to definition, preparation and design:

Stage 0: Strategic Definition
Stage 1: Preparation and Brief
Stage 2: Concept Design
Stage 3: Developed Design
Stage 4: Technical Design
Stages 5–7 cover post-design activities: Construction, Handover and Close-out, and In Use

The plan notes that it embraces the principles of sustainability, and throughout the document sustainable design processes are integrated with the stage-by-stage activities. The plan recognises the continuous nature of the building cycle and promotes improved briefing processes and encourages the use of beneficial feedback from completed projects to inform decision-making. Although specific design activities are listed at different stages, the plan notes that it might be necessary to vary the process or have an overlap between certain stages, depending on aspects such as the procurement route or the timing of the planning application (RIBA 2013).

Defined work stages of this type represent a standard framework that provides the client with an awareness of the process, together with the expectations for the outcome at each stage. They also provide key points for payment of staged fees for the design team and signing-off approvals (Makstutis 2010). It should be recognised that the sequence of the work stages or the content may vary, and they could overlap to suit different procurement methods (RIBA 2013). The intention at each stage is to develop the team's design proposals to a position whereby as many of the considerations as possible have been resolved. Following completion of each stage, it is normal to seek client approval of the proposals and consent to proceed to the next stage. Traditional programmes have a sequential process through the design stages to procurement, construction and occupation, with formal client sign-off at each design stage. Business imperatives have meant that 'fast-track' programmes have become commonplace, and there is often pressure to overlap stages to achieve earlier completion.

2.4   Brief development

2.4.1   The nature of the briefing process

An essential activity for the whole team at the beginning of the project process is to liaise with the client to develop the brief. From the building services perspective, the brief should be developed to a sufficient level so that the full extent of the client's functional requirements, and the key performance criteria influencing the engineering systems, can be determined. The briefing criteria should ideally be agreed to the most detailed extent possible during the appointment stage. However, in reality, there is usually a more formal development of the brief during Stages 1 and 2 and usually some ongoing development during subsequent stages. It is inevitable that some of the detail of the brief will continue to evolve as the client's needs change. However, to facilitate design in the most effective way, the key factors should be recorded and agreed to as early as possible.

In many cases, a client might be unsure about the specific functional performance that they need or may not be able to articulate it clearly. It is important for the design team to critically examine and question the initial brief in a rigorous way. This is so they can get to the root of the need, rather than accept and adopt a pre-ordained or preferred starting point that implies a particular solution that might, in reality, be inappropriate. This aspect of client collaboration will be essential to achieving a sustainable solution.

An important source of information for the briefing process is feedback from existing buildings. This includes feedback from the users and on the operational performance. This real-life information should be a primary reference and is, in many ways, more valuable than theoretical design guidance. Chapter 4 describes how post occupancy evaluation from similar buildings can provide a useful input for the design briefing and help designers to achieve low-carbon building performance in practice.

It should be expected that the design brief will continue to evolve as the project moves forward. In reality, it is often continually developed right through the staged design and approvals process until the design can be said to be in an agreed and finalised format. Thereafter, the nature of development is primarily related to information production for use in the procurement process (RIBA 2013).

2.4.2   Standard and institutional design criteria

The design of many engineering systems can be progressed, within reason, by reference to standard published criteria by professional institutions, such as the Chartered Institution of Building Services Engineers (CIBSE). These criteria will provide the normal design parameters for generic or common types of spaces. More specific guidance on criteria and acceptable design solutions for particular building types can be found in publications of the relevant public-sector bodies for health, education and other public buildings. In the private sector, many large organisations – such as large hotel or airport groups – have their own standard design briefing publications. There are also publications by specific property-sector bodies, such as the British Council for Offices (BCO), which covers the offices sector. Many client organisations, particularly investment organisations, will often defer to such established 'institutional criteria', as these are seen as the norm for the industry and therefore represent a safe and reliable way to protect the value of their asset. However, every building is unique to some degree, and there is a requirement of the client to state specific objectives and criteria in order to describe the desired functional performance of the building and its engineering systems.

2.4.3   Specific briefing development

Building types that are well established and that do not deviate markedly from the norms for their type might require little in the way of brief development. However, building types, or parts of buildings, that are inherently complex in nature or that tend to have uniquely defined criteria are likely to require considerable brief development. Such buildings might include laboratories, research centres, data centres, factories, entertainment centres, exhibition centres, conference centres, museums, media centres, industrial facilities, archives, distribution and storage buildings. Some of these buildings contain spaces that are primarily dedicated to equipment or materials with minimal occupancy levels. They might have design criteria that are markedly different from those for spaces that are primarily for occupancy and whose criteria largely derive from human comfort parameters. It is therefore most important to establish any non-standard criteria prior to commencement of the design. Typical aspects that might be clarified during the briefing process are described in the following sections.

Numbers and types of occupants

The anticipated numbers of occupants for the whole building, and individual spaces within the building, together with any transitory aspects, should be confirmed so that concurrent occupancy needs can be determined. The types of occupants will also be relevant, as this will indicate their activity and hence the related comfort criteria and any special considerations (such as H&S or quality of provision). The nature of activities will also determine the functional requirements of services, for example, power and data provision for workstations.

Functional performance of spaces

Each type of space should be defined according to the performance requirements. This might be directly in terms of design parameters or indirectly, as a description from which an appropriate design parameter can be proposed. It is normal to summarise the criteria on a room data sheet for each space. This would record engineering design parameters for thermal, lighting and acoustic performance, together with facility requirements for power, communications, public address, security and so on.

Special equipment or facilities

It is often the case that a client will require certain items of proprietary equipment to be incorporated within the building. This could be process equipment, production equipment, computers, audio-visual facilities or other specialist or bespoke equipment. The technical details, connection facilities and so on should be defined.

Facilities management and operational engineering requirements

So that the engineering systems may lend themselves to successful operation and mainte-nance throughout their life – and hence provide sustainable operational energy efficiency – it is essential for the design to consider the client's operational and maintenance regimen. This aspect of the brief development is likely to require extensive engagement with the client's facilities management or operational engineering staff. It should include the propos-als for energy management. It should also include the requirements for such access, facili-ties and spaces that will be necessary to allow effective testing and commissioning; effective monitoring and data logging of performance; planned preventative maintenance or other maintenance regimen; plus inspection, operation and periodic partial or full replacement of equipment. This will include metering, monitoring, data collection and automatic controls, as described in Chapters 3 and 10.

Phasing

The extent to which the construction work, and associated activities, might be undertaken in more than one phase must be described. This is often the case on large projects. It is some-times necessary to provide enabling works or temporary facilities as a prior contract to effect conventional or phased construction.

Expansion and flexibility

This relates to allowances that must be included within the design to accommodate future expansion of the facility and flexibility in the way the building and the individual spaces will be used. This is sometimes called 'future-proofing', although the term has much wider connotations.

Environmental target

The environmental assessment methodology should be agreed on. The target should be agreed on, and the design team should work with the client to assist in this process. The client should be encouraged to set a high target (CIBSE 2007). This is an important step in providing clear objectives for the design and also when creating a building with identifiable sustainable credentials (see Chapter 4).

One example of where it is necessary to develop a highly specific brief is when determining the level of resilience for engineering systems for buildings housing business-critical facilities. Certain engineering systems will require a particular level of 'resilience' or 'redundancy' (CIBSE 2008a) to provide the desired level of reliability or availability of operation. This is particularly the case for facilities where continued operation is essential for security or business reasons. This would include data centres;communications centres;transport control centres; and certain industrial, security and defence facilities. This is different from the criteria for life-safety facilities that are defined in statutory regulations. Systems' resilience is usually defined in terms of either a numerical reliability or availability level or using the well-known N-based redundancy statements, such as 'N+1', 'N+2' and '2(N+1)' (CIBSE 2008a). The definition of such statements, as understood by the client, is all-important and should be agreed to explicitly. It is preferable to use industry-standard definitions, such as the tier levels described by the Uptime Institute or other institutional body. The resilience definition should cover not only the primary function requiring continued operation – such as electric power for data processing facilities and equipment – but also the relevant support systems, such as cooling and controls, upon which the primary systems' function will also be reliant for continued operation.

The importance of defining and agreeing to resilience criteria in the design of these facilities cannot be over-emphasised. It will have a fundamental impact on the numbers, sizes, arrangement and types of equipment such as transformers, switchgear, uninterruptible power supply (UPS) modules, generators, cooling equipment and other ancillary equipment. This will, in turn, have a fundamental influence on the plant space requirements (and potentially the shape and size of the building), planning issues and capital and running costs. It will also have a major influence on operational energy performance due to the usually lower overall energy efficiencies of multiple parallel items of the plant sharing the load. It will also have a carbon impact due to the increased influence of the embodied energy in relation to multiple parallel items of equipment and their operation and maintenance demands. System resilience criteria can therefore fundamentally influence the whole-life carbon impact of certain types of buildings.

2.5   Design objectives for building services engineers

2.5.1   The essential aspects

Typical design objectives that might be appropriate at each stage are discussed in Sections 2.5.2 to 2.5.5. However, it is probably more important to understand the *essence* of the overall design intention and what is trying to be achieved, rather than just thinking in terms of specific activities at each stage. The key behaviour that should be established from the outset is for all disciplines to work as a team in a collaborative way. This will allow the design to develop in a way that addresses mutual objectives and aspirations, rather than as an outcome of different interests pulling in different and contradictory directions. In particular, the three principal design disciplines of architect, civil/structural engineer and building services engineer need to work in harmony as a team with a common understanding and mutual respect.

Figure 2.4 shows some of the key objectives for the design team to achieve whole-life performance in the widest sense and not just in relation to sustainability and energy. The architect will be seeking a rational spatial arrangement and logical interrelationships for the different functional spaces required to meet the brief and will be aiming for a satisfying aesthetic outcome that works in a wider sense, including planning acceptability. The civil/

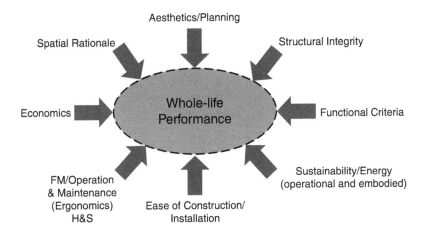

Figure 2.4 Key objectives for design team interdisciplinary collaboration

structural engineer will be seeking to enable the architect's ambitions while providing integrity of the structural solution. The building services engineer will be seeking to provide functional performance of the spaces to satisfy the brief and to do this in a sustainable way with minimal environmental impact, mindful of both operational and embodied energy. All the designers will want to achieve ease of construction and installation, as will the contracting organisation(s). Similarly, all the designers will want ease of operation and maintenance, as will the facilities team, mindful of ergonomics for their operatives and satisfying health and safety matters. The cost consultant – and the design consultants – will be aiming to provide an economic solution for the client (in whichever way that may have been defined). But more than this, all of these objectives are interrelated, so all parties (client, design team, contractors, FM team) are, to a greater or lesser extent, involved in and trying to achieve a combined outcome that achieves mutual success in all of these objectives.

It is important to understand that building design is a highly iterative process involving the development of concepts, ideas and proposals from the different disciplines, and there is regular re-working and refining of those proposals as the design evolves toward an integrated and coordinated resolution. Figure 2.5 shows a greatly simplified view of the design process, emanating from the need, which should be clearly identified through the brief development process. Ideas develop through concepts to a form for the enclosed spaces. There will be division of the form into zones – designated spaces or areas. This leads to proposals for treatment of spaces and selection of appropriate systems. System design includes decisions on the numbers, types and arrangements of terminal devices, with suitable duties, capacities or ratings to cover the relevant zones or spaces. Design integration covers building services and their coordination with other elements. As the design continues through integration to the greater level of detail required for eventual design resolution, it will include integration of terminal devices with architectural features and finishes to satisfy aesthetic criteria. In a more general sense, the services design throughout is intended to be robust, discreet, reliable, safe, economic and energy-efficient.

Risk management is an important feature of construction projects. The design approach should continually address risk management aspects as a matter of course. This can include

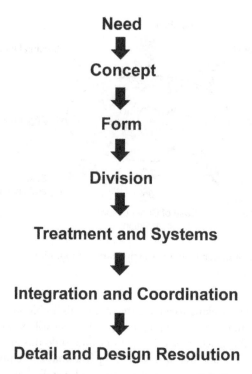

Need

⬇

Concept

⬇

Form

⬇

Division

⬇

Treatment and Systems

⬇

Integration and Coordination

⬇

Detail and Design Resolution

Figure 2.5 A simplified view of the design process

regular, recorded activities to identify risks; create and maintain risk registers; and reduce risks through decision-making as the design develops.

There is an opportunity to bring together holistic influences that can create buildings that will perform in a sustainable way and also meet the architect's aesthetic aspirations at the brief and concept design stages, making these key periods in design. The opportunities for influencing both the design outcome and the whole-life performance reduce sharply with time. The potential for change diminishes and the resistance to change grows, along with the cost of change (CIBSE 2008a).

As outlined in Chapter 1, there is a compelling need for carbon mitigation throughout the whole building life cycle, as well as a need for the design to address adaptation to future climate change. The generic measures for carbon reduction through energy efficiency are outlined in Chapter 3 and expanded upon in later chapters. In simple terms, the key considerations for energy efficiency can be represented as three interrelated elements, as shown in Figure 3.3: the passive elements of the building fabric; the active elements of the mechanical and electrical engineering systems; and the whole-life operation, which will be determined by the management strategy for the building. As such, early design considerations should include durability, buildability and commissionability. Taken together, these can be considered to address the energy efficiency aspects of sustainability that have the primary carbon influence, while accepting that sustainability is a wider objective, embracing many other factors.

From the outset, the building services engineer should be involved in influencing the conceptual development of the shape, form, orientation and performance of the building

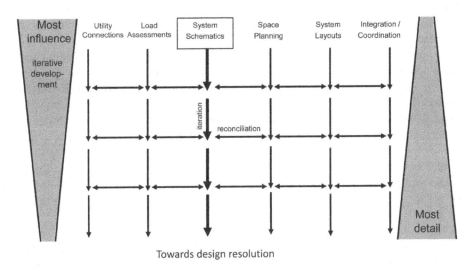

Figure 2.6 Concept for iterative development of the building services design

envelope in order to achieve suitable performance from the passive elements to minimise the need for active engineering systems and their energy usage. As the conceptual development of the active systems commences in line with the energy strategy, it will proceed through iterative development, generally as shown in Figure 2.6. The most important drawing for each system is the schematic, as this identifies the engineering system components and their interrelation. In essence, each system comprises the central or distributed plant, or a mixture of both; an infrastructure of distribution elements from the plant to the relevant branches or circuits; and terminal devices that treat the space or provide a functional requirement or life-safety feature. Alongside these are the controls and components for data collection and monitoring, which will be required to enable successful energy management.

The schematic diagram is the primary reference for the system design; therefore, as any significant change occurs, it is the series of schematics that should be amended first. The design development should ensure that several aspects are addressed concurrently and reconciled at each stage, with the system schematics as the reference: load assessments, which will in turn affect the utility connections required and the space planning for plant and equipment; system layouts in relation to the architectural model or drawings; and the integration and coordination of building services elements with architectural and structural elements, including the interfaces for terminal devices. As each iteration in the design occurs, the same re-working and reconciliation take place across these aspects of the system designs as they move toward design resolution. It cannot be over-emphasised that the earliest design stage is the most influential, but with little detail, while the later design stages have much less influence, but much more detail.

There is widespread adoption of environmental assessment methodologies, such as Building Research Establishment Environmental Assessment Method (BREEAM) and Leadership in Energy and Environmental Design (LEED), which are described in detail in Chapter 4. Under these methodologies, an assessor is appointed and works with the design team throughout the design period to provide continuous focus and guidance on how individual elements of the design might be developed to achieve the credits that contribute to the target rating.

There is an increased focus on health and wellbeing in buildings, with the use of relevant standards (such as the WELL standard as described at Chapter 5) also influencing aspects of the design.

Dynamic simulation modelling is widely used as a design tool to simulate building performance under different scenarios to inform decision-making. It is also often used to demonstrate compliance with Building Regulations Approved Document Part L2A (HMG 2016), as described in more detail in Chapter 3.

Typical areas of involvement are outlined here against each design stage, focusing on the intended outcome rather than the software tools, digital methodology and information exchange that might be used. For convenience, these have been set out against the RIBA Plan of Work stages (RIBA 2013). However, this is a list of selected activities that might be useful and may be relevant to some projects and not to others, and the relation to each stage is fairly loose. It is strongly emphasised that this is not intended to represent in any sense an interpretation of formal building services activities or duties under the plan of work stages, or any agreement or appointment criteria. Instead, it is intended simply as a useful list of building services design actions, loosely related to stages, with a particular emphasis on achieving an energy-efficient solution, together with some wider aspects of integrated design across disciplines. It should be recognised that, in practice, design is not always carried out with strict adherence to the formally defined stages. Equally, design is not necessarily a linear process and so it does not often follow a direct linear path through the stages of development activities on a project (Makstutis 2018). Certain design activities might move from one stage to another – for example, for different procurement routes, or if a decision has been made to submit a planning application at an earlier or later stage than normal (RIBA 2013).

2.5.2   *Stage 0: strategic definition*

This stage involves appraisal and definition of a project in a strategic sense to allow creation of a detailed brief in the next stage. It covers aspects such as the client's strategic brief/business case, assembly of the team and creation of the initial project programme. A key activity that can provide benefits to the sustainable objectives is gathering feedback from similar projects to help inform the brief. A strategic sustainability review is developed, covering aspects such as the client needs and the potential sites that could be available. This early review of needs and sites could involve considering re-use of present buildings, facilities or materials, or other alternative solutions that do not require new construction. This stage seeks to properly consider the key factors relevant to the project scope to allow commencement of the preparation and briefing in Stage 1 (RIBA 2013). For certain types of projects where their specialist insight could provide useful guidance, the building services engineer might be involved in supporting the architect in relevant aspects. This could involve some of the activities outlined in Stage 1, depending on factors such as the project type, scale and complexity.

2.5.3   *Stage 1: preparation and brief*

At this stage the structure of the design team and the roles of the individual design disciplines will usually be established and the team assembled. The design team leader will be identified – usually the architect – and the design responsibilities, interrelationships and lines of reporting will be confirmed. It is likely that any necessary specialist sub-consultants will be identified. The initial briefing will include identifying the client's needs and objectives for the building's function. The team will develop project outcomes along with aspirations for sustainability, based on the initial brief, budget and programme (RIBA 2013).

The client's initial brief should identify their business case in general terms, together with any key factors that might constrain the development. It should also include the client's initial views or aspirations for environmental or sustainable performance, normally in terms of a score or category for an environmental assessment methodology, and a specific aspiration for carbon performance. The design team should encourage the client to aim for a high target. The sustainable performance is likely to include a target for life expectancy of the facilities. The energy strategy should consider the future climate criteria and adaptation. Appropriate feasibility studies will be undertaken and options assessed so that there is sufficient information to allow the client to decide on whether to proceed with the project. As part of a holistic approach, a strategy should be created for the handover activities, which should include aspects such as the type of post-completion activities, which is highly relevant to the building services engineer.

A significant focus is on obtaining a standard range of background documentation from the client in relation to key aspects of the brief and the site. This will allow the identified needs to be developed into a design brief. (See Section 2.4 for an overview of key aspects of brief development.)

For the building services engineer, two site aspects are particularly important for identifying key requirements and constraints: the existence, extent and capacities of utility supplies and physical site restrictions that might impact upon the design proposals. The standard utility supplies are electricity, gas, water, sewerage and telecommunications. The specific utility arrangements will vary from country to country and location to location, e.g. some regions will not have a mains gas network, or in some circumstances, the availability of district heating (or, much less commonly, district cooling) might influence the energy strategy.

Where the client is the occupant of the site, it is likely that they would have record drawings available in relation to the site that will show the nature and locations of existing utilities on the site. If the client does not have this information, it would need to be obtained from the utility companies. This would include the sizes of the infrastructure elements, such as pipes and cables, and the terminations within substations, gas meter rooms and similar utility-owned buildings, rooms and enclosures. Such record information showing the presence of both 'live' and disconnected underground services is also of vital importance to discharging the designers' duties under H&S management, as described in Section 2.7.3.

While the concept design development takes place in Stage 2, it can be useful to develop initial ideas for the energy strategy to help inform the commencement of concept design across the disciplines, specifically with a view to ingraining energy efficiency into the design thinking at the outset. To develop the initial concepts for the energy strategy, as outlined in Chapter 3, it will be necessary to analyse the site opportunities. These would cover aspects such as the annual solar path variations and high and low sun angles; the prevailing wind pattern and direction; and the possible locations for an energy centre, as shown in Figure 2.8. As the early concepts evolve, it is beneficial to undertake simple load assessments for heating, cooling and power (perhaps using 'rules of thumb') that will allow initial enquiries to commence with suppliers for gas and electricity, together with water, drainage and communications connections, as shown in Figure 2.9. This would be particularly important if on-site generation is being considered for projects where there are constraints on the network(s). It is beneficial for the designer to make initial enquiries to the relevant suppliers and network operators to establish the up-to-date scenario regarding available service capacities to feed the development's needs. This should confirm any recent developments to the situation shown on the record drawings, and can provide information on planned or proposed

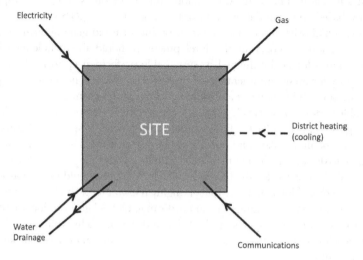

Figure 2.7 Utility connections

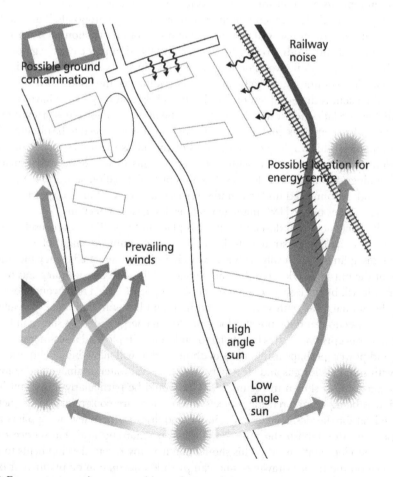

Figure 2.8 Energy strategy: site opportunities and constraints

Source: Reproduced from CIBSE Guide L (2007) with the permission of the Chartered Institution of Building Services Engineers

infrastructure developments in the vicinity that could influence the level of availability for the proposed building or site. It must be emphasised that the availability of services connections of the required capacities can have an influence on energy strategy, and could even determine whether the building is a feasible proposition. Where the existing infrastructure is insufficient, the relevant network supplier(s) should be able to advise on the level of local reinforcement required to provide the capacity of connection necessary, for which there could be considerable cost implications.

Many physical constraints could impact upon the engineering systems design, including access limitations, noise, relative site exposure and shelter, wind, electromagnetic interference (EMI) and air quality. Any access limitations might restrict the maximum sizes of components that can be delivered to the site for the initial construction (or subsequent replacement) and might influence the equipment selection in the design or methodologies for construction. High ambient noise levels might require acoustic features to be incorporated into the site landscape layout, building envelope and heating, ventilation and air conditioning (HVAC) systems to provide attenuation to satisfy internal acoustic criteria. Similarly, a site in an area with very low ambient noise levels might require higher levels of acoustic attenuation than normal in the engineering systems to minimise noise influence from the new development. Sites in different locations can vary widely in their levels of exposure to the elements – from low-level city centre sites which are relatively sheltered, to sites on higher ground and which might have very high levels of wind and precipitation, to marine environments with a saline atmosphere. The ambient air quality might be affected by heavy traffic or industrial activity, again influencing the selection and disposition of HVAC system equipment.

Many of these factors will be relevant to planning matters and the feasibility assessment for utilising renewable energy and might require special features and materials selection in the design proposals. For example, sites close to sources of EMI (such as railway lines or power lines) or radio frequency interference (such as radar facilities) might require special features to minimise any undue influences within the building to achieve electromagnetic compatibility.

2.5.4   Stage 2: concept design

During this stage the team will develop the overall concept design from the initial, fairly loose and sketchy proposals to an outline design of meaningful shape and form. This will include outline proposals for the MEP solutions. Strategies will be developed for the project's sustainability proposals, together with the proposals for maintenance and operation. Along with proposals for the handover activities, these strategies are likely to have a considerable influence on both the overall concept design and the MEP solutions. For the design to be developed to the required level, the building services engineer is likely to visit the site to obtain detailed information on existing services and features. It is also likely that there will be meetings with client representatives for management of the facilities, and any relevant surveys will be undertaken. As the initial concepts for the form evolve, the building services engineer should be involved in contributing to discussions about energy performance and the nature of passive features to minimise energy and carbon impacts, as outlined in Chapter 3.

Communication with authorities is likely to have been fairly minimal prior to this stage, but should now involve more detailed consultation on the key principles affecting the engineering systems, particularly those for energy performance and life-safety (which would be related to the fire strategy and 'means of escape' routes).

While the initial concepts for systems may have been considered at the previous stage, it will be necessary to consider alternative ways in which different systems might satisfy the requirements. The building services engineer should critically examine those aspects of the brief, budget and programme that will have the most impact on the energy performance and engineering services design. The programme should be examined and assessed in terms of practicality for the extent of design work and studies required under the scope of duties in the appointment. For complex projects that are likely to require non-standard solutions, sufficient design time will be required for the design team to undertake the necessary research and specialist studies and to develop and review a range of solutions. Similarly, for sites with inherent complexities – such as congested inner-city 'brownfield' sites or sites of irregular shape, on difficult terrain, in remote areas or with limited access – sufficient design time will be required to assess the impact of these restrictions on possible design solutions.

The building services engineer may move forward with an initial Part L assessment, based on the energy strategy and outline proposals. This is likely to include the HVAC systems approach covering the variations in seasonal conditions. The proposals will be summarised in an outline design which the architects will use, along with outline design information from the civil and structural consultants, to prepare concept proposals. The proposals will represent an integrated, but not especially detailed, solution covering all disciplines. The overall environmental impact of the design proposals is likely to be assessed, covering the construction approach and the options for building materials. As part of the sustainability strategy, the resilience of the design proposals to future climate change are likely to be assessed, which would include the principles being adopted for adaptation. The design concepts should include a strategy for the practical aspects of construction and installation and this should include consideration of the ease of 'buildability' during construction (RIBA 2013).

The previous discussions on the design brief should be continued and concluded. The intention is that, while some further minor development of the brief is inevitable, in effect, it should be considered as concluded at this stage.

There should be some form of preliminary cost plan compiled by the project cost consultant. This is likely to show all the costs associated with the project activity, covering not just the construction value of the building but also aspects such as land purchase, design team fees, local authority charges and fees for statutory authority services connections. There is usually a requirement to produce basic building services cost information at this stage. This might be a simple cost estimate based on unit costs, either $£/m^2$ or $£/$room, or another useful unit basis, or some other form of advice. Although the cost consultant normally has responsibility for the overall cost management (including the costs for building services), it is important to recognise that any cost information produced by the building services designer can be a useful contribution to discussions in relation to this element of the cost plan.

2.5.5 Stage 3: developed design and Stage 4: technical design

For convenience, Stages 3 and 4 are considered together in this section. As with all the design stages, there is, in any case, inevitably some overlap, and some aspects of the technical design may also overlap with the construction in Stage 5.

In Stage 3 the team will provide updated design proposals, including outline specifications. The design work will involve the development, checking and completion of spatial coordination exercises between disciplines (RIBA 2013). To achieve this there is likely to be much iteration, reconciliation and re-working through design exercises and workshops. At the end of Stage 3 the coordinated design is costed, with alignment to the budget. Planning applications may be made at the completion of this stage, so there is development of MEP matters that will affect planning issues. Other activities at this stage are development of a full sustainability assessment and an interim Building Regulations Part L assessment, and there should also be an assessment of resources and waste management opportunities to reduce impacts.

In Stage 4, technical aspects of the design are developed further and completed, along with all relevant information. This provides technical definition of the proposals. Specialist MEP sub-contractor design work proceeds and concludes, with involvement from the building services engineer as required, some of which may be concurrent with construction. There will be an input and review to the design development from the lead designer. At this stage the team should address conditions attached to a planning consent, if relevant, and conclude the Building Regulations Part L submission, with an updated carbon and energy declaration. At conclusion, the design will be complete in its entirety, although some of the sub-contractor design elements might be concurrent with construction. Other information that is prepared includes the logbooks and aspects such as the strategy for conclusion and handover, which should address details of testing and commissioning. Relevant strategies are reviewed and updated, including for construction and H&S (RIBA 2013).

A client's decision to give approval for the design team to proceed to these later design stages is usually an indication that the client has a considerable commitment to proceed with the project, for the process is primarily one of adding more detail to the concept proposals and therefore incurring further design costs (Makstutis 2010).

At this point, the programme is usually of a simple form and should indicate the design, procurement, contractor mobilisation, construction, handover and occupation. It is essential for the team to determine the client's imperatives for the completion date, which might have absolute limitations related to immovable dates or might have an inherent degree of flexibility. It is also important to identify any further complexities, such as a requirement for staged completion with partial occupation.

The development of proposals for the incoming utility services with the relevant providers is key and will receive significant attention. This will involve estimated load (or demand) requirements, as shown diagrammatically in Figure 2.9. Load assessments for thermal and electric power are described at Chapters 11 and 12, respectively. The discussions are likely to cover aspects such as allocated capacities, services sizes, terminations and demarcation, and metering arrangements. Equally important will be the provision of spaces for the service providers' equipment, which will require dedicated rooms or enclosures in accordance with their standard criteria. The nature of space planning for utility providers is different from that for the other building services equipment, as it will require:

* external connections
* external access, usually on a 24-hour basis
* possibly wayleaves or similar reserve rights over the external services routes.

The integration of these spaces forms a fundamental part of design during these stages (see Chapter 13).

For the design of civil and structural engineering to progress, it will be necessary to provide details of the key structural loads related to MEP equipment. This will require information showing the location of the heaviest equipment, with footprint dimensions and mass for each item. The equipment items of most relevance are large water-filled items such as tanks, thermal storage vessels and large pipes, together with large and heavy individual items of equipment such as chillers, boilers, transformers, generators, switchgear, cooling towers and UPS modules.

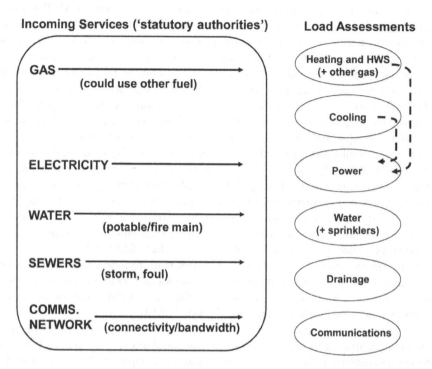

Figure 2.9 Load assessment and incoming services

The structural engineer may allocate a uniform loading density throughout all plant areas to cover a normal arrangement and distribution of MEP equipment. However, the heaviest items may require additional structural features, such as thicker floor slabs that will impact upon the structural design and cost. The structural engineer might impose limitations on the locations of heavy equipment, and once these have been agreed to, there is limited scope for changing the locations of equipment.

The load assessments for heating, cooling and electric power will be further developed as the design detail progresses. As the architectural proposals for the envelope – facade and roof – are developed further, the HVAC system proposals will become more detailed so that preliminary sizes can be allocated for components such as boilers, heat pumps, air handling units (AHUs), fans, pumps and chillers.

It is likely that the application will be made for planning permission, with the building services engineer providing supporting information to the necessary level of detail (see Section 2.7.1). The Building Regulations submission, including Part L, might be prepared and submitted during Stage 4 (see Section 3.8).

The design can now be developed sufficiently to allow a good level of coordination of the various components and elements. This will involve detailed calculations for all items of equipment and detailed development of all spatial information. While the earlier design would have shown the generality of the design for all spaces, the developed and technical designs are likely to have evolved to also to cover all particular and non-standard aspects. This will normally include each terminal device (diffuser, grille, luminaire, socket, radiator, chilled beam, etc.) and branch or final circuit from the on-floor distribution point. Schedules can be created with the key design and performance parameters to summarise the important information for items of equipment.

The design coordination up to this stage will have required the incorporation of the principal builder's work features into the architectural and structural designs. The typical scope of the builder's work is described in Chapter 13.

Relevant information should be prepared for statutory requirements, including information necessary for the management of H&S.

There is a need to create design information of a sufficient level of detail for the tender package(s) that will be used in the procurement process. This will become the formal set of design material from which the tendering contractors will be able to understand all requirements of the systems and submit tender bids accordingly. The specific sets of design information created for the tender package(s) will depend on the procurement route selected and the agreed upon information exchange format, including BIM/digital information.

2.6   Provision for testing and commissioning

It has previously been noted that a key design objective is to facilitate whole-life performance. Testing and commissioning comprise the conclusive stage of the construction process in which the various installed systems are separately tested and then set to operate in the required integrated operational mode to achieve the intended design conditions. As outlined in Section 2.4.3, the planning for the testing and commissioning process should commence at the briefing stage. The testing and commissioning activity will establish plant operation in accordance with the design parameters and set the systems to maintain the conditions for the likely range of load conditions and to the tolerances that have been defined. For this activity to be successful, the systems will require suitable features to allow ease of commissioning. The engineering systems designs must therefore include all necessary features required to effect testing and commissioning to satisfy the design parameters, achieve optimum performance and allow for ongoing adjustment to maintain the design criteria. This will include standard features, such as:

- for air systems: dampers, test points
- for hydronic systems: regulating valves, pressure-tapped valves, commissioning stations
- for electrical systems: meters, instrumentation, control panels, adjustable settings (protection relays).

It might also include non-standard features required for more complex testing and commissioning procedures and sequences, such as temporary connections and test loads (fixed or temporary). All of these features should be identified within the design material. Information should be provided on all system parameters to be achieved during the testing and commissioning process.

2.7   Legislation, regulations and consents

There are several areas of approvals or legislation in which building services engineers are involved to a significant extent. The specific requirements will depend on the location and jurisdiction. In the UK this includes planning consent; building regulations, or building control approval; and designers' duties for health and safety. The nature of the involvement for these aspects is briefly outlined here. The regulatory framework and processes will vary between countries, but it is likely that many countries will have regulatory frameworks or national or local codes with similar objectives.

There are also, of course, numerous standards (such as British Standards) relevant to most technical aspects of buildings and engineering systems. These are often referred to in codes and regulations.

2.7.1 Planning

Planning relates to the appropriateness and suitability of a proposed construction project in relation to its location. A project should not proceed to construction unless the required planning permission or consent has been obtained from the relevant planning authority, such as a local municipality or council.

In the UK, the processes for achieving planning and building regulations approvals are separate. Both are based on submission of drawn information, together with other design material, as required. Planning approval is sought first, with a series of drawings of a relatively simple nature, but suitably informative about the nature of the proposals. Building Regulations approval is applied for at a later stage, using more detailed technical information covering most disciplines. In some countries, it might be normal for both applications to be made at the same time (Makstutis 2010)

For most locations in most countries, there will be some type of policy for planning approval so that the process for development of a neighbourhood, village, town, city or region can be managed. In essence, planning is a method of regulating land use in its widest sense (Makstutis 2010), so that only certain types of use considered suitable and appropriate can be undertaken in any particular location. Policy can be at a local municipality level or for an area, district or region. For some developments of a particular type, scale, potential impact and significance, the approval process might be at the national level. In other cases, a city or other locality might have its own unique set of outline and detailed planning requirements, usually in the form of policy guidelines. The intention is that these allow for growth and development of the built environment, but in a sensible and properly managed way. They also provide a degree of control of the visual impact of development. The policies seek to ensure that the distribution and pattern of types of buildings is suitable, along with their scale and uses. They also seek to protect and conserve buildings or areas of historic character and provide built development that is in the interests of the locality and those who may be affected by development (Makstutis 2010). Planning policy sets objectives in a general sense and provides some opportunity for negotiation.

Planning can be a complex issue covering many aspects of the proposed development, including its context, function, appearance, scale, impact and the transport issues related to goods, people and vehicle movements. It will also consider zones, heights, noise, views and overlooking of other properties. For a building services designer, the main aspects that will normally require particular attention are:

*   plant locations and appearance, where they might have a visual impact
*   visible features on the elevations, such as louvres and flues
*   noise breakout from the plant that might have an impact on ambient noise levels
*   fumes or other emissions from the plant
*   energy strategy, which is increasingly a consideration in planning policies, and ventilation statements, which are increasingly required for planning.

The planning approval process consists of submitting a range of material that describes the intended development. For outline planning approval, the information does not usually need to be in detailed form and tends to be fairly general. The objective is to provide suitable information so that the authority can appreciate the key features of the development and make a decision accordingly (Makstutis 2010). A wider range and level of information are likely to be required for a larger scale of development. A more comprehensive set of information will be required for detailed planning approval.

The planning departments of local authorities seek to ensure that proposals comply with the relevant policies governing the nature of the built environment as it develops in their locality. Most urban areas will have a unitary development plan (UDP) or similar that allocates zones for particular types of development, such as residential, industrial, commerce, retail, etc. The objective is to protect the interests of residents and other parties that might be affected by changes taking place (Makstutis 2010).

Planning can be a lengthy process for certain types of proposals and will often involve public consultation, meetings and negotiations. In some cases, buildings might obtain approval without going through the formal procedure of a full planning process. It should be noted that planning only considers the nature and acceptability of the proposed development in relation to the prevailing policies for the area. It does not consider any safety aspects, which are covered in Building Regulations.

2.7.2   *Building Regulations*

'Building Regulations', or 'building control', relates to the compliance of the specific design for the building and its engineering systems in relation to the relevant codes or regulations, such as the Building Regulations approved documents in England and Wales. In contrast to planning, Building Regulations are standards for the performance of the construction works.

The Building Regulations applicable in England and Wales exist principally to protect the public by ensuring the H&S of people in and around buildings. They also cover access to and around buildings. The regulations apply to most new buildings and many alterations of existing buildings, whether domestic, commercial or industrial. The regulations cover most aspects of a building's construction, including its structure, energy conservation, fire safety, sound insulation, drainage, ventilation and electrical safety. Details of certain elements of the design have to be submitted for approval. The construction on-site is monitored by a local authority official or approved inspector, previously often known as the 'building control officer'.

For building services designers, the most important aspects will normally be:

Part B: Fire safety
Part F: Ventilation
Part G: Sanitation, hot water safety and water efficiency
Part H: Drainage and waste disposal
Part J: Heat-producing appliances
Part L: Conservation of fuel and power (see Chapter 3)
Part M: Access to and use of buildings
Part P: Electrical safety.

2.7.3   *Designers' duties for health and safety*

An important duty for all professionals involved in the design of construction projects is to manage the H&S implications of the structures or systems they design. The construction industry has, unfortunately, given rise to a significant number of injuries and fatalities and, historically, its record has been poor in comparison with other industries. It is useful to consider situations in terms of 'hazard' (something with the potential to cause harm) and 'risk' (the resultant likelihood that harm will occur from the hazard, taking account of the controlling measures introduced). Certain specific aspects of the construction process have easily identifiable potential hazards – for example, activities involving craneage manoeuvres; working at a high level, particularly in exposed areas such as roofs; and working in trenches and tunnels. These are generic hazards relevant to the designs of all disciplines.

For those designing mechanical and electrical systems in buildings, however, the nature of the fuels, systems and equipment involved means that they have specific inherent hazards that need to be properly managed and controlled. Specific H&S hazards could potentially arise from a variety of sources: electricity, gas, combustion, water, drainage, sewerage, steam, hot surfaces, rotating plant, noise and vibration, fumes, fibres, particulate matter, confined spaces, chemicals, compressed air and many others. The nature of the construction process provides a sense of impermanence – the creation of a one-off, short-term, in situ activity, where the 'workplace' is undergoing continual change and can have a high degree of work-force density and on-site mobility. This inevitably makes it more difficult to manage risks when compared with industrial activities in surroundings of a more permanent and easily controlled nature.

It is important to have a whole-life approach to H&S management, and the recent legislative developments have recognised this, with the emphasis on managing risks from commencement (HSE 2015a). In some ways, it parallels the life cycle approach required for sustainability by looking beyond just the delivered and completed construction to cover lifetime usage and, ultimately, dismantling or demolition as well. The building services designer must therefore consider the potential hazards that could arise during the complete life of the building:

1   Construction and installation: H&S hazards arising from the delivery, assembly and installation of systems and equipment as part of the overall construction process.
2   Testing and commissioning: H&S hazards arising from the process of testing equipment and systems and making adjustments during the commissioning process, as well as when setting systems to work in their final operational state.
3   Usage: H&S hazards arising from usage of the building by occupants.
4   Operation and maintenance: H&S hazards arising from the activities required to operate and maintain systems and equipment throughout their operational life.
5   Dismantling, replacement and demolition: H&S hazards arising from the activities at the end of its useful life for systems, equipment and components.

While items 1, 2, 4 and 5 relate to activities undertaken by competent personnel, the hazards in item 3 relate to building occupants who are not competent with MEP systems, and these obviously require careful consideration. Building Regulations are largely concerned with this aspect of H&S.

Normal good practice is for most potentially hazardous items of MEP plant and equipment to be located in segregated plant rooms and risers, so that they are only accessible by competent personnel. The selection and arrangement of spaces for the plant therefore needs to take into consideration relevant H&S matters. This aspect of building services design is covered in Chapter 13.

In the UK, regulations are in place to improve the management of H&S related to the construction process and the use of the building throughout its life – the CDM regulations

(HSE 2015a). The primary aim of the regulations is to integrate H&S into the management of a project and to encourage everyone involved to work together. This is so the team can plan the work in a sensible way and manage the risks involved from commencement through to completion, including:

- improve the overall planning and management of H&S considerations on construction projects from the outset
- identify potential hazards at an early stage in order to eliminate, or at least reduce, them at the design stage, which will allow risks to be reduced and the remaining residual risks to be communicated to others and managed properly.

The regulations place certain specific duties directly on the principal designer and designers, as well as the principal contractor, contractors and workers. Clients also have defined duties.

Designers (organisations or individuals) have duties to eliminate, reduce or control foreseeable risks that could occur during construction, maintenance or usage and a duty to provide information to other project team members, so as to assist them in fulfilling their own duties (HSE 2015b).

Principal designers (organisations or individuals) have duties in the pre-construction phase of a project to plan, manage, monitor and coordinate H&S. This includes identification, elimination or control of risks that are foreseeable and ensuring designers undertake their duties. There are also duties to prepare and provide information to others and to liaise with the principal contractor during the construction phase (HSE 2015c).

In order to discharge these duties, a competent building services designer will need sufficient knowledge of how their system components are delivered and assembled, together with experience of the construction and installation process.

In the recent past, designers often sought to discharge their duties under the CDM regulations by creating formal risk assessments for each identified hazard, in which they made a numerical or relative evaluation of the initial and residual risk. This risk assessment approach has been considered to be unsatisfactory. This is because, while it created a considerable amount of 'audit trail' paperwork, it was questionable as to the real impact it had on reducing H&S risk, which is the whole purpose of the regulations. Rather than creating paperwork, the main focus of the designer should always be on eliminating the hazard from the design. If this is not economically practical, they should always seek to reduce the hazard and communicate in explicit terms the nature of the residual hazard to those who must manage it during the construction stage and during later stages, where this is appropriate. The principal contractor is the construction organisation appointed by the client to coordinate the construction phase of a project involving more than one contracting organisation (HSE 2015a).

Following elimination of as many hazards as possible, a 'Residual Hazards' list can be passed to the principal contractor. This will enable them to properly plan and cost the construction phase, considering measures that will be necessary to address the residual risks. This list could include aspects such as the nature of the activity, specific hazard, and people potentially exposed; design measures taken; and explicit information communicated to raise awareness of the residual hazards.

An example of a design feature that can eliminate or reduce hazards is a pre-fabricated plant room. These can be provided for MEP equipment, such as boiler houses, generator rooms and switchrooms. This approach allows construction to be undertaken off-site in a factory environment where H&S and quality control measures can be better managed than in a construction site environment. In situations where plant rooms are located at roof level – as is often the case – the use of a pre-fabricated enclosure complete with the required MEP equipment can significantly reduce the extent of activities in this potentially hazardous location. It will primarily require lifting the pre-fabricated plant room into location by a crane, followed by fixings and connections to the relevant distribution elements. Such an approach might also be preferred for its cost-effectiveness, reduced time involved and (potentially) lower embodied energy of materials through good materials management; but in the context here, it should be clearly recognised as a good solution to reduce H&S risks. This is a good example of systems thinking and holistic design, which are about mutual benefit and concurrently achieving multiple objectives.

2.8   Quality management for designers

It can be seen that the design stage for construction projects is a complex process involving many individuals of different disciplines, usually across a number of different organisations. For a consistent and successful outcome, the design stage must be undertaken as a professionally controlled and managed process. So that a design can be developed in any discipline, it is essential that all relevant design-related information is stored, updated and maintained in a properly structured manner. This allows easy access for all the designers and others to the relevant information. It is also necessary so that it is formally defined in relation to provenance – authorship, date, revision and cross-references to other material. A formal quality management (QM) procedure is required for design in the same way as for many other quality-controlled aspects of business and industry. Most design organisations hold QM certification for their design activities from an independent accreditation body, in accordance with criteria such as British Standards or the International Organization for Standardization (ISO). This requires a comprehensive set of principles and procedures to be in place to ensure that quality can be maintained throughout the design process. This normally includes a filing management regimen that covers aspects such as:

- appointment and brief
- recording any subsequent client instructions and changes in a sequential manner, which will cover all confirmed changes to the brief and the associated correspondence
- recording design stage approvals and any client conditions and directives that need to be addressed in subsequent stages
- complete sets of design calculations and evidence of the checking regime, which would typically include manual calculations; computer calculations together with the associated input data and the interpretation of the output; sketches and diagrams; and formal drawings
- registers of the issue history of drawings and updates from the designer and the receipt of those of other designers with whom the design has been integrated and coordinated
- registers of the issue history of equipment schedules
- details of all design reviews, usually at an interim stage and at output stage, immediately prior to tender issue for procurement (see later)
- registers of the issue history for all specifications, studies and reports.

An important part of the QM process is the undertaking of design reviews. Careful consideration should be given to the strategy for the review process at the outset, so that it is appropriate for the project's scale and complexity. It is normal for all calculations and drawings to be checked independently, usually by a competent colleague. Most projects have independent design reviews, usually at an interim or scheme design stage and at the output stage, prior to the information being issued for tender and procurement. Where the project involves a non-standard brief or design approach involving innovative principles or technologies, it is important to have a detailed peer review of the concepts prior to proceeding. The design review activity will normally involve checking the developed design against the original brief and changes to the brief, together with checking adherence to (and acceptability in relation to) standard engineering principles and the relevant statutory and industry standards and regulations. It will usually result in marked-up comments related to calculations, drawings and specifications recording the reviewer's recommended aspects for attention. There should be a mechanism through which the designer absorbs and acknowledges each comment and records that this has been done. Large, important and technically complex projects might have many design reviews, both within the design organisation and externally. Clients sometimes appoint a separate consultant as a monitoring or 'checking consultant' for this purpose.

One important objective of the QM process is to create an 'audit trail'. This means to retain all necessary information to demonstrate how the final design came about such that this could be inspected by a third party. An audit trail is essential should there be any future litigation in relation to the design. Project files, including the design and calculations files, are usually archived for a number of years in accordance with the conditions in the designer's appointment agreement with the client.

With the growing use of modern common electronic information exchange systems for projects, the emphasis has shifted to auditable electronic filing.

2.9 Summary

To undertake their role successfully, building services designers must have a clear understanding of their own duties and the duties of other professionals within the design team. It should be recognised that building and engineering systems design is an iterative and interdisciplinary process, requiring cooperation, integration and negotiation. Teams should work together in a collaborative manner. The design normally proceeds to a programme through clearly defined stages, with cost reporting on the proposals and client sign-off before approval to proceed to the next stage. A key early activity is the development of the brief to confirm the specific aspects influencing the building services concepts. Sustainability matters and targets should be considered from the outset by all disciplines. In order to influence these aspects, building services engineers should be involved in the decision-making at the earliest concept design stage. The design proposals eventually form a detailed set of information that is issued for tender and procurement of a contractor. The building services designer must adhere to relevant legislation, policies and codes, and has particular duties in relation to the H&S management of the systems designed, as well as an involvement in submissions for planning consent and building control compliance. Designers' work should be undertaken within a QM process to maintain the quality of the design outcome and delivered material and to create an audit trail of the design decisions and information exchange that resulted in the final design.

3 Generic design strategies for energy-efficient, low-carbon buildings

3.1 Introduction

This chapter outlines strategic approaches to reducing energy consumption and carbon emissions in buildings. The focus is on identifying appropriate generic design considerations that can be addressed in priority order as practical steps to promote energy efficiency. Specific technical aspects of active engineering systems are not covered in depth here, as most of these are explored in Chapters 6–10. Reference is made to selected aspects of the approved documents to the Building Regulations in England and Wales to provide some context of regulatory frameworks for energy conservation. It should be noted that this chapter only covers strategies related to energy efficiency and carbon mitigation. It does not address the many other issues relevant to sustainability in buildings – such as water, drainage, materials management, 'circular economy', recycling and biodiversity – which should also be considered as part of a wider strategy for sustainable design, as well as approaches to health and wellbeing, as outlined in Chapter 5.

An important point to emphasise is that this is a period of great change for the those involved in the planning and design of building energy performance, with the unprecedented challenge to make deep reductions in carbon, at a rapid pace, to achieve the objective of many nations to achieve 'net zero carbon buildings' by 2050. National regulatory frameworks, municipality development rules and relative metrics on carbon emissions for different fuels are all likely to be going through a period of transition over the coming years, both in the UK and elsewhere. It is clearly not possible to provide firm strategic guidance in such a changing scenario with so many variables. Instead, this chapter provides general guidance, referring to existing approaches, likely trends and changes. Engineers will need to make astute judgements on strategies for energy use and carbon reduction as new regulations, guidance and metrics come into place.

3.2 Developing a focused approach

A useful approach to achieving low-carbon buildings is shown in Figure 3.1 and involves a sequence of planning, designing and managing. The typical range of interdisciplinary design collaboration during project stages was described in Chapter 2. The early stages will involve appraisal of the site to assess the physical features and the potential passive and renewable energy opportunities, and hence the options available to satisfy the development's requirements with a minimal use of energy and carbon. The site appraisal can influence the locations and layout of building(s) on the site to provide an optimum arrangement. It should also identify the key factors for availability of utility supplies of sufficient capacity. This is becoming

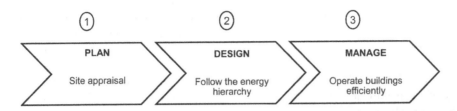

Figure 3.1 Path to low-carbon buildings

increasingly important with the need to decide on fuel for heating, as the use of electricity will necessitate an overall supply capacity that may not be available (and this has wider implications at local, regional and national levels). The next step is to design the building(s) and systems based on a logical energy hierarchy, taking note of the prevailing and anticipated scenario for the carbon content of different fuels. The third step is for the building(s) to be operated in an efficient way throughout their life while meeting the functional requirements.

In Chapter 1, a conceptual model was presented of an approach to reduce carbon emissions in buildings. This proposes that the focus should, first of all, be on reducing energy demand. The energy demand is a function of both necessity – to achieve comfort and functional criteria – and demand management. The energy demand will depend on many factors and will vary with the building type and function, but for most building types it can be influenced significantly by the characteristics of the envelope – primarily the thermal performance of the external walls, glazing (in particular the frames) and roof. The initial focus is therefore on optimising the envelope, building form, orientation and massing and incorporating such passive features that might be appropriate to beneficially utilise available climate conditions and thereby reduce the need for active energy. Another aspect of demand reduction is to put in place such management regimes and controls and monitoring facilities that will reduce the usage of active energy to only that amount required to satisfy the occupancy criteria, adjusting accordingly as requirements and occupancy patterns vary. The intention is always to do this without reducing the 'enjoyment' or the functional performance of the occupied space. This aspect of demand reduction is a function of technology (data collection, automatic controls, metering and monitoring), the management regime for operation and maintenance and occupancy behaviours.

An important starting point, therefore, is appropriate engagement with the client at the briefing stage and to seek an opportunity to communicate with the operational and maintenance (facilities management) staff who will be responsible for running the building's services and managing energy usage (to the extent that it is within their control – for example, they may only have limited influence on 'unregulated' small power usage). This will provide an opportunity to explore features that can be incorporated to allow ease of operation, maintenance and energy management consistent with their preferred approach. It will also provide an opportunity to review any assumptions about occupancy behaviours that might be amenable to review, with the intention of adopting a less energy-intensive approach. These measures will help to reduce the need for active energy. Such early engagement may not always be possible, however, particularly for speculative buildings.

The second priority is to address the active energy provision process, which can be considered as comprising all of the processes for energy generation, supply, distribution and the active energy-using systems in the building. The emphasis is on zero or low-carbon energy supply and efficient distribution, along with energy-efficient systems. This might include measures that have traditionally been best-practice engineering design considerations. Systems should be designed to be inherently energy-efficient and incorporate the simple steps that can be taken to minimise carbon emissions, such as recovery of heat in thermal systems.

The use of renewable energy systems has, until recently, been considered the logical third stage of the hierarchical energy strategy. This is likely to be reviewed in many countries in light of the changing scenarios for the relative carbon content of fuels, availability levels, relative costs and so on. Provision of renewable energy might be subject to specific local authority planning policy, requiring a proportionate on-site contribution. However, renewable energy should be seen in the context of other carbon reduction measures, so that the relative effectiveness and potential limitations are understood. In many cases renewables may not be an appropriate choice compared with investment in more rigorous energy efficiency measures. There is no real sense in incorporating renewables where they would be merely offsetting carbon emissions arising from the use of inefficient systems; the passive and active systems should, instead, be made more efficient. Simple energy efficiency measures may not be glamorous and may have no visible presence to occupants or the public, but are often more effective than some of the more 'token' renewable technologies that are often – incorrectly – seen as the essential hallmarks of sustainability.

A sensible order of priority for addressing carbon mitigation through design would be:

- Optimise the building envelope by utilising passive design strategies to reduce the energy demand. 'Passive design' means utilising the building itself to assist in creating the desired internal environment, through suitable design and selection of the shape, form and orientation, and the thermal and light transmission characteristics of relevant building elements.
- Further reduce the demand through energy efficiency measures in the active systems.
- Supply energy efficiently.
- Incorporate suitable renewable energy technologies, where appropriate.

Renewable energy technologies generally come under the term 'low and zero carbon' (LZC) technologies.

In the UK, a useful example of an energy hierarchy of this type is the London Plan of 2016 (GLA 2016). This can be expressed in simple terms as:

- Be LEAN: be energy-efficient (i.e. reduce the demand)
- Be CLEAN: incorporate low-carbon energy sources (i.e. supply energy efficiently)
- Be GREEN: incorporate renewable energy sources.

This forms part of the Greater London Authority (GLA) planning policy for London and placed particular emphasis on promoting connections to district heating schemes and the use of site-wide energy networks. Figure 3.2 shows an energy hierarchy based on this approach. However, it should be noted that it is anticipated that this approach will be altered, with a focus on net zero carbon and monitoring and reporting on energy performance.

By undertaking passive design and energy-efficient design for the active systems, the energy demand will have been reduced to its 'leanest' level and will make the 'clean' energy supply measures most effective. Incorporating an appropriate level of renewables can further reduce the level of carbon emissions.

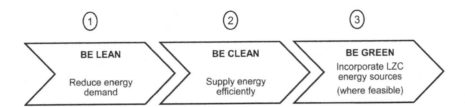

Figure 3.2 Energy hierarchy

Source: London Plan, GLA 2016

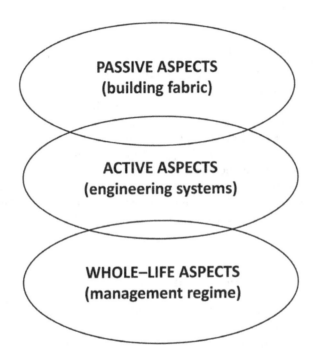

Figure 3.3 Key considerations for energy efficiency

In essence, the factors affecting energy efficiency can be broken down more conveniently into passive aspects, active aspects and the arrangements for the whole-life operation, as shown in Figure 3.3. These overlap to some extent and cover both design and operational aspects, but need to be considered together in the decision-making at the design stage. Sections 3.4, 3.5 and 3.6, respectively, cover generic strategies for passive aspects, active engineering systems (including renewables) and whole-life aspects for operation.

The design must, of course, comply with any relevant statutory and/or local codes and policies related to conservation of energy, fuel and carbon reduction. So a design strategy can be established as outlined here, while also ensuring that the design achieves compliance with prevailing regulations, which are likely to include similar considerations to achieve similar objectives. Section 3.9 provides a context for regulatory criteria, using the Building Regulations–approved documents in England and Wales as an example.

It is likely that many countries will soon be adopting legislation and guidance to promote carbon reduction in buildings as part of their commitment to meeting the immense challenge set out in the Paris Agreement: global carbon emissions to be almost halved by 2030 and then eliminated by 2050. In the UK, there is an intention to adopt net zero carbon targets for buildings. This is likely to profoundly influence regulatory and planning legislation, which will have a significant impact on approaches to energy strategy, so it is a changing scenario. The UK Green Building Council has published a definition in their document 'Net Zero Carbon Buildings: A Framework Definition' (UKGBC 2019). The framework has two approaches, one based on construction and one based on operational energy. It is a first step, aiming to achieve net zero carbon based on the whole life of a building. This framework is initially intended as guidance, with industry involved in the further developments required over the next decade to add detail.

3.3   Energy strategy reports

The need to address whole-life building performance requires a strategic approach to the energy proposals. The development of a strategy should be sufficiently broad to capture all the relevant influences and opportunities and have a creative approach to potential solutions, including aspects such as solar and wind exposure, daylight, embodied energy and relevant life cycle issues (Moe 2008). The best way to encapsulate an appropriate approach is to set this out as a formal energy strategy report. A report of this type is required for obtaining planning consent by a number of local authorities in the UK and should ideally be an integral part of the early-stage design collaboration outlined in Chapter 2. It is good practice to include a suitable energy strategy statement for all large planning applications (CIBSE 2007). An energy strategy report should provide recommendations on the principles affecting the energy and carbon performance (Bateson 2009), including:

- site factors
- available energy infrastructure and capacities, including for community energy schemes
- estimated thermal and power loads and their dynamic nature
- statutory legislation and other standards and regulatory criteria
- benchmark data or building-specific energy analysis based on modelling
- initial energy assessment outcomes
- initial proposals for shape, form, orientation, envelope and other passive design features
- fuel selection, energy generation and distribution proposals
- energy efficiency measures for active energy systems
- ambient energy potential and incorporation of renewable energy technologies, where appropriate
- approach to compliance with regulations, codes and planning, as well as targets for environmental assessment methodologies
- initial data, metrics and predictions for energy consumption; carbon emissions; and the associated cost implications
- correlation of the energy strategy with wider objectives for sustainability and whole-life matters, including embodied energy.

The following sections introduce the key generic strategies. Where appropriate, each refers to other chapters where the particular aspect of design is covered in detail.

3.4   The building envelope and passive design measures

3.4.1   Concepts for passive design

The operational energy performance of a building is, to a very large extent, determined by the location, shape and form of the building and the properties of the envelope in contact with ambient conditions: the roof, external walls and relevant floor slab(s). Using a passive design approach, the envelope can be designed to minimise the amount of energy usage and carbon emissions from active energy systems and help to optimise energy performance. The site's potential for renewable energy should be assessed at the same stage and should form part of the energy strategy report. Computer modelling of likely operational energy usage can provide early assessment to guide the conceptual development and, through iterative development of architectural and structural designs, can and should provide a direct influence on the evolving building morphology (Moe 2008). This should be considered in relation to embodied energy (and carbon) also, so that the overall concept developed has good performance for the whole life of the building.

Figure 3.4 shows the basic concept for a passive approach in very simple terms in relation to reducing the demand for cooling. An inappropriate approach is shown on the left, in which there is no solar control for the windows, resulting in overheating. With this approach, to maintain an acceptable indoor climate, an extensive mechanical cooling system will be required, resulting in a major operational energy demand, which might use carbon-intensive

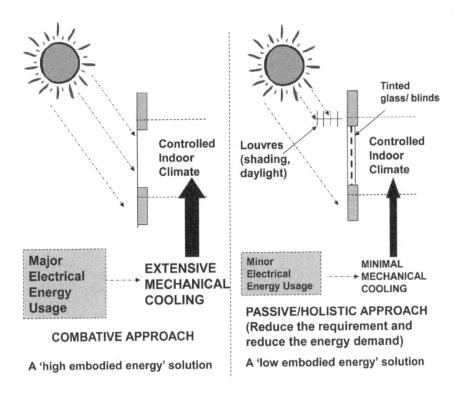

Figure 3.4 Developing a passive approach: concept

electrical energy. This is also a high 'embodied energy' solution for the building services equipment required; needs space and ongoing maintenance for the equipment; and, furthermore, is very likely to involve refrigerants with potential impacts in terms of global warming and ozone depletion. A passive or holistic approach is shown on the right, in which solar control features are provided in terms of louvres and carefully specified glass and/or interpane blinds. This will minimise the solar gain in the space, requiring no or minimal mechanical cooling (and refrigerants) and greatly reduced operational energy usage. This will also result in lower embodied energy (for the louvres and blinds) compared with that for an extensive mechanical cooling system. This approach is all about reducing the demand using passive features so that the requirement for active energy systems and the associated operational (and embodied) energy is reduced.

The complex subject of building physics and passive design, encompassing detailed analytical modelling and assessment of the energy performance of building elements, is beyond the scope of this book. However, all building services designers should recognise that the performance of the structure and fabric should be the initial focus for their attention.

There has been a renewed focus on the need for research and education in building physics to address the relatively limited guidance previously available, as outlined in the 2010 Report from the Royal Academy of Engineering (King 2010a). We describe passive aspects here in a fairly simplistic sense as an introduction to some of the main considerations. For detailed guidance, reference should be made to other publications covering building physics, building science and information on the performance of building materials (Hopfe and McLeod 2015).

In order to create an acceptable envelope, it will be necessary to work closely with the architect from the outset to explore and evaluate site considerations and the practical options for the location and orientation of the building(s). These issues are complex, because all aspects are interrelated and cannot be viewed in isolation. Computer modelling is used extensively as a tool to inform design and/or check compliance with Building Regulations. The specialist area related to the interdisciplinary design of the external facade is an evolving discipline sometimes termed 'facade engineering'. The objective is to achieve an optimum balance for the envelope's energy performance, alongside achieving a satisfactory outcome for non-energy aspects, such as aesthetics, planning, acoustics and fire performance. The intention, from an energy perspective, is for the envelope to modify the climate towards that desired for comfort through moderating and discouraging those aspects that are undesirable and enabling and encouraging those aspects that are beneficial (CIBSE 2004a). This will usually include:

- providing good levels and quality of daylighting
- providing effective natural ventilation, where practical
- minimising heat losses from fabric and infiltration during cold periods
- minimising solar gains during hot periods (and gains from fabric and infiltration, where necessary)
- maximising beneficial solar gains during cold periods.

The specific intentions will depend upon the geographical location (and hence the prevailing climate conditions) and whether the building type means there will be a predominant need to limit or attract solar intrusion. However, it is worthwhile listing key features individually, as outlined next.

3.4.2   *Location, shape and form*

If the size and nature of the site provide some freedom to select the location of the building, use can be made of features that might assist the energy performance. The main features that can be beneficial are shown in Figure 2.8. Existing trees or buildings can provide a shelter belt to minimise exposure to prevailing winds, as well as providing visual delight, a sense of wellbeing, better air quality and biodiversity. The opportunity to benefit from trees may be reduced in urban settings. Where there is a significant slope in the ground, the opportunity for partial burying could be explored, as it can provide energy benefits. The use of ambient conditions to provide renewable energy is discussed briefly in Section 3.5.13 and in more detail in subsequent chapters.

The form of a building is an important starting point, as the requirements for other aspects of the nature and physical characteristics of the building envelope will follow on as consequences from the three-dimensional shape (CIBSE 2004a). In simple terms, desirable features that should be sought are:

- Seek a shape that minimises the surface area in relation to the volume, so that the heat losses (and gains) from outside air are reduced to the minimum for each unit of occupied space. This tends toward seeking a form that is compact, rather than long and narrow.
- Minimise elongated exposed blocks (or elements) protruding from the main body of the building, as they will have relatively higher heat losses (or gains) per unit volume.
- Avoid exposed overhanging floor slabs, which will give rise to unwanted additional heat losses (or gains) compared with floor slabs in contact with the ground.
- A narrower floor plan will assist daylight penetration and improve the potential for natural crossflow ventilation, as outlined in Chapter 6.
- Include or encourage internal courtyards, light wells or 'streets' to promote daylight and natural crossflow ventilation.

3.4.3   *Orientation*

Where a building is likely to require cooling during summer periods (such as an office building), seek a shape that is orientated on an east–west axis. This will minimise solar gains from low sun angles in the early morning and late afternoon. Solar gain from higher sun angles in the middle of the day can be more easily controlled by horizontal shading.

3.4.4   *Thermal insulation*

To minimise static heat losses (and heat gains, where appropriate) from heat flow due to external design temperatures, the overall envelope should be designed so that the proportion of glazed areas is minimised compared with non-glazed, or opaque, areas, and both opaque and glazed areas should have the lowest U-values that might be economic and practical for these elements. However, it is usually necessary to seek a balance, as some solar gain is beneficial, and it will, of course, be necessary to take account of other considerations, such as the desirability of providing good levels of daylighting. It is often the case that walls will require high levels of insulation. Compared with other desirable features of the envelope, this can be achieved in a relatively easy way through selection of suitable fabric 'sandwich' construction for walls related to materials, air gaps and thicknesses. There is a tradition in cold northern climates to have 'super insulation' at levels well above those that have been the norm in the

UK. However, highly insulated spaces with low levels of air infiltration can potentially give rise to condensation problems.

3.4.5 Daylighting and glazing

The locations, orientations, shapes, sizes and characteristics of glazing – whether windows or glazed panels within cladding on facades or incorporated as rooflights or similar – will have a fundamental impact on the internal environment and energy performance. Their methods of shading will also have a major impact, as outlined in Section 3.4.6. From a thermal point of view, glazed areas will always have higher U-values than opaque areas, so they are major areas of static heat loss and heat gain, as well as solar gain. This is normally undesirable, but can be desirable in circumstances where solar space heating in winter is a viable proposition. From a lighting point of view, of course, glazed areas provide the natural daylighting that is an essential amenity requirement for occupied spaces. Providing a suitable quality and quantity of daylighting that satisfies illumination needs for the majority of the time when activities take place is an essential feature of a successful overall lighting solution. It would also be a major contributor to the health and wellbeing of occupants, as described at Chapter 5, and suitable provision can, of course, significantly reduce the energy and carbon required for artificial lighting. From an acoustic perspective, glazed areas are a weak point in the envelope and will often have a detrimental impact on the acoustic integrity of a space, particularly where there are exacting noise criteria.

Daylight design is involved with planning the positions and sizes of glazing and related surfaces so that daylight can penetrate a building in a beneficial way. The daylight can enter directly through glazing or indirectly via reflections from other surfaces. The daylighting requirements will be related to the nature of the occupants and activities and the types of spaces. For a successful daylighting outcome, ideally, the activities should be primarily in spaces with good levels of daylight levels and should be no more than 5 metres from a window-wall (Hopfe and McLeod 2015). Daylight factor (DF) is a basic metric for the quantity of daylight in a space. DF is the ratio between the external and internal horizontal illuminance under a fully overcast sky as a percentage, and for rooms where the majority of activities take place and are lit from the side, a DF of more than 2% is usually an indicator of good daylight potential (Hopfe and McLeod 2015). It should be noted that in the guidance for achieving Part L compliance for lighting energy efficiency, there is a definition for 'daylit space' and related guidance – see Section 10.4.6. The quality of daylighting will be a function of the actual locations, sizes and characteristics of glazing, rather than just the DF figure. There is more detail about the health and wellbeing aspects of daylighting in Chapter 5.

So that daylight penetration can be maximised, floor plates should be narrow and glazing should extend towards the upper parts of walls to make the best use of ceiling reflectance. It is necessary to have suitable proportions of glazing with good light transmittance characteristics to achieve a high level of daylight in occupied spaces. Achieving good daylight levels also requires pale-coloured wall, ceiling and floor finishes with high values of reflectance. The selection of suitable colours for internal finishes will also, of course, benefit the artificial lighting levels and is therefore a key criterion where the lighting designer can influence the architect. This is an important area of design coordination for energy efficiency that is often overlooked. The selection of the glazing proportion and characteristics will, of course have a major impact on the thermal performance of spaces, as well as on the acoustic performance. In most cases, improving the level of daylight will have a negative impact on the thermal and acoustic performance. So in seeking to maximise levels of daylighting, there is inevitably a compromise to seek an appropriate balance with the thermal and acoustic performance. There is also a need to ensure the design satisfies the requirements for compliance with relevant regulatory criteria. Selecting a building shape with inner courtyards or light wells can be highly beneficial in this regard. For spaces where both the quantity and quality of light is of particular importance, glazing arranged to provide light from the north can be beneficial. This 'north light' provides a 'cool' colour source, without the detrimental impact of direct sunlight and solar gain. There is growing use of climate-based daylight modelling to estimate daylight under different building geometries and climate scenarios.

There are, therefore, conflicting positive and negative impacts from glazing. Inevitably, a balance is usually required between the appearance and the thermal, visual and acoustic issues. The design of the glazed components of a facade is one of the key areas requiring a cooperative and integrated approach with the architect in order to reach an optimal solution. It is one of the main aspects of design resolution that will impact upon the passive energy performance, and hence the requirements of the active engineering systems. Among the characteristics of glazed components that should be considered are:

•   U-value, light transmission index, G-value and emissivity
•   body tints that might be useful in certain circumstances, but these should be considered with care, as they can have a detrimental impact on daylight and cause spaces to appear gloomy
•   for retrofits in existing buildings, light tubes can provide a useful way of introducing daylight into areas that would otherwise have little or no daylight from glazing.

In seeking a balance, there should be a particular emphasis on achieving good levels of daylight, as this has such a significant impact on creating a satisfying interior with a sense of wellbeing. A robust solution is to select low-emissivity double glazing with a thermal transmittance similar to triple glazing and an 80% light transmission factor (CIBSE 2004a). As the glazing performance will be relevant to thermal and acoustic aspects, as well as daylighting, there is inevitably a trade-off to achieve a balanced solution. For double glazing with argon filling, typical U-values are about 1.1 W/sq.m.K, with typically a high 80% light transmittance factor. For triple glazing with argon filling, typical U-values are about 0.6 W/sq.m.K, but the light transmittance factor often reduces to a lower figure of about 70%, so the significant benefit in improved thermal performance is offset by a 10% loss of performance in daylight transmission. The guidance for achieving Part L compliance defines 'daylit space' as having glazing light transmittance of at least 70% (HMG 2016).

There are likely to be changes in the approach to daylighting design in the UK with the proposed adoption of a new European standard, EN17037. This would be the first pan-European standard for daylighting in buildings. It will provide new daylight targets in terms of illuminance, in addition to DF. This will allow designers to achieve the standard by using either a climate-based or DF approach. A significant change will be adjustment of the DF targets for a given location. The targets in the standard are expressed at three levels – high, medium and low – rather than the previous single target. The standard also provides targets for sunlight exposure, with low, medium and high exposure targets (Waskett 2019).

3.4.6   *Solar and glare control*

While selecting a suitable orientation and facade geometry can limit the total amount of solar radiation on the facade of a building, further reduction must be achieved through control features on the facade that will shade the glazed areas and thus reduce the transmission of solar heat gain through the glazing. Solar control features should be considered alongside the glazing characteristics and interpane blinds as part of an overall approach to minimising the impact of solar gains. In each case, the key determining factor is the sun's path at the building's location in relation to the orientation of the facades. It will also be necessary to have some form of glare control to protect occupants from discomfort arising from glare entering their field of view directly from the sun, external reflections or the sky. An ideal solution would be to incorporate control features that provide both solar and glare control, but this is not always practical.

There is a wide variety of features, some of which can only be integral to the architectural treatment (and the resulting geometry) of the facade and others that can be considered more as discrete control features. Features that include local manual adjustment can provide particular benefit by allowing fine-tuning to satisfy personal preferences under varying conditions. A selection of control features follows:

- Overhanging roofs and eaves that provide shading to glazed areas.
- Balconies and other overhanging or protruding structural elements arising from the geometry of the facade that provide shading to lower levels.
- Window reveals or balconies with inset walls that are deep enough to provide shading, both horizontally and vertically.
- Horizontal slatted louvres that provide interception to the sun's rays, but which will allow some daylight to penetrate at particular angles. The louvres can be in the horizontal or vertical plane. The most effective (optimal) blade spacing and angle geometry can be selected through computer modelling. It is possible to incorporate motorised horizontal louvres that will adjust the pitch of the blades as the sun's angular position changes to optimise solar control performance. However, practical experience has shown that the motors and actuators require considerable maintenance due to weather impact. The best solution is nearly always the simple solution of fixed-position louvre blades.
- Louvred shutters.
- Vertical fins that can be narrow, but can be placed on east- and west-facing facades to provide useful shading from the lower angles of the sun's path earlier and later in the day; interpane blinds within glazing; or, less ideally, internal blinds.
- A natural solution that can bring wider benefits is to locate deciduous trees close to the south facade. This solution can provide solar shading in summer, when the trees are in full leaf, and allow beneficial solar gain to penetrate into the building when the leaves have been shed in the winter.
- Awnings and external blinds – which allow manual adjustment so as to control the solar cut-off angle – can be useful, provided they are not exposed to windy situations and suitable maintenance is undertaken.

3.4.7   *Thermal mass (external and internal)*

While the (area-weighted) thermal transmittance of envelope components will determine the static heat flow, the thermal mass will determine the dampening of temperature changes due to thermal inertia, and hence the extent to which the structure can act as a thermal buffer. Buildings and parts of buildings can be considered more 'heavyweight' or 'lightweight', depending on the amount of thermal mass. External and internal walls (and floor slabs) can be thermally massive due to the specific heat capacity of their materials and their thickness. This will cause these parts of the building structure to absorb and store a proportion of the heat (or coolth), rather than letting it pass through, resulting in a thermal time lag which evens out, or smooths, the internal temperature variation. By selecting suitable arrangements of thermal mass of appropriate specific heat capacity, the time lag can be such that heat (or coolth) can be stored when not wanted and released when beneficial. For example, when utilising solar space heating, heat can be stored in well-insulated floor slabs and walls through exposure to solar radiation during the daytime and can be released in the evening, when required, as the external temperature falls. Similarly, where cooling is required during the day, it can be achieved by 'night-time cooling' through ventilation, i.e. by introducing naturally ventilated air at a high level to exposed soffits at night to cool them down. This is so that when the external temperature rises during the day, the internal temperature swing is reduced (see Chapter 6).

A useful arrangement when seeking to minimise solar gains is for the east and west walls of a building to be provided with significant thermal mass, where practical in relation to the architecture. This reduces the effect of the prolonged periods of lower sun angles early and late in the day, and thus limits the temperature levels resulting from the solar influences later in the day. Solar shading can be provided on the south facade to shade against the high-level sun during the middle of the day.

The thermal effects of the multitude of structural elements within buildings of different thermal transmittance and mass can only be assessed in detail through dynamic modelling. Chapter 11 describes the principles of thermal load assessment to demonstrate the impact of thermal inertia.

3.4.8 Control of air infiltration

It is necessary to control the level of air infiltration to avoid additional heat losses in cold periods and heat gains during hot periods. This requires good sealing and attention to architectural details for elements of the building envelope. Controlling air infiltration can be more problematic when the air pressure differential increases, for example, in tall buildings and buildings in exposed locations. Controlling air infiltration can also contribute to satisfying acoustic criteria by reducing noise infiltration. Section 3.9 describes the limiting design criteria in Part L2A. Chapter 6 outlines the design approach to air infiltration in relation to natural ventilation.

3.4.9 Arrangement of internal spaces

The allocation and disposition of spaces within buildings should be planned so that, where possible, their comfort criteria can be achieved in the most energy-efficient way. This goes beyond passive design for the envelope; it is about appropriate adjacencies for spaces and the suitability of locations for areas with different functions. Of course, it must always work alongside architectural aspirations for the internal planning and circulation and the functional interrelationships between spaces. In many cases there might be limited scope to achieve this, but it is usually worth exploring. Examples could be locating areas which are more sensitive to temperature variation, and which require good daylighting, on the north side; locating space which is transient with low environmental requirements in more problematic perimeter areas; and grouping spaces with similar environmental requirements together, providing the potential to utilise the local plant as a thermal control zone for heating, ventilation and air conditioning (HVAC) systems.

3.4.10   *Influence on architectural form*

The need for energy demand reduction through passive design is likely to have a continuing influence on the appearance of contemporary buildings. There is now much more evidence of solar control features, sometimes incorporated within more complex facades, that are a fundamental aspect of the architectural expression. There is likely to be a move toward well-insulated, thermally heavyweight construction and much less evidence of lightweight buildings with continuous facades of unshaded glazing. This trend is likely to continue.

Figures 4.7, 4.8 and 4.9 show a contemporary three-storey office building with a range of design features that provide an energy-efficient approach. The building's orientation and envelope have been designed to provide optimum energy performance. The wide horizontal brise soleil extending from the south facade and the deep overhang roofs on the east and west are prominent solar control features and are becoming more commonplace on modern buildings. The brise soleil shields the glazed facade from high sun angles in the summer, but allows lower-angle solar gain when it is beneficial during the heating season. The building has a typical depth of 16 metres to benefit from the prevailing wind direction so that it can use natural ventilation. This is also used at night-time for pre-cooling the office spaces in conjunction with the thermal mass of the exposed concrete soffit. The narrow depth also provides good daylighting levels, which aid health and wellbeing, as well as reducing lighting energy consumption. The building also uses building-integrated renewable energy systems: photovoltaics (integrated into the brise soleil) and solar water heating. Chapter 4 includes a post occupancy case study on this building and has more details about the design features.

3.4.11   *Passivhaus*

'Passivhaus' is a particular approach to the design and construction of buildings with significant passive features that have minimal requirements for active heating and cooling and good indoor air quality. The concept is based on a set of principles developed by the Passivhaus Instutute (PHI) in Germany. Passivhaus buildings require a high level of detail and a rigorous approach to both design and construction. There is a highly specific quality assurance process that can provide certification (Passivhaus Trust 2019).

The design objectives are based on achieving thermal comfort primarily through heat provided by solar gain and occupants, plus appliances and heat recovery from extracted air. The heat recovered heats the supply air that has a flow rate calculated to provide the amount of fresh air necessary for good indoor air quality. This can suffice for maximum heating loads below 10 W/sq.m of living space. This avoids having to recirculate air, which would reduce air quality (Passivhaus Trust 2019).

The key features to achieve the Passivhaus concept and standard include high levels of insulation ('superinsulation'); windows with insulated frames to avoid thermal bridging; a compact form to minimise surface area in relation to volume; a building that is airtight to avoid uncontrolled leakage; optimal orientation, plus glazing performance and ratios, to provide good daylighting and beneficial solar gains; and mechanical ventilation with high-efficiency heat recovery (MVHR) (Hopfe and McLeod 2015)

As the heat losses are so low, it is likely that only a small amount of heating (or cooling) will be required to supplement what is provided via the mechanical ventilation supply air in order to satisfy the peak heating (or cooling) load.

An important aspect of achieving the low-energy objectives in practice, and operational performance in line with the design and standard expectations, is that construction standards are exacting. An essential feature is that the building must be extremely airtight.

3.5 Active elements: engineering systems

3.5.1 *Key energy-consuming systems*

Having explored the potential for utilising passive design to reduce energy demand (and minimise the need for active systems), the next requirement is to look at the strategic approach for the required active engineering systems. It should be noted that there is an overlap between features that minimise energy demand (i.e. by reducing the magnitude and/or duration of the load) and those that improve the energy efficiency of the system itself. The measures are noted generically here, and most are covered in detail in other chapters. Many of the features listed are interrelated to some extent. To provide a focus for measures to reduce carbon, it is useful to examine the systems that account for the highest emissions and which would benefit from the most attention. This will vary with the building type, so it is useful to look at the typical breakdown by end use in different sub-sectors of the built environment.

Figure 3.5 shows the breakdown of final energy consumption in commercial office buildings in the UK, by system, in 2012. It can be seen that delivered energy consumption in

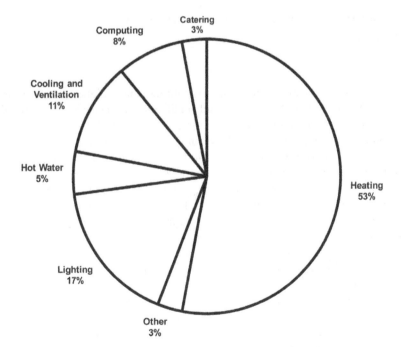

Figure 3.5 Breakdown of final energy consumption in commercial office buildings

Source: DECC (2012)

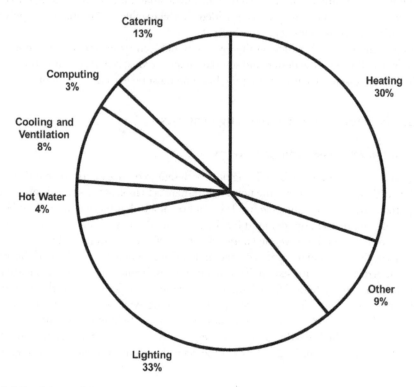

Figure 3.6 Breakdown of final energy consumption in retail buildings

Source: DECC (2012)

offices is dominated by heating. The figure for lighting is next highest, at about one-third the value for heating, followed by cooling and ventilation. A process load – computing – is higher than the next fixed building service – hot water services. As a comparison, the final consumption breakdown by system for the retail sub-sector in 2012 is shown in Figure 3.6. Here the highest energy consumption is for lighting, which is just above the figure for heating. A process load – catering – is next highest, followed by cooling and ventilation. It should be noted that Figures 3.5 and 3.6 represent an aggregate for all buildings in the relevant sub-sector. As such, they are representative of the building stock as a whole and do not therefore represent contemporary standards and design approaches or specific new building types. For both sectors, the energy consumption under 'other' is not defined, but it is likely that a considerable proportion relates to lifts and escalators. As would be expected, this is higher in the retail sector – with high levels of inter-floor traffic throughout the day – compared with the commercial offices sector.

   The pattern for the hotels and catering sub-sector is different. Based on 2012 figures, the proportion of end-use energy consumption for heating (at 33%) is not too far ahead of cater-ing (at 26%), and hot water (at 17%) is above lighting (at 14%), while for warehouses, heating (at 52%) and lighting (at 20%) together account for nearly three-quarters of the overall con-sumption (DECC 2012). In the UK residential sector (which represents the majority of the building stock), energy consumption for existing buildings tends to be dominated by heating, followed by hot water and lighting, although the specific pattern will depend upon the nature of the accommodation, building characteristics and occupancy arrangements. While these figures are general for 2012, they still provide a useful indication of how energy usage patterns differ across building sectors.

Examining system contributions by carbon emission (rather than final energy consumption) will present a different picture. For example, using the 2013 Building Regulations for England and Wales carbon fuel factors, consumed electricity would have a carbon impact about 2.4 times that of gas. Using these factors for the system contributions shown in Figure 3.5, the carbon emissions due to lighting would be about 77% of those from heating (assuming heating to be gas-fired). For Figure 3.6, the carbon emissions due to lighting would be about 2.6 times those due to heating (again assuming heating to be gas-fired). However, these carbon fuel factors are likely to change to better reflect the decarbonisation of the electricity grid and real-life seasonal carbon factors of the grid, so the relative carbon impacts noted earlier would change accordingly. Similarly, if the heating were from heat pumps (rather than gas-fired), the comparison would again be different.

A further consideration for carbon mitigation strategies is to minimise 'unregulated' or 'uncontrolled' and 'process' loads. The terminology is not always consistent, but 'unregulated' loads usually covers loads such as electronic equipment in offices, data processing or computer installations, audio-visual, catering and other equipment that can be deemed to be part of the business or activity processes, rather than fixed building services systems 'servicing' the occupied spaces in the building. Where the loads are specific systems or installations such as data centres or catering facilities, they are often called 'process loads'. Where there are miscellaneous loads fed from the small power system, they are often called 'plug loads' (particularly in the United States). These loads receive little attention in regulatory frameworks. While this can help to focus attention on the fixed building services systems, it can be seen from Figures 3.5 and 3.6 that some process loads (computing and catering) can be significant. Moreover, unregulated loads will contribute to the internal gains in spaces, and hence potentially the energy demand for cooling and ventilation. To be fully effective, energy and carbon reduction strategies (especially management regimes) should also seek to address uncontrolled and process loads.

3.5.2   *Energy strategy, carbon fuel factors and fuel selection*

The main features of an energy strategy are outlined in Section 3.3. In order to choose the most appropriate fuels for the different energy-using services, it is necessary to understand many interrelated factors. These will include aspects such as anticipated loads (and the dynamic nature of their load profiles, in particular, the frequency of low load conditions); availability within the local infrastructure to meet the capacity; access, space, delivery and storage (where required); and the environmental impact (particularly for carbon) and costs.

The load or demand assessment aspects are described in Chapters 11 and 12. At the earliest design stage, it is useful to develop simple demand profiles for heating, cooling and power (CIBSE 2007), which will aid the decision-making for fuel types and provide a focus for the main areas requiring attention. Figure 3.7 shows typical 12-month demand profiles. As the design concepts develop, it is likely that dynamic simulation modelling will be used as a design tool on many projects, providing a more meaningful representation of load scenarios, which can further influence decision-making for energy strategies.

In the UK, the preferred choice of fuel for heating is undergoing a fundamental change. For a long time, natural gas has been the preferred choice for heating, where a suitable

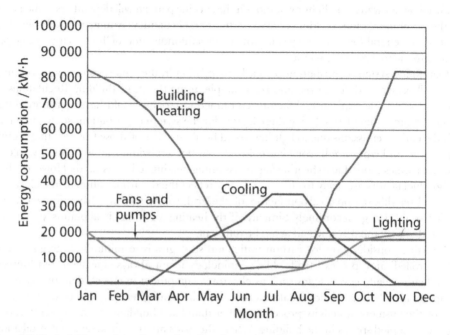

Figure 3.7 Typical 12-month heating and cooling demand profile

Source: Reproduced from CIBSE Guide L (2007) with the permission of the Chartered Institution of Building Services Engineers

capacity of mains supply is available, due to its lower primary energy carbon factor when compared to electricity, and the use of electricity for heating purposes was previously considered to be inappropriate, except in special cases. However, with the decarbonisation of the UK electricity grid that has already been achieved (and which is continuing with further growth in offshore wind production), it is likely that the carbon content of the electricity grid will continue to reduce, so serious re-consideration is necessary, and the 'decarbonising' or 'electrification' of heat is considered to be essential to meet climate targets. The anticipated reduction in Part L carbon fuel factors means that electric heating using heat pumps is likely to be a viable low-carbon and economic solution for many buildings. A further factor supporting the widening of electricity usage more generally is its versatility – the majority of energy-using equipment can utilise electricity – and its ability to be derived from a variety of renewable or low-carbon sources. However, the move towards widescale use of electric heat pumps will have impacts on the available capacities in the electricity supply infrastructure (which will, in any case, require significant development to provide the anticipated charging for electric vehicles and accept the continuing growth in renewable electricity generation). This might limit its adoption in some locations at some points in time, so the practicality is not yet clear.

A key factor in the selection of fuel type for heating is the carbon factor for electricity and its comparison with the carbon factors for alternatives (primarily natural gas in the UK). There has been considerable frustration that the factors used in the Building Regulations Part L in England and Wales (see Section 3.9) have not changed since 2013. These factors are 0.519 $kgCO_2$/kWh for electricity and 0.216 $kgCO_2$/kWh for gas. These provide a ratio of 2.4:1 between electricity and gas and are considered by many to be outdated and not representative of the major beneficial changes that have taken place in decarbonising the grid. The factors used in Part L are not due to be changed until the next update, anticipated for 2020. The Standard Assessment Procedure (SAP) used for new homes in the UK has a new version, SAP10, in which the carbon factor for electricity has been reduced by about half, to 0.233 $kgCO_2$/kWh, which brings it close to parity with gas; however, this is not due to come into effect until Part L is updated. In London, the GLA has been an early adopter of the SAP10 figure in their energy assessment guidance.

An important aspect of carbon factors is that they should properly reflect the real-life scenario that is likely to prevail for the assessment being made. For comparative decisions between fuels and systems to be meaningful, for the anticipated life expectancy of the system, fuel carbon factors should relate to the expected seasonal variations and the likely future variations over the relevant period of time. Seasonal variations should reflect the different electricity grid operational scenarios, particularly during winter (when the electricity grid demand is likely to be at a consistently high level). As key decisions will be primarily for heating proposals, the heating season scenario is all-important. The seasonal aspect will be influenced by changing weather due to climate change. Another dynamic of relevance is the changes in the generation mix of electricity grids over the relevant period of time.

In the UK and other countries where the decarbonisation of the grid has been taking place (and will continue further), the previous significant benefits from using combined heat and power (CHP) as a low-carbon and economic solution are unlikely to continue. The clear trend is towards electric heating via heat pumps. So, CHP is only likely to be worthwhile in countries or locations where decarbonisation has not taken place and the electricity to gas carbon factor ratio is sufficiently high and is likely to remain at that level for a significant proportion of the plant's life to justify its viability. However, much will depend on the relative carbon factors and the availability of the electricity infrastructure to support electric heating. Heat pumps for heating are discussed in Section 3.7 and Chapters 8 and 9. CHP is discussed in Section 3.8.

It is clear that, in many locations, the key factors influencing energy strategies for buildings are undergoing fundamental change at a rapid pace and are likely to continue doing so, causing uncertainty for decision-making. During this period of transition, engineers will need to use astute judgement to determine appropriate strategies for their projects.

3.5.3   *Energy-efficient plant and distribution arrangements*

A key consideration for any active engineering system is the extent to which the plant will be centralised or dispersed, as this will have a considerable impact on the energy losses due to distribution, and hence the overall energy efficiency. The basic approach should be similar for power, heating and cooling distribution, namely to locate the plant close to the main load centre(s) to minimise distribution losses in relation to the anticipated annual load pattern. The plant arrangement (i.e. the numbers and sizes of items of plant to meet the required installed capacity) should be selected to provide the best overall performance through the range of load scenarios. The norm might be a part-load condition well below full load, so diurnal and seasonal variations in energy efficiency should be considered. This will often require selection of multiple equal-sized modules or components to provide the best opportunity for optimised energy performance for the anticipated load pattern and for future load growth. A modular plant arrangement of this form, with facility to add future modules, could also aid adaptation for climate change load increase (or reduction), as outlined in Section 1.9

in Chapter 1. The selection should consider base-load capacities and shared part-load efficiencies. The need for energy-efficient distribution relates not only to primary distribution (or infrastructure) but also to the distribution elements of systems, such as cabling, ductwork and pipework. The intention is always to design distribution elements so that they are inherently energy-efficient by reducing lengths and resistance per unit length.

Spaces that are remote or used on a highly intermittent basis should be served by a separate localised plant that can be sized accordingly and run independently of the central plant. In such cases, it is usually better to run a smaller, slightly less efficient plant, rather than running a central plant with significant distribution losses and/or running in an inefficient mode for long periods.

It will also be necessary to minimise standing losses in all systems and equipment. This applies most obviously within thermofluid systems to reduce energy losses from bypasses and redundant legs. It also applies to electrical systems, including lift systems, multiple parallel uninterruptible power supply (UPS) systems and standby loads for process equipment.

3.5.4   Energy-efficient HVAC systems

The systems for controlling the indoor climate are pivotal to the overall building services design strategy and the energy performance. Suitable HVAC systems should be selected to satisfy the comfort criteria for the spaces served and to provide a healthy indoor climate. This will include considerations of the characteristics of the emitters or terminal devices and their locations within the spaces, so that the conditions are maintained effectively within the occupied zones. It will also include the methods for generating heating and cooling power, the types of thermofluid circuits and their operating modes and the arrangement and characteristics of fans or pumps for circulation. Treated spaces should be subdivided to allow separate control in a meaningful and representative way, so that spaces with different conditions or criteria act differently. It should be possible to match the particular energy demand with fast response, through suitable controls, so that energy consumption is no more than necessary to meet the requirements.

Systems should minimise the creation of waste heat and maximise the beneficial re-use of any waste heat that is unavoidably created, including heat recovery and recirculation of 'valuable' treated air, where this is viable. Designs should seek to minimise unwanted heat gains into cooled spaces and similarly maximise unwanted heat gains into heated spaces, where this is practical (CIBSE 2004a).

The design of energy-efficient HVAC systems forms a significant part of this book and is covered in Chapters 6–9 and their controls aspects are covered in Chapter 10.

3.5.5   Explore the design parameters

Design parameters, particularly those related to comfort conditions, should be explored and challenged, where appropriate, to see whether there is some scope to relax them – or perhaps to seek a partial relaxation for certain times and circumstances and thereby reduce energy demand (CIBSE 2007). This relates primarily to thermal and illumination criteria, which have a fundamental influence on energy demand. It also relates to acoustic criteria, which can influence attenuation requirements in ventilation systems, and hence fan energy (and the associated embodied energy). The opportunity to relax criteria will depend, to some extent, on the client's flexibility. However, the designer may be able to identify, as part of the brief development, the order of carbon (and cost) benefit versus the perceived shortfall in conditions for specific parameters to promote discussion so that an informed choice can be made.

3.5.6　*Motive power for fans and pumps*

The main motor-driven components within the HVAC systems are circulation pumps for heating or chilled water systems and fans for mechanical ventilation systems. There will also be motors for the compressors in heat pumps, where these are used for heating. The motors that drive pumps and fans can consume a considerable amount of energy. This can be optimised through selection of the motor and drive equipment, as outlined in Chapter 10, and through selecting suitable parameters for the hydronic or ventilation systems, as outlined in Chapters 6–9. Variable-volume systems should be used where appropriate for the dynamics of the load to allow matching and reduce energy usage for fans and pumps.

3.5.7　*Coolth generation and heat rejection*

The energy consumption and carbon impact related to the generation of cooling power will depend upon the types of equipment, choice of refrigerant and system parameters. This is covered in Chapters 7–9.

3.5.8　*Lighting*

Artificial lighting systems should be selected so that they maintain appropriate lighting of the spaces, utilising natural daylighting wherever practical. There is a particular need to minimise lighting energy usage in spaces that require cooling, as this can represent an internal gain that will require additional cooling energy to offset. The design approach for internal and external lighting is covered in Chapter 10.

3.5.9　*Unregulated and process loads*

Unregulated and process loads can represent a significant level of energy consumption in certain types of buildings. This can, in turn, increase the energy consumption for cooling and ventilation in many buildings. However, there is often considerable wastage of energy from small power systems and process loads, so it is one of the main areas requiring attention by the management regime (see Chapter 10).

3.5.10　*Hot water services*

Hot water services usually account for a relatively minor proportion of energy consumption and carbon emissions in most commercial and public buildings, when compared to HVAC systems and lighting, but it varies with building type. The relative proportion will be higher in buildings with high water usage, such as hospitals, hotels, leisure centres and certain types of residential accommodations. The primary focus should always be to reduce demand through incorporation of water-efficient devices, such as low-flow showers, to limit the amount of storage or heat-up energy required. There are also potential useful energy reduction measures through wastewater heat recovery for residential or shower blocks, and possibly for certain types of hotels. For buildings that have fairly continuous high-density usage, solar hot water can be a viable alternative, as outlined at Section 9.4.

3.5.11   *Lifts*

In some types of buildings there can be high levels of lift passenger traffic, either due to the disposition of the occupants by floor or the nature of the occupancy activity, or both. In such cases, lifts can account for significant levels of energy consumption. Various measures can be taken to optimise energy performance (see Chapter 10).

3.5.12   *Controls, monitoring and data collection*

All of the active engineering systems will require suitable controls to ensure that regulation takes place to maintain the functional performance at relevant times, while minimising energy usage. Metering and monitoring will also be required for the ongoing involvement and attention from the management regime. An essential feature of energy management will be the collection of data, and usage of information derived from analysis of the data, to inform actions and interventions to reduce energy consumption. The provision of controls, monitoring and data collection can be considered an aspect of both the active engineering systems and the management regime, as outlined in Section 3.6 and in Chapter 10.

3.5.13   *Renewable energy technologies*

The potential for utilisation of a site's available ambient or renewable energy should be considered within the energy strategy report. The main site considerations for renewable energy are the solar path and the wind pattern, as shown in Figure 2.8, together with ground (and possibly water) conditions for ground source energy systems. Solar and wind aspects will depend on the location and site factors, such as latitude, elevation and exposure. As stated earlier in this chapter, renewable energy technologies should not be considered the first step in a low-carbon energy hierarchy. They should instead be a lower-priority consideration and assessed in the context of the relative benefit of other available carbon reduction measures. Renewable technologies should only be considered viable if they provide a cost-effective solution or if their environmental or social benefits can be deemed to outweigh any additional costs when compared to alternative measures (CIBSE 2004a).

The range of renewable technologies that could be considered would typically include the following (see subsequent chapters):

- wind power
- photovoltaics
- solar thermal energy for hot water
- ground energy sources.

The key to selecting appropriate renewables is to understand the site potential and make a careful assessment of the likely, and most appropriate, renewables contribution that would be sensible within an overall coherent energy strategy. All renewable technologies will have a capital cost and an energy 'cost' related to embodied energy, together with some operational and maintenance costs. All renewables could involve some of the issues listed here, to a greater or lesser extent, which basic energy efficiency measures may not require:

- capital cost justification and/or client approval
- grants, to make an economically viable case

- planning permission, particularly where equipment is visible or where water courses or groundwater might be affected
- novel or untested technologies
- maintenance
- complexity and risk related to integration with the site, building and active engineering systems
- additional space allocation within the building or on the site, which has a cost impact and might limit other design aspirations.

The cost comparison will vary with location, load pattern and other factors. It is likely that the cost assessment will be subject to regular change due to variations not only in the capital and operational energy costs but also in the availability of grants and (where applicable) the tariffs for the sale of exported energy. It is always necessary to distinguish between cost per unit of carbon reduction (which relates to primary energy) and cost per unit of delivered energy reduction. It should be emphasised that the cost assessment, and hence the viability, will be highly dependent on location. In the UK and other countries, with the decarbonisation of the electricity grids, payback periods and carbon savings for renewable electricity generation are likely to extend and decrease, respectively.

3.6 Whole-life operation: management regime

As outlined previously, the realisation of good energy performance will, in practice, depend upon the way in which the building is occupied and managed. We can consider the 'management regime' to encompass both the formal arrangements for operating the building and the behaviours of the occupants – which can, in turn, be influenced to a considerable extent by the management regime. The desired outcomes can only be achieved through the combined involvement and commitment of both the operational staff and the occupants.

In order to encourage the occupants to adopt the most appropriate behaviours, it will be necessary to get their commitment and engagement, which can be assisted through providing education and information on the building's energy performance. This should directly relate the energy and environmental impacts in a way that can be understandable in relation to occupancy behaviours, management and personal intervention (CIBSE 2004a). The psychology of behavioural engagement is well beyond the scope of this book, but it should be recognised that creating a sense of ownership and involvement can be an essential factor in the success of the building.

The management regime should include an active strategy for energy management (CIBSE 2008a). This would be in the form of a comprehensive energy management policy that forms an integral part of (or is developed alongside) the overall plan and regime for operation and maintenance (CIBSE 2004a). This is also a function of plant space allocation to ensure maintenance effectiveness, as outlined in Chapter 13. A monitoring plan should form a key part of the policy and include a data collection strategy to provide information that will allow continual monitoring and control of the energy consumption. This will require a mixture of automatic and manual controls and metering and monitoring facilities that jointly provide the human–machine interface (HMI). The concept is shown in Figure 3.8 Appropriate automatic and manual control features should be incorporated within designs to facilitate matching system usage against actual requirement. This can include exploiting the operational variables in systems so that the mode selected is the most appropriate for the changing usage pattern of the building (see Chapter 10).

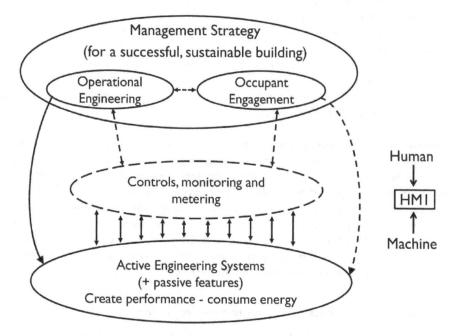

Figure 3.8 Concept for controls, metering and monitoring

The active engineering systems should be amenable to initial testing and commissioning in an integrated way, allowing the building and systems to perform in the manner intended by the design (CIBSE 2004a). This should include demonstrating achievement of design conditions for the anticipated range of scenarios. It is also useful to provide independent validation of the results and to seek optimisation through extended involvement during the initial period of occupation, as outlined in the BSRIA Soft Landings Framework. It is essential that sufficient time is allocated within the construction programme to allow for the full and integrated testing and commissioning of systems and for separate seasonal testing, where appropriate.

The systems should also be amenable to subsequent operation, adjustment and regular fine-tuning throughout the building's life, so that optimal energy performance can be achieved and maintained as the norm. This can be achieved through close agreement with the client and operational staff and the adoption of a suitable strategy for planned maintenance, usually including schedules of maintenance activities for all building services equipment, which will contribute to its energy efficiency. For example, to reduce fan energy consumption, there should be a planned activity for cleaning filters on a regular basis and replacing them periodically, so that the pressure drop does not increase. Because the occupancy arrangements and functional needs will change with time, it is inevitable that some periodic fine-tuning will be necessary. This will include adjusting the systems so that they match more closely the changing needs. Certain systems might require occasional partial re-commissioning from time to time so that the range of likely changes can be accommodated. This could include changes of usage; changes of occupancy types and levels; re-arranged layouts; or new departments, functions and associated equipment resulting in changes in internal gains (CIBSE 2008a).

3.7   Heat pumps for heating

In the UK there has been rapid progress in the decarbonisation of the electricity grid, as outlined earlier, primarily from the growth in supply from wind and solar photovoltaic sources and the reduction in supply from coal-fired power stations. Similar developments have been happening in some other countries. These developments are expected to continue in the UK as a key contributor to achieving net zero carbon emissions by 2050. This means that the effective carbon impact of grid electricity has been reducing significantly and will continue to reduce over the coming years, closing the gap in carbon impacts when compared to natural gas. This is changing the scenario for low-carbon heating provision for buildings in the UK. There is growing adoption of heat pumps for space heating and water heating in both new-build domestic and commercial buildings, and this is likely to become much more widespread with forthcoming changes to building regulations and adjustments to carbon fuel factors. The use of heat pumps for heating is described inChapters 8 and 9.

The anticipated growth in use of heat pumps for heating will have an impact on electrical loads for individual buildings and sites and will have a major impact locally and regionally at the distribution level and nationally at the grid level.

3.8   CHP and the impact of grid decarbonisation on viability

The viability of CHP in the UK is undergoing change due to the decarbonisation of the electricity grid achieved to date and the anticipated further decarbonisation and changes in carbon fuel factors and regulations. As the carbon fuel factor for grid-derived electricity reduces, there is a reduced case for CHP based on the environmental (and potentially economic) benefits over the expected lifetime of the installation. There is also growing concern at the emission of pollutants from CHP engine exhausts, particularly in urban locations where air pollution is a serious public health issue. It is likely, therefore, that the usage of CHP will only be considered viable in certain countries or locations where electricity generation remains dominated by conventional thermal power stations and where decarbonisation of the electricity grid has not happened to any great extent and is not expected to a significant extent during the anticipated life expectancy of the CHP plant. The basic concepts for CHP are outlined next in terms of appropriate locations and applications, which can be beneficial in environmental and/or economic terms and be deemed as viable; but this technology may feature in a more limited way, and for very specific applications, as electricity grids decarbonise in many parts of the world.

The basic principle and logic for co-generation of heat and power has been outlined briefly in Chapter 1. Conventional heat engines used for electricity generation only convert about one-third of their fuel (primary energy) input into mechanical energy, which is then available for conversion to electrical energy via an alternator. In a CHP plant, the engine is fitted with heat exchangers to capture the heat that would normally be rejected (e.g. from exhaust, jacket cooling, lubrication oil) and use it for applications such as space, water or process heating. Heat and power are therefore generated simultaneously, which provides a significant increase in overall energy efficiency. The specific efficiency will depend on the engine type and rating, the load cycle and so on. Typically, up to 80% (and possibly more) of the fuel input can be converted into useful energy (IET 2008; CIBSE 2009a). Figure 3.9 shows an energy balance for a typical small gas-fired CHP plant. Where it is viable, utilising CHP can reduce the relative emissions of carbon dioxide and operational energy costs (and may also reduce the relative levels of other pollutants), but

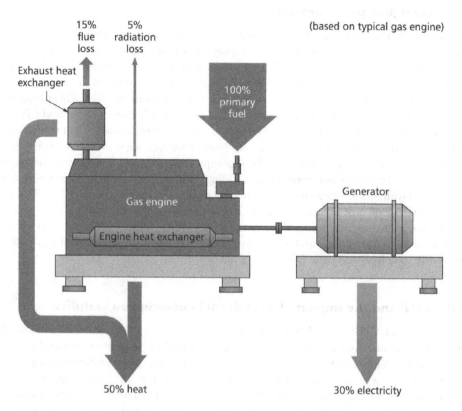

Figure 3.9 Energy balance for a typical small gas-fired CHP plant

Source: Reproduced from CIBSE KS14 (2009a) with the permission of the Chartered Institution of Building Services Engineers

the benefit will depend on the particular scenarios for carbon and other pollutants that are being displaced.

The prime mover for CHP systems used in buildings is normally a gas engine. It could also be a micro gas turbine. These are reciprocating engines fuelled by natural gas. Heat is recovered from the engine's exhaust and the cooling water jacket. The low-grade heating requirement for space and water heating allows a high proportion of heat recovery from otherwise wasted heat. However, it must be recognised that CHP is only viable in certain situations and applications and is highly dependent on the carbon fuel factor (and cost) for the grid-derived electricity that is displaced. For CHP to be viable, it usually requires specific site load criteria, namely daily demand for heat and power for either the whole or a considerable part of the year. The thermal demand would preferably be for applications that require elevated temperatures of low-temperature hot water (LTHW) for a building's heat load. If the thermal demand is not coincidental with the power demand, there should be a facility to store heat to suit the usage pattern. The main factor to consider in selecting a CHP size is the building's base heat load, so that the heat utilisation can be maximised. The most favourable sites will have heat demand throughout the year, while a good guide for economic viability is that the plant should operate at, or close to, full load for more than 5,000 hours p.a. (CIBSE 2009a)

The CHP plant is normally sized so that the daily heat output is equal to the daily base heating load, so the presence of a base-load heating demand is the key. The control system should ensure that the CHP plant always operates as the 'lead' heat generator (IET 2008; CIBSE 2009a). The electric power available is used when the CHP plant is running to meet the thermal demand and operates in parallel with the grid supply. When planning the sizing and integration of a CHP plant into the main heating and power systems, there are important design considerations on both the thermal and electrical sides (CIBSE 2004a). On the thermal side, there are issues concerning temperatures, pressures and thermal storage in relation to the heating demand and the operating characteristics of other items of the heat-generating plant. On the electrical side, there are considerations related to parallel operation, controls, protection, earthing and the load characteristics, including harmonics.

To allow a CHP unit to generate electricity to a level where the corresponding heat output is more than required at that time by the system and/or thermal store, a 'heat dump' facility would have to be included to reject the excess waste heat. If the plant is oversized, at times there will be an excess of heat that has to be dumped. It is obviously undesirable to have an operational cycle that involves significant dumping of heat, as it will be less beneficial in environmental and economic terms (IET 2008).

CHP has widespread applications in industries where there are continuous process thermal loads. In terms of buildings, potential applications could include:

*   swimming pools, because of the continual requirement to heat pool water and for space heating
*   leisure centres and large hotels, because of the continual requirement for hot water
*   hospitals and similar institutions, because of the continual requirement for hot water.

The assessment of viability for CHP should be based on detailed profiles for the thermal and power loads. These should be both realistic and reliable so that they represent diurnal and seasonal load variations with a high level of confidence (Carbon Trust 2010; IET 2008).

3.9 Regulatory context: Building Regulations Approved Document Part L in England and Wales

3.9.1 Background

Many countries have legislation in place to regulate the conservation of energy in buildings. Building services designers must create and submit their design proposals to the relevant authorities in the required format to achieve compliance in accordance with the prevailing legislation, both at the design stage and post-completion. In England and Wales, the Building Regulations Approved Document Part L 2013 (HMG 2016) is the document covering the conservation of fuel and power. There were minor non-technical amendments in 2016. The Approved Document provides practical guidance on ways in which compliance can be achieved for the relevant energy efficiency requirements of the Building Regulations. Reference is made here to selected aspects only of Part L 2013 (with 2016 amendments) for the purpose of providing some context of regulatory frameworks and how they influence the design approach. However, it should be noted that revised Part L regulations are anticipated in 2020. It is expected that these will change the design approach fundamentally.

The 2013 regulations were a development of earlier versions in which regular improvements had been made to reduce environmental impact. The 2006 regulations represented a significant tightening of the carbon emissions criteria compared with the previous version (2002), with typical reductions in target levels for non-residential buildings in the order of 28%, and about 20% for residential buildings. The strategic objective for Part L 2010 was to further reduce carbon emissions from the building stock and to improve compliance by closing the performance gap between that predicted at the design stage and the actual emissions as measured at completion. The stated aim was to achieve an overall reduction in carbon emissions of 25%, relative to the 2006 regulations, as cost-effectively as possible. The objective was to achieve this in two ways: first, as a flat 25% reduction for domestic buildings; and second, as an aggregate 25% reduction for non-domestic buildings. The 2013 regulations had an objective to deliver 9% carbon dioxide savings across the new non-domestic building mix, relative to Part L 2010.

It is usual for the primary energy fuel factors used within the CO_2 emissions calculations to change with each update. In the 2010 regulations, the carbon emission factor for grid gas increased by 4% to 0.198 $kgCO_2$/kWh, and for grid electricity by 22% to 0.517 $kgCO_2$/kWh, compared with 2006. In the 2013 regulations the carbon emission factor for grid gas increased by 9% to 0.216 $kgCO_2$/kWh, and for consumed electricity by 0.4% to 0.519 $kgCO_2$/kWh, compared with 2010. The 2013 regulations also use a figure of 0.519 $kgCO_2$/kWh for electricity displaced (e.g. by CHP or photovoltaics). These carbon factors are considered to be outdated and not representative of the present or future grid scenarios. It is anticipated that the carbon factor for electricity will be reduced by about half for the next update, which would make it close to the figure for gas.

With the urgent need to reduce emissions and the anticipated further tightening of criteria in the near future, it is appropriate to see the Part L criteria as the minimum acceptable performance standards, rather than the actual preferred target. They represent a set of overall regulations for carbon reduction against which designers need to demonstrate compliance, but that does not prevent designers from aiming for further improvements in performance, some of which may not be too difficult to achieve. This aspect of the project brief should be discussed with the client as part of environmental target-setting during the brief development stage. Indeed, other aspects of environment benchmarking may override Part L targets (such as targets set by an environmental assessment methodology).

The 2013 regulations cover new buildings and changes to existing buildings for both domestic and non-domestic buildings:

Part L1A Domestic buildings (new)
Part L1B Domestic buildings (existing)
Part L2A Non-domestic buildings (new)
Part L2B Non-domestic buildings (existing)

Because of the relatively higher energy impact per unit area of non-domestic (i.e. commercial and public) buildings, the following sections mainly relate to Part L2A as an example of the requirements.

3.9.2   *Criteria for compliance in Part L2A 2013*

For new non-domestic buildings, the regulations have five criteria, which have a sequential relationship. Of these, Criterion 1 is a regulation and is mandatory, while Criteria 2–5 are for guidance.

Criterion 1 relates to the building's calculated emission rate for CO_2, called the building emission rate (BER). This figure is required to be no greater than the target emission rate

(TER) for the building. In order to achieve compliance, a target-achieving building has to be presented in the design submission. The regulations do not provide a direct target figure for CO_2 emissions. Instead, the building design has to be input into approved software. This creates a notional building with specific properties that is geometrically similar to the proposed building. The TER figure is the CO_2 emissions rate from the notional building. This provides a flexible approach, as a BER which is equal to or less than the TER can be achieved using any mix of design factors for the envelope, active engineering systems and LZC technologies, as long as the building meets the limits on design flexibility in Criterion 2.

Criterion 2 relates to the performance of individual elements of the building fabric and the fixed building services, i.e. the fixed systems for heating, hot water services, air conditioning or mechanical ventilation and internal lighting systems (but excluding emergency escape lighting or specialist process lighting). The regulations require that these should be no worse than the stated minimum energy efficiency standards in the Non-domestic Building Services Compliance Guide (HMG 2013) for systems installed in new buildings. The intention is to place limits on design flexibility, which is statutory guidance, rather than a regulation. However, in reality, it is likely that many aspects will need to be considerably better than the minimum standards in order to meet the TER figure under Criterion 1.

A few selected examples of the guidance for Part L2A 2013 are:

- The system efficiencies of the fixed building services must be equal to or better than the minimum efficiencies in the Non-domestic Building Services Compliance Guide (HMG 2013).
- For air distribution systems, there are limitations on fan power per unit of air flow rate. The specific fan power (SFP) in watts per l/s is the total design circuit watts of all the fans that supply and exhaust air in the air distribution system, divided by the system's design air flow rate. It includes losses in switchgear and controls. For example, there is a limiting SFP of 1.6 for a central balanced mechanical ventilation system that includes heating and cooling, or 1.5 with heating only. There are allowances for components such as additional fine filters and heat recovery devices. The guidance suggests that fans rated at more than 1100 W should be equipped with variable-speed drives.
- There are minimum efficiencies for heat exchangers for dry heat recovery in air distribution systems, e.g. 50% for a plate heat exchanger, 60% for a heat pipe, 65% for a thermal wheel and 45% for a run-around coil. However, it should be noted that the notional building in Criterion 1 utilises heat recovery, with sensible efficiency of 70%, in zones with mechanical supply and extract.
- There are design limits for fixed internal lighting, with two options to demonstrate compliance. The first option relates to 'effective lighting efficacy'. For general lighting in office, industrial and storage areas, the effective lighting efficacy should be not less than 60 luminaire lumens per circuit-watt. This is the average for all luminaires in the relevant areas of the building, divided by the total circuit watts. Luminaire lumens is (lamp lumens × LOR), where LOR is the light output ratio. For general lighting in other types of space, the limit is 60 lamp lumens per circuit-watt. For display lighting, the limit is 22 lamp lumens per circuit-watt. Control factors may be applied to reflect the energy reduction arising from controls. For example, for a luminaire in a daylit space with photoelectric switching or dimming control, a control factor of 0.9 can be applied. The second option is for the whole lighting system to have a lighting energy limit lower than that indicated in the Lighting Energy Numeric Indicator (LENI). This is described in more detail in Chapter 10.

- For comfort cooling systems, chiller performance is expressed as an energy efficiency ratio (EER), which is defined by calculation. For example, for a vapour compression cycle water-cooled chiller with a capacity greater than 750 kW, the minimum EER is 4.7, whereas for a vapour compression cycle air-cooled chiller with a capacity greater than 750 kW, the minimum EER is 2.65. There is also a cooling system seasonal energy efficiency ratio (SSEER). This takes account of distribution losses and fan energy associated with heat rejection.
- For heat generators, there are seasonal coefficients of performance (SCOP).
- For elements of building fabric, the limits on design flexibility have not changed since 2010, as shown in Table 3.1.

The design limit for air permeability relates to the airtightness of the building envelope that encloses 'treated' spaces. It is the air leakage rate per hour for each square metre of envelope area (i.e. the total area of all floors, walls and ceilings bordering the internal volume) at the reference pressure differential of 50 Pa. The regulations provide a design limit of 10 m³/(h.m²) at 50 Pa. While it is allowable for air permeability to be as high as 10.0 m³/(h.m²) at 50 Pa, the notional building assumes lower air permeability rates; therefore, using an air permeability of the design limit level is likely to make it more difficult to achieve the target BER.

Criterion 3 relates to control measures to limit solar gains. It requires that appropriate measures are provided to limit solar gains through glazing to each space that is occupied or mechanically cooled, with the intention to reduce the requirement for mechanical cooling. The purpose is to limit solar gains to reasonable levels during the summer period, so that the need for, or the installed capacity of, air conditioning systems is reduced (HMG 2013).

Criterion 4 is about measures for ensuring that the performance of the completed building is consistent with the predicted performance of the BER assessment. This includes a requirement for testing the achieved permeability, the air leakage rate from ductwork and the commissioned fan performance.

Criterion 5 relates to the operation of the building. It requires that certain provisions are put in place to enable the building to be operated in an energy-efficient way. This includes provision of information in a logbook, including the data used in the TER/BER calculation and the Energy Performance Certificate recommendations.

It is emphasised that the brief outline of the regulations provided here is selective and has been greatly simplified to provide a few specific examples only. To appreciate in full the detailed implications of the regulations and guidance, reference should be made to the original source and supporting documentation.

Table 3.1 Part L2A: limiting U-value standards (W/m²K)

Element	U-value (W/m²K) (worst acceptable standard based on area-weighted average for all elements of that type)
Wall	0.35
Floor	0.25
Roof	0.25
Windows and rooflights	2.2

Source: Adapted from Part L2A 2013 (HMG 2016)

3.10 Summary

In this chapter we have seen the need for a focused approach to energy efficiency and carbon mitigation based on a logical hierarchy of measures. The starting point should be to reduce demand by optimising the building envelope with respect to heat loss, heat gains and daylight. An introduction has been provided to the wide range of passive measures that can be considered as part of the early-stage concept design to allow development as part of the interdisciplinary design collaboration. The next priority should be to incorporate energy efficiency and demand reduction measures in the energy supply and the active engineering systems. The decarbonisation of the electricity grid in the UK (and elsewhere) will have an increasing influence on energy strategies and system selection. It is likely that the 'decarbonisation of heat' for buildings will be largely achieved through heat pumps. The use of CHP is likely to be limited to energy scenarios where the electricity to gas carbon ratio remains high. A brief introduction has been provided for the generic energy efficiency considerations of the most relevant mechanical and electrical systems, all of which are covered in some depth in subsequent chapters. Incorporation of renewables should generally be a lower priority. It is essential to address all the technical issues related to renewables, particularly the likely energy yield and carbon savings in practice, so that the viability can be assessed in the context of alternative carbon reduction measures and regulatory criteria.

An essential aspect of the approach for delivering energy-efficient buildings is the adoption by the client of a management regime that will be committed to running the building in an optimal way. Some of the key considerations for effective whole-life operation have been outlined, including commissionability, data collection to provide information for management reporting and improvement actions and a planned and structured approach to operation and maintenance.

While building services designers can identify a suitable hierarchical approach to energy performance, they must, as a minimum, ensure that their designs comply with the relevant regulatory requirements. The Building Regulations Approved Document in England and Wales Part L2A 2013 has been briefly described as an example (noting that it is anticipated that it will be changed fundamentally in 2020).

4 Post occupancy evaluation for optimal energy and environmental performance

4.1 Introduction

This chapter outlines the methods of carrying out a post occupancy evaluation (POE) of a building. It continues from the previous chapter, which outlined how to address sustainability at the design stage to outline how to assess whether sustainable design strategies have been implemented and operated successfully. The POE is of practical importance in the improvement of existing buildings and also generally, as it can help to inform design decisions for future buildings by highlighting common pitfalls and successes to be avoided/implemented.

With the introduction of Energy Performance Certificate (EPC) regulations and Display Energy Certificates (DEC) for most buildings in the UK and other European countries, the investigation of design and operational energy performance is compulsory (or it will be soon) for most buildings; the investigation of environmental performance and user satisfaction seems to be the next step. As in Chapter 3, this chapter only covers carbon mitigation and not the many other issues relevant to sustainability in buildings, such as water, drainage, materials, recycling and biodiversity.

4.2 The European Energy Performance of Buildings Directive

In Europe, as elsewhere in the world, there is a strong demand to reduce energy use, both to mitigate CO_2 emissions and to strengthen security in supply. In the case of buildings in European Union member countries, this is managed through the Energy Performance of Buildings Directive (EPBD), its recast and targeted revision (EPBD Directive 2002/91/EC, 2010/31/EU and 2018/844/EU), which require member states to apply minimum requirements covering the energy performance of new and existing buildings. It also addresses nearly zero-energy buildings, building renovation and indoor environmental quality. The EPBD therefore plays a major role in forming building energy policy in the EU. The directive requires:

- a common methodology for calculating the integrated energy performance of buildings
- minimum standards on the energy performance of new buildings and existing buildings that are subject to major renovation
- systems for the energy certification of new and existing buildings and, for public buildings, prominent display of this certification and other relevant information (certificates must be less than five years old)

- regular inspection of boilers and central air conditioning systems in buildings and, in addition, an assessment of heating installations in which the boilers are more than 15 years old.

The common calculation methodology must include all aspects that determine energy efficiency and not just the quality of the building's insulation. This integrated approach should consider of aspects such as heating and cooling installations, lighting installations, the position and orientation of the building and heat recovery, etc.

Some countries have based minimum standards on a reference building approach in which a specified minimum improvement over an equivalent 'reference building' is required. However, this has led to anomalies that can result in air conditioning being used even when it is shown to be far more energy intensive than equivalent passively cooled buildings providing the same level of comfort. Some countries have avoided the reference building approach and based requirements on an actual energy target, irrespective of the means by which this is achieved. In such cases an allowance or 'fictitious' cooling amount is often factored into the energy design calculation to discourage a low-energy solution that subsequently requires air conditioning to meet comfort needs. In other words, if a design is shown to not reasonably fulfil summer thermal comfort requirements, an allowance for subsequent cooling energy must be factored into the energy estimate, even if a cooling system is not included in the initial construction. Evaluation is largely based on approved calculation methods followed by energy monitoring of the actual building once constructed (Liddament 2009).

This chapter does not cover the requirements and methodologies to produce EPCs and DECs required by the implementation of the EPBD, as these are regulated by Building Regulations (mainly Part L in the UK) of different countries in Europe.[1]

4.2.1   Case study: the present implementation in the UK

EPCs are now required in the UK for all buildings whenever built, rented or sold. The certificate records how energy-efficient a property is as a building and provides A–G ratings. These are similar to the labels now provided with domestic appliances such as refrigerators and washing machines. An EPC is always accompanied by a recommendation report that lists cost-effective and other measures (such as low- and zero-carbon-generating systems) to improve the energy rating. A rating is also given showing what could be achieved if all the recommendations were implemented. EPCs are produced by accredited energy assessors.

DECs show the actual energy usage of a building, the operational rating and help the public see the energy efficiency of a building. This is based on the energy consumption of the building as recorded by gas, electricity and other meters. The DEC should be clearly displayed at all times and be clearly visible to the public. A DEC is always accompanied by an advisory report that lists cost-effective measures to improve the energy rating of the building. Figure 4.1 shows the DEC of one of the buildings at Brunel University for 2018.

At present in the UK, DECs are only required for buildings with a total useful floor area over 1,000 m² that are occupied by a public authority and institution providing a public service to a large number of persons and therefore visited by those persons. They are valid for one year. The accompanying advisory report is valid for seven years. The requirement for DECs came into effect on 1 October 2008.

Display Energy Certificate
How efficiently is this building being used?
⊛ HM Government

Brunel University London
MARY SEACOLE
Brunel University London
Kingston Lane
UXBRIDGE
UB8 3PH

Certificate Reference Number:
9646-1067-0711-1304-2421

This certificate indicates how much energy is being used to operate this building. The operational rating is based on meter readings of all the energy actually used in the building including for lighting, heating, cooling, ventilation and hot water. It is compared to a benchmark that represents performance indicative of all buildings of this type. There is more advice on how to interpret this information in the guidance document *Display Energy Certificates and advisory reports for public buildings* available on the Government's website at: www.gov.uk/government/collections/energy-performance-certificates.

Energy Performance Operational Rating

This tells you how efficiently energy has been used in the building. The numbers do not represent actual units of energy consumed; they represent comparative energy efficiency. 100 would be typical for this kind of building.

More energy efficient

| **A** 0-25 |
| **B** 26-50 |
| **C** 51-75 ◀ **54** |
| **D** 76-100 |

•••••••••••••••••••••••••••••• 100 would be typical

| **E** 101-125 |
| **F** 126-150 |
| **G** Over 150 |

Less energy efficient

Total CO₂ Emissions

This tells you how much carbon dioxide the building emits. It shows tonnes per year of CO_2.

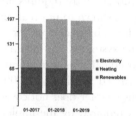

▪ Electricity
▪ Heating
▪ Renewables

01-2017   01-2018   01-2019

Previous Operational Ratings

This tells you how efficiently energy has been used in this building over the last three accounting periods.

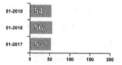

01-2019   54
01-2018   56
01-2017   53

0   50   100   150   200

Technical Information

This tells you technical information about how energy is used in this building. Consumption data based on actual meter readings.

Main heating fuel: Natural Gas
Building environment: Heating and Mechanical Ventilation
Total useful floor area (m²): 4210
Asset Rating: Not available

	Heating	Electricity
Annual Energy Use (kWh/m²/year)	76	57
Typical Energy Use (kWh/m²/year)	174	92
Energy from renewables	0%	0%

Administrative Information

This is a Display Energy Certificate as defined in the Energy Performance of Buildings Regulations 2012 as amended.

Assessment Software:	i-Prophets Energy Services, digitalenergy, v3.1
Property Reference:	661743640014
Assessor Name:	Jonathan Lee Cranefield
Assessor Number:	LCEA141129
Accreditation Scheme:	CIBSE Certification Limited
Employer/Trading Name:	Digital Energy Ltd
Employer/Trading Address:	Manchester Science Park, Unit 10 Enterprise House, Lloyd Street North, Manchester, M15 6SE
Issue Date:	22-03-2019
Nominated Date:	01-01-2019
Valid Until:	31-12-2019
Related Party Disclosure:	Not related to the occupier.

Recommendations for improving the energy performance of the building are contained in the associated Recommendation Report - .
You can obtain contact details of CIBSE Certification Limited at cibsecertification.com.

Figure 4.1 Example of a DEC in the UK taken from one of the buildings at Brunel Campus, Uxbridge

Source: Courtesy of Brunel University London

4.3   Why do we need POE?

Energy assessment (already covered by Building Regulations in Europe, as outlined in the previous section and elsewhere) is not the only issue to be addressed by a POE, which is wider to include environmental conditions and user satisfaction.

4.3.1   Environmental assessment tools

A number of environmental assessment tools have been developed, with BREEAM and LEED being the most well-known.

BREEAM (BRE Environmental Assessment Method) was developed by Building Research Establishment (BRE) in the UK in the 1990s. Credits are awarded in nine categories according to performance and combined together using the weighting of each category to produce a single overall score. A building is then awarded a Pass, Good, Very Good, Excellent or Outstanding depending on its overall score. These categories include management, health and wellbeing, energy, transport, water, materials, waste, land use and ecology and pollution. There are BREEAM schemes for a range of building types.

LEED (Leadership in Energy and Environmental Design) was developed by the US Green Building Council (USGBC) for the US Department of Energy. The first version was launched in August 1998. In LEED version 3 (2009), there are 100 possible base points plus an additional 6 points for innovation in design and 4 points for regional priority. Buildings can qualify for four levels of certification, which are Certified (40–49 points), Silver (50–59 points), Gold (60–79 points) and Platinum (80 points and above). The rating system addresses six major areas, which are sustainable sites (16 possible points), water efficiency (10 possible points), energy and atmosphere (35 possible points), materials and resources (14 possible points), indoor environmental quality (15 possible points), innovation in design process (6 possible points) and regional priority (4 possible points).

Similar to BREEAM, different versions of the rating system are available for various building types. Other countries have been developing environmental assessment tools as follows:

CASBEE (Comprehensive Assessment System for Building Environmental Efficiency) started in Japan in 2001. According to CASBEE website, it is a joint industrial/government/academic project initiated with the support of the Japanese Ministry of Land, Infrastructure, Transport and Tourism (MLIT). There are two spaces involved in CASBEE, internal and external, which are divided by a hypothetical boundary defined by the site boundary and other elements. Therefore, two factors are considered, namely:

- Q (Quality): Building Environmental Quality and Performance: evaluates 'improvement in living amenity for the building users within the hypothetically enclosed space'.
- L (Loadings): Building Environmental Loadings: evaluates 'negative aspects of environmental impact which go beyond the hypothetical enclosed space to the outside'.

CASBEE deals with four assessment fields, namely energy efficiency, resource efficiency, local environment and indoor environment.

NABERS (National Australian Building Environmental Rating Scheme) is the Australian government's initiative to measure and compare the environmental performance of buildings in Australia. NABERS provides ratings for buildings based on their measured operational impacts on the environment. NABERS rates the waste and indoor environmental quality for offices, as well as the water and energy use of hotels, offices and homes. The higher the NABERS star rating, the better the actual environmental performance of a building. The ratings are from 1 to 5 (poor to exceptional) with an increment of 0.5. The Green Building Council of Australia's Green Star rating is different from NABERS: the focus of Green Star is on the potential of design features in new buildings to reduce a range of environmental impacts, whereas NABERS is focused on the actual environmental impact of existing buildings over the previous 12 months.

GBTool has been under development since 1996 and is used in the Green Building Challenge, which is an international collaborative effort to develop a building environmental assessment tool that exposes and assesses the controversial aspects of building performance and from which participating countries can draw ideas to incorporate into or modify their own tool. It has evolved considerably, and over 25 countries are now involved in the system. The latest version, GBTool 2005, is developed by the International Initiative for a Sustainable Built Environment (iiSBE). There are three factor levels when assessing a building using GBTool: the high level 'Issues', the second level 'Categories' and the third level 'Criteria'. The top level consists of seven general performance issues, 29 categories are included in the second level, whereas the third covers 109 criteria. The weighting of the scores at the lower levels is used to derive the assessment scores, i.e. category scores are obtained by aggregating the constituent criteria-weighted scores. The weighted scores of issues are then used to obtain the overall score of the building. The weighting value, from the lower levels to the overall building, is a total of 100%.

Hong Kong Building Environmental Assessment Method (HK-BEAM) was first launched in December 1996 with funding from the Real Estate Developers Association of Hong Kong. HK-BEAM is used to measure, improve, certify and label the whole-life environmental sustainability of buildings. It assesses buildings in terms of whole-life site, material, energy, water, indoor environment and innovative aspects. Improvements are identified during assessment, and buildings are labelled Platinum, Gold, Silver, Bronze or Unclassified accordingly. It integrates the following aspects such as land use, site impacts and transport, hygiene, health, comfort and amenity, use of materials, recycling and waste, water quality, conservation and recycling, energy efficiency and conservation. The overall grade is based on the level of applicable credits gained, as well as on a minimum percentage of indoor environmental quality.

Thus, in the last 20 years, a number of environmental assessment tools have emerged that are useful for POE assessments. But how did POE start?

4.3.2   History of POE

A literature review on the subject carried out (Pegg 2007) indicates the following:

1   Theory followed on from 'operational research' – post Second World War.
2   Initial POE findings appear in the late 1950s and early 1960s, focused on mass house building projects.
3   Part of original RIBA – Royal Institute of British Architects (1965) Plan of Work (Stage M), later omitted (1973).
4   The field of environmental psychology grew in the 1970s, which led to a desire to understand how buildings affected people.

5   This was coupled with the energy crisis, out of which came a movement for 'green' buildings.
6   Concerns arose as sick building syndrome was observed in a number of deep plan air-conditioned buildings, leading to a wider debate about the purpose of buildings.
7   Probe Studies (UK) performed POEs on 20 UK low-energy buildings, 1995.
8   UK green building effort steps up with revisions to Building Regulations, planning policy and the adoption of BREEAM.
9   Federal Facilities Council (US) reported that they systematically review all new buildings (2003).

A study in the late 1990s found that clients of the UK construction industry were frequently dissatisfied with the completed product, while the industry reported low profitability. One of the reasons for this dissatisfaction is the lack of a natural feedback route for designers (and contractors) to learn how their buildings are working and what the users really think of the solutions. This process is indicated diagrammatically in Figure 4.2.

Indeed, it is not only designers who lack feedback. There is a trend to outsource facilities management in large corporations; this means that the parent organisation is not closely connected to the operations of the building and is not able to improve briefing on subsequent buildings.

However, feedback was recognised as being a useful component of design and was included as Stage M of the RIBA Plan of Work (this is a document that describes the responsibilities of architects at each stage of a building project) in 1965. However, it was not sustained, being removed in 1973. This is possibly because architects did not receive fees for reviewing their work.

One reported problem is that benefits are split between the client (current and future), design team and construction team and, therefore, no one believes that they should fully fund the process of POE. Of course, other barriers exist, such as the reluctance to discover negative aspects of the design (potentially leading to costs) and the fact that a designer appreciates that decisions made within a project are a result of a variety of mitigating factors (personalities, costs, time, etc.) and therefore feels POE can make little difference.

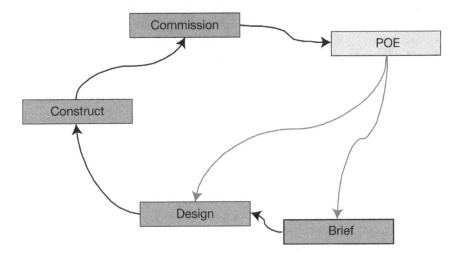

Figure 4.2 The role of POE in the building design and use cycle

Barriers to POE can be summarised as:

- designers do not get paid for reviewing their work, as the current client rarely benefits
- decisions are made in a growing team, responsibility is diluted and therefore it can be difficult to use the feedback
- litigation is becoming more common, and therefore the POE can highlight problems that would have been ignored in the past
- managing knowledge and capturing context in large organisations, and within teams, are difficult (the possibility of information overload)
- problems are more commonly reported than successes, which can demotivate staff
- there's a danger of only reporting what is easily measurable, rather than what represents performance.

Despite these barriers, there are a wealth of benefits reported, such as:

- rapid feedback leads to quicker reaction (rather than other research routes)
- is closely linked to design and therefore can become easily actionable
- improves the ability to predict and therefore minimises risk in the future
- good results can be marketed effectively
- decisions can be made more quickly if relevant evidence is available
- the alternative is to continue to build in ignorance of the outcomes.

In addition, user satisfaction is a very important consideration when operational buildings are investigated. A building with low energy consumption alone does not indicate a successful project. From early on it has been stated that buildings are for, and should go some way to, supporting objectives and activities carried out. We must therefore make some reference to the users and their needs within a building to ensure that a balanced evaluation of a building has taken place.

The link between people and their environment has been the study of extensive research in offices by environmental psychologists. Satisfaction in the built environment is commonly assessed by rating scales of satisfaction for tangible environmental variables, such as acoustics and thermal comfort. It is the purpose of these questionnaires to gauge how satisfactory a building or a space is, with the intention that designers can know what they should improve next time. An example is described in the following section.

4.4   POE methods

A historical literature review on the subject (Pegg 2007) identified the POE methods described in this section. As mentioned before, the purpose of a POE is always to understand more about how a building performs, which suggests that objective POE requires a method, metric and benchmark.

The key elements that these fields have in common are the recognition that the subject concentrates on the interactions between the users and the built environment and subsequent connection of that information with the building design process. The feedback process can be summarised in Figure 4.3.

Some of the earlier studies looked at attempts to meet housing demands in the 1930s and 1940s to assess the effectiveness of architects' attempts at providing functional spaces

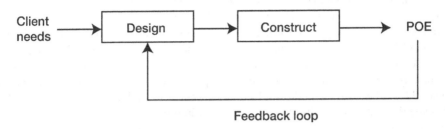

Figure 4.3 Basic concept of feedback within building design

around society. These studies found that architects' assumptions about how these societies would function were naïve and in error. The studies were grounded in the social sciences, using interviews and observation to determine the successes and failures of estates. The work looked at how spaces affected socialisation and practical living, all compared to explicit assumptions and hypothesis.

Studies in the early 1970s used a variety of techniques to determine the needs and assess the current practice of building comprehensive schools. They viewed the problem from first principles and developed a representative model of how organisations within buildings operate, determining the influence of the environmental systems on the activities of the users. The studies used a variety of methods, including examining current design practice, using surveys and taking physical measurements in buildings.

The work identified methodologies and spatial planning guidance that affected future buildings, highlighting some workable concepts for planning a building (such as the perimeter on plan ratio, a comparison of the actual perimeter of the building compared to a cylinder of equivalent floor area).

In North America the theoretical foundation of studying buildings after completion was discussed in studies, which viewed the work from an environment–behaviour point of view and used multi-methods to look at how the building affects the users. The work pointed out the conflicts between airtight 'energy-efficient' buildings and health-related ventilation problems.

The prevalence of open-plan air-conditioned offices saw much dissatisfaction. Sick building syndrome was one of the key problems in these open-plan offices, and comprehensive studies were undertaken to provide more knowledge about this effect. Most studies found little direct linkage between a particular building type and heating, ventilation and air conditioning (HVAC) technology, but did find that the ability to exert control over one's environment was associated with symptoms.

The interest in the environment, and specifically energy usage, was investigated in a range of POEs of buildings incorporating passive solar features. Known as the Solar Building Studies carried out in the 1990s, they monitored the energy consumption of a number of building types and passive approaches.

More recently, the Probe studies (see the next section) reviewed over 20 buildings previously featured in the *Building Services Journal* and used techniques to assess the build quality (air pressure test), energy performance (audit) and user satisfaction (questionnaire). Key findings showed that designers' estimates of consumption were often half the actual consumption (due to unrealistic assumptions) and that chronic problems persisted to some extent in most of the buildings.

One will begin to note that the topic of POE is incredibly diverse in terms of methods used and focus of the studies. This highlights the numerous activities and objectives that buildings support and the ever-growing list of stakeholders involved in buildings. It is clear that no POE can claim to be truly holistic; however, do we consider a simple building energy audit a POE? It seems that taking some researchers' belief that buildings are constantly designed as users adapt their environment to support their needs, then it appears that we should. This is because the audit is a feedback mechanism leading to a decision based on balancing the cost and benefits of energy efficiency improvements.

Yet in order to frame the field and tentatively suggest that to become manageable POE should rest alongside design, we suggest that the studies should seek to evaluate the outcomes of design within a recognisable project environment. What is more, we should *also* consider that the outputs should inform future design, with the intention that a body of knowledge is generated and tested.

While the work should inform future design, it should concentrate on the improvement of building performance from the point of view of the building stakeholders and seek to develop robust and measurable performance indicators that demonstrate progress.

This consideration has implications for future POE practitioners, as it means that work carried out must attempt to understand and describe the original context of the building design or disseminate to a field where the context is shared. The idealised process of POE is outlined in Figure 4.4. The figure shows that the outcomes of the design process are a building with high environmental impacts and moderate 'building performance'. The POE process adds to the 'knowledge' of the design organisation and feeds into the next project; the cycle is repeated with improved outcomes from the next building.

As is suggested by the nature of the studies highlighted in the previous section, POE studies are diverse in nature. The types of studies carried out vary from single-issue studies, using fairly focused methodologies, such as field thermal comfort studies; to wider focused studies, such as the sick building syndrome studies, which utilise a variety of methods to elicit influencing factors; and finally 'holistic studies', which attempt to cast a

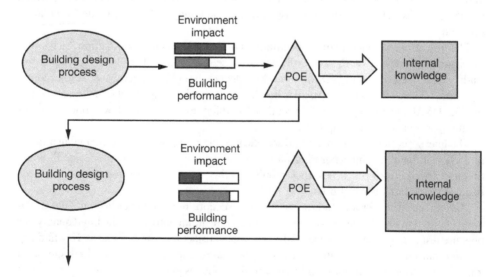

Figure 4.4 Conceptual representation of the POE system in design

wide net to learn about a wide range of performance parameters. There are three types of studies:

- *technical* – looking at the measurement of physical parameters to describe the environment generated by the interaction of building services, building fabric and human behaviour
- *behavioural* – looking at what links the occupants' satisfaction with the physical environment
- *functional* – determining how the building directly supports the activities carried out within the building.

The question arises: What is possible to measure in a building? Parameters that have established methods of measurement are:

- *user perception about their environment* – usually through questionnaires and comparison with established benchmarks.
- *energy* – usually addressed by prescribed methods in national Building Regulations in European countries. In the UK, approved methods are included in Part L of the Building Regulations. Most assessments of operational energy have been developed from a method described in CIBSE TM22 (CIBSE 2006) that uses measurements and an audit to determine the overall energy usage of each system compared to benchmarks. Benchmarks have been published in TM46 (CIBSE 2008b).
- *objective environmental measurements* – usually carried out for thermal comfort, indoor air quality (IAQ), noise, daylight and luminance, airtightness and in some cases thermographic survey.
- *changes to original design* – usually evaluated by inspecting the original drawings and carrying out a survey of current use of space and services. Consultation with designers and facilities managers helps in the interpretation for the need of such changes.

4.5   A method developed in the UK: the PROBE study

In the 1990s, a UK study investigated operational buildings and established a methodology which is followed today. The study was called 'Post occupancy review of building engineering' (PROBE).[2] The methodology had two parts: building investigation and data analysis.
   The building investigation includes the following:

- design and construction
- energy consumption
- occupant questionnaires
- management interviews
- maintainability
- control issues
- review of performance
- changes made.

The data analysis includes the following issues:

- comparison with original design
- energy assessment

- occupant survey analysis
- key messages.

The report includes the following sections:

- study report
- response from design team
- response from building occupier
- publication.

The outcome, as mentioned, helped to:

- improve industry practice
- inform further research by identifying
  - common/repeated mistakes
  - specific aspects that lead to good performance.

As an example, some results are presented here of a study on the Elizabeth Fry Building, which is a learning resource centre at the University of East Anglia, UK[3] and one of the buildings investigated. The building was also revisited in 2011, and the results of this investigation were published in the *Building Services Journal* (Bordass and Leaman 2011).

The building was designed on low-energy-design principles; has a four-storey, 3,250 m^2 floor area; and includes a hollow-core ventilation system. This is an activated thermal mass strategy where hollow-core floors enhanced access to the thermal capacity of the structure to reduce cooling and heating demand. Heating is satisfied by three 24-kW domestic condensing boilers.

Energy consumption results are shown in Figure 4.5 in terms of normalised energy consumption and carbon dioxide emissions. Normalisation, usually per treated floor area, makes comparison with other buildings and benchmarks easier. Comparison with benchmarks is important to show the performance of the building in comparison with others. In the UK, publications exist to facilitate this, and in the case of the examined building are identified within Figure 4.5. They show that the building performs very well both in terms of both energy consumption and CO_2 emissions. During the 2011 revisit, a slight increase of the annual CO_2 emissions was found of approximately 10 kg/m^2 treated floor area, and this was mainly attributed to heating and hot water due to the change to constant hot water and the appearance of some additional electric heaters (Bordass and Leaman 2012).

User satisfaction results are shown in Figure 4.6, where it can be seen that occupants are happy with their environmental conditions and mostly with the controls provided.

One important issue at the time of the study was airtightness, now regulated by Building Regulations Part F, in which airtightness targets for domestic and non-domestic buildings are specified and airtightness testing is mandatory. The building was tested in 1994 (before handover) and also in 1998 as part of the PROBE study; the result was an air leakage index of 6.5 m/h, which placed it at below very airtight buildings in the UK. Tests were repeated in September 2011, and the result was 5.3 m/h air leakage index, which is better than 1998. This was attributed to changes in the building, such as the removal of the catering kitchen and its ventilation plant (Bordass and Leaman 2012).

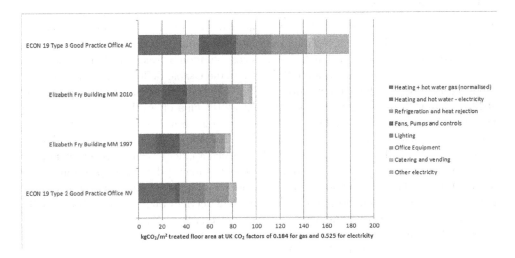

Figure 4.5 Energy consumption results

Source: Redrawn from Bordass and Leaman 2012, after *BSJ* March 2012

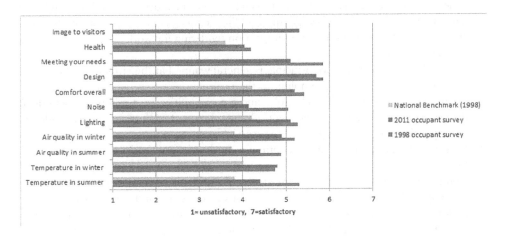

Figure 4.6 User satisfaction results

Source: Redrawn from Bordass and Leaman 2011, after *BSJ* March 2012

All investigated parameters were summarised to key design lessons, which included:

- energy performance
- ventilated hollow-core slabs
- construction supervision
- controls
- aftercare
- occupant comfort.

4.6   Soft Landings

The term 'soft landings' refers to a strategy adopted to ensure the transition from construction to occupation, with the aim of realising optimal operational performance of the building. The methodology was developed in the UK as a collaboration between the Building Services Research and Information Association (BSRIA) and the Usable Buildings Trust (UBT) who produced the Soft Landings Framework in 2009. The framework was further updated in 2014 to align with the project stages of the RIBA 2013 Plan of Work. The Soft Landings Framework serves as a methodology, available for anyone to use and tailor to suit the particular circumstances of an individual project. The Soft Landings Framework was adopted as Government Soft Landings (GSL) in the UK, which applies to central government procurement. The framework is supported by a range of guidance documents providing more detail. Of particular use for building services engineers is CIBSE's DE9: Application of Soft Landings and Government Soft Landings in Building Services Engineering (CIBSE 2018), which has mapped the Soft Landings Framework to RIBA Work Stages. The Soft Landings Framework is built upon 12 core principles, and its relevance to building services engineers is outlined here as presented in (CIBSE 2018):

1   Adopt the entire process: This is an important aspect of Soft Landings usually adopted before the engagement of the building services engineer. This is because Soft Landings is designed to start at RIBA Stage 0, which usually includes the employer and lead designer as contracted parties. As the building services engineer is one of the most affected by Soft Landings, this may impact on tendering for work, as additional activities are required.
2   Provide leadership: This core principle is generally aimed at the employer, who should appoint a 'Soft Landings Champion' to the project.
3   Set roles and responsibilities: The roles and responsibilities are traditionally set out by the employer. For building services consultancy, a named engineer is recommended to be responsible for each project.
4   Ensure continuity: This is about continuity of process and is closely connected with the first principle.
5   Commit to aftercare: This is the most obvious and yet revolutionary aspect of Soft Landings, which ensures that the project runs as designed and should help close the performance gap between the design and in-use phases. It is suggested that troubleshooting and fine-tuning of the building be done by the original design team and that commissioning be protected from time and budget constraints.
6   Share risk and responsibility: The concept of sharing risk and responsibility is rarely achieved in practice due to adversarial contractual forms and ingrained business practices. A new contractual approach is needed.
7   Use feedback to inform design: The concept of using feedback and previous project data to inform the design of a new project has its merits and is the concept that would help mostly to close the performance gap and improve practices. However, all variables between the two projects must be considered. The recommended 'pit-stopping' meetings give an opportunity to reflect on the design and design principles used, along with lessons learnt from previous projects and their applicability in each case.
8   Focus on operational outcomes: The process of reality-checking or pit-stopping has been developed to help design and construction teams think about the project from the perspective of the facility operator and its inhabitants.

9   Involve the building managers: Involve those that will actually run the building in the design and construction processes, so as to gather information about how that work could be done more efficiently and designing for that. However, some methods of building may make this difficult, for example, in a speculative office build.

10 Involve the end users: This is a desirable feature of every good design. In addition, informing the users of how the space is designed to operate will help them acclimatise when they move into the space. As before, there are instances where this is not possible, such as speculative office buildings.

11 Set performance objectives: This would provide real and achievable targets for the building, but clear definition around client and end-user management responsibilities is needed.

12 Communicate and inform: This helps all processes, including Soft Landings, although there may be some contractual barriers, depending on contract type and procurement route. It is recommended that all communications with clients and end users be in plain and clear language and free from jargon.

4.7   One example of POE from a European study

As mentioned, operational energy studies are now part of the regulatory framework in Europe and elsewhere. The next step is to carry out studies that include monitoring of achieved environmental conditions and user satisfaction surveys. Such a study was conducted under the European programme, Intelligent Energy Europe.

One studied building from the UK is Red Kite. The building is situated in a climatic region with a moderate heating and cooling load. An external photo of the building is shown in Figure 4.7.

Figure 4.7 External view of the Red Kite House building

Red Kite House is a three-storey office building in southeast England with a total floor area of 2,500 m². Each floor is mainly an open-plan office area with some meeting and other rooms. The total number of staff is about 250, some of whom are permanently stationed in the building; others spend a proportion of their working time away from the office and are only intermittently present, using a 'hot-desking' arrangement. Occupied weekday hours are between 8.00 and 18.00.

The form of the building is designed to ensure that an effective natural ventilation strategy can be achieved and, as a consequence, it has a relatively long, narrow plan on an east–west axis, with a typical depth of 16 m. The limited depth ensures good use of natural lighting. A brise soleil (Figure 4.8) is situated at roof level on the south facade and is designed to provide

Figure 4.8 The brise soleil incorporating photovoltai ccells in Red Kite House

Figure 4.9 Open-plan office showing natural and artificial lighting

protection from direct solar gain in the summer months but to allow useful heat gains in the heating season. The brise soleil incorporates photovoltaic cells that reduce the building's electricity demand on the conventional grid supply. Roof-mounted thermal solar collectors provide hot water for washrooms.

The building is naturally ventilated by automatically controlled high-level windows on each floor of the main facades. Larger, manually operated windows are also available. The ceiling of each storey is exposed concrete (Figure 4.9). This thermal mass is used in conjunction with night-time ventilation to reduce peak internal temperatures in summer. In addition to the ventilation strategy, described in more detail later, the building incorporates a number of other sustainable features, including a rainwater storage system, which collects surface water from the roof and recycles it for washroom flushing, and sustainable drainage measures, e.g. permeable paving, permeable gravel beds around the building and grass landscaping.

The building achieved an excellent rating based on the UK BREEAM assessment method.

4.7.1 Ventilation strategy

Except for simple mechanical extract ventilation for the toilets and meeting rooms, the building is fully naturally ventilated by openable windows on the north and south facades. High-level top-hung lights on each floor are automatically opened by motorised links. Larger top-hung opening lights are provided at a lower level, and these can be manually operated by the occupants.

The open-plan arrangement allows free movement of air across the building. Even where partitions are provided – for instance for meeting rooms – these are not carried to the full ceiling height of 3.2 m, again allowing crossflow of air.

There are no suspended ceilings, and the concrete soffit is exposed on each floor. This provides substantial thermal mass which, combined with night-time ventilation cooling, acts to minimise peak temperatures during the occupied period and obviates the need for air conditioning.

Operation of the high-level windows is controlled by local temperature measuring devices. When the temperature rises above a set point, determined by the building management team, the windows open and remain open until either the temperature falls or rain is detected by a roof-mounted sensor.

4.7.2 Evaluation of energy and environmental conditions performance

Energy performance

The monitored energy consumption data for 2006, normalised on a unit-treated floor area basis, yields annual consumptions of 66 kWh/m^2 for heating and 127 kWh/m^2 for electricity (Figure 4.10).

These may be compared with benchmarks current at the time of construction given in the UK Energy Consumption Guide 19: Energy Use in Offices (ECG19) for naturally ventilated, open-plan offices (see also CIBSE TM46). For typical practice, the benchmarks are 151 kWh/m^2 for heating and 81 kWh/m^2 for electricity. The good practice values are 79 kWh/m^2 for

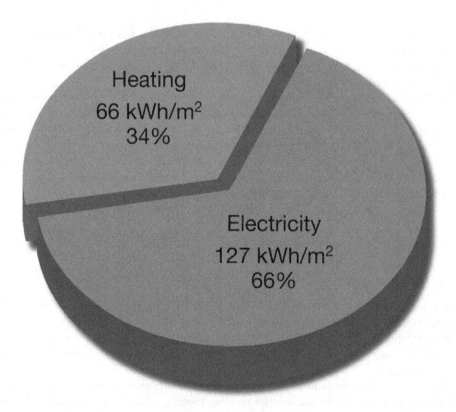

Figure 4.10 Heating and electricity measured energy consumption

heating and 54 kWh/m² for electricity. These are shown in Figures 4.11 and 4.12, together with similar benchmarks for a standard air-conditioned building.

Red Kite House has an excellent heating consumption, below the good practice benchmark for naturally ventilated open-plan offices, and substantially lower than for air-conditioned offices. The electricity consumption, however, is higher than both typical and good practice for naturally ventilated open-plan offices. This is likely to be a result of the Red Kite House's relatively high density of occupation and the high computer and office appliances not reflected in current benchmarks. It was not possible to check this, as sub-metering of the electricity circuits was not available.

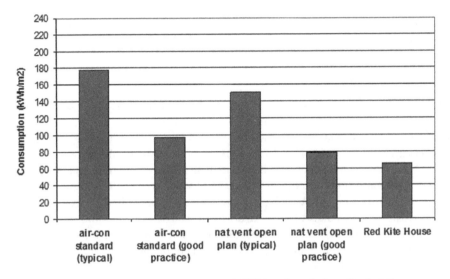

Figure 4.11 Comparison of heating consumption with UK benchmarks for office buildings

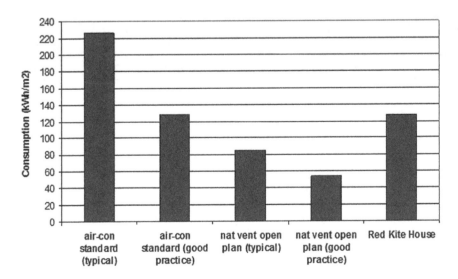

Figure 4.12 Comparison of electricity consumption with UK benchmarks for office buildings

Indoor climate

Thermal: Temperature measurements were made at six locations during the period 15 March to 15 September 2006. These provided an initial overall assessment of performance, the results of which are summarised in Table 4.1, indicating the proportion of occupied hours for which the temperature at five locations exceeded 28°C.

Ventilation: Continuous measurements in two locations in open-plan office areas over a two-week period in June/July 2008 showed low concentrations of carbon dioxide: always less than 800 ppm, with 700 ppm being exceeded for only three hours. Measurements made in a two-week period in October 2008 yielded slightly higher rates with mean values of 590 ppm and 660 ppm during office hours and occasional peaks above 1,000 ppm. Higher concentrations were measured in December 2008, but these may have been atypical because of relatively low indoor temperatures resulting from a problem with the heating system.

Occupant assessment of performance

About 55% of occupants responded to a questionnaire survey to assess satisfaction with the indoor environment. The results are summarised in Table 4.2.

In general, a significant majority of occupants are satisfied in summer with the overall indoor environment together with specific aspects, including thermal comfort, air movement and IAQ. In winter, satisfaction with the thermal environment is reduced, although detailed analysis of responses indicates that this appears to relate to a local problem with building operation rather than design. As with many open-plan buildings, there is some dissatisfaction with internally generated noise. Dissatisfaction with external noise in summer is likely to have arisen from a nearby temporary construction site. There is a very high level of satisfaction with the natural and artificial lighting. This is a benefit of the open-plan design of limited depth with windows provided on two opposing facades.

Table 4.1 Proportion of occupied hours that air temperature exceeds 28°C

Ground floor	West wing Level 1	East wing Level 1	East wing (south) Level 2	East wing (north) Level 2
1.81%	2.79%	1.76%	2.30%	1.42%

Source: Bateson (2008)

Table 4.2 Summary of occupant assessment of the indoor environment

	Summer %	Winter %
Overall indoor environment is acceptable	82	69
Thermal environment is acceptable	77	61
Indoor air quality is acceptable	93	90
Acoustic environment is acceptable	51	65
	Natural %	Artificial %
Lighting is acceptable	97	84

Design lessons

Occupants indicated a high degree of satisfaction with the indoor environment, but designers of future buildings using the same principles might consider including monitors within the Building Energy Management system to allow windows to be opened if carbon dioxide concentrations reach a set limit. Control algorithms need to be carefully designed, where window opening is controlled by several variables (such as temperature, rain and carbon dioxide) to ensure optimal operation.

With the increasing use of computing equipment in offices, in order to manage energy use, it is useful to meter lighting and other building-related electricity consumption separately from other uses.

Key points concerning the design

- Night-time natural ventilation in common with a thermal sink provided by exposed concrete ceilings limits peak temperatures in summer.
- The combination of brise soleil and solar photovoltaic array both limits unwanted solar gains in summer and provides a useful supply of renewable electricity.
- The deliberately limited building width, combined with the open-plan design and orientation, provide for efficient natural ventilation and daylighting, both of which increase the satisfaction of the occupants.

A simplified, subjective, easy-to-do environmental checklist suitable for people starting POE studies is included in the Appendix.

4.8   Summary

In this chapter the usefulness of POE studies to complete the design cycle was outlined and some methods and examples of POEs were presented.

- The history of POEs in the last 50 years indicates that valuable lessons can be learnt for better operation of buildings and also for input to future designs.
- Apart from operational energy evaluation (which is now compulsory in many European countries), a POE should include evaluations of environmental conditions to confirm design intentions and fine-tune systems and a user satisfaction evaluation conducted to confirm that the building performs according to occupants' expectations. User satisfaction evaluation could play an important role in future design aspirations.
- Soft Landings.
- Two case studies of low-energy buildings with POE evaluations were presented.

Appendix

Environmental building appraisal forms

Stage 1 – first appraisal: total environment

Using the 7-point scale provided, rank the following factors associated with the total environment (1 least favourable, 7 most favourable).

Use this space to elaborate on your impressions and define the points on the scale.

Table A1 Sheet 1 (total environment)

	1	2	3	4	5	6	7
The external environment in which the building is placed							
The visual character of the building							
The social setting							
The approach to the building							
The site itself; space, levels, vegetation							
The building exterior							
The building interior							
The effect of the building on the locality							

Stage 2 – detailed appraisal: organisation of spaces

Using the 7-point scale provided, rank the following factors associated with the organisation and design of spaces within the building (1 least favourable, 7 most favourable).

Table A2 Sheet 2 (organisation and design of spaces)

	1	2	3	4	5	6	7
Consider the distribution of the total space horizontally and/or vertically. This may be determined by: • function • space available • daylighting policy • policy decisions (e.g. prestige) • economic considerations • assumptions about social organisations How appropriate are the decisions which have been made?							
Consider the communication between spaces – the access, ease in locating individuals and comprehending the spatial organisation. How effective is the spatial communication in this building?							
How 'efficient' has the planning been? What is the ratio of usable to service space provided?							
Consider the status of the building within the hierarchy of its type. What is the standard of space provision in relation to people, functions, furniture and equipment accommodated?							
Consider the use of space. Is the space fully used? Does it include provision for expansion/reduction of activities?							
Consider the shape of the space. Is it appropriate, for example, to the function, the space available and the daylighting?							
Consider the height of the space, real or apparent, together with the effect of light and colour. How suitable does the height of the space appear to be?							

Stage 2 – detailed appraisal: visual and lighting factors

Using the 7-point scale provided, rank the following factors associated with the visual environment of the building (1 least favourable, 7 most favourable).

Table A3 Sheet 3 (visual and lighting factors)

	1	2	3	4	5	6	7
Consider the appearance of the interior of the building in terms of both natural and artificial light. Consider also the appearance of persons and objects seen in the interior, e.g. modelling, colour rendering. How effective is the visual design in facilitating the performance of visual tasks, e.g. inspection or reading?							
Daylight design. Consider the quantity and consistency/variability of daylight illuminance. Consider also the quality – brightness distribution, glare, reflection, shadow, solar penetration, suitability, character. Consider the view from the windows – distance, interest, obstruction. How effectively has daylighting been used?							
Electric lighting design. Consider the quantity and consistency/variability of electric lighting illuminance. Consider also the quality – brightness distribution, glare, reflection, modelling, shadow, suitability of luminaires and other fittings and installation, as well as the appearance of luminaire when illuminated and when not. How effective has the electric lighting design been?							
Consider the surfaces and finishes: floors, walls and ceilings. How do the quality, condition, colour and texture affect the visual environment?							
Consider the furniture and equipment. How do their suitability, quality, condition, colour, textures and visual organisation affect the visual environment?							

Stage 2 – detailed appraisal: heating and air conditioning

Using the 7-point scale provided, rank the following factors associated with the heating and air conditioning environment of the building (1 least favourable, 7 most favourable).

Table A4 Sheet 4 (heating and air conditioning)

Private	1	2	3	4	5	6	7
Consider the following air comfort factors: • the air temperature within the building • experience of radiation (ceiling heating, sunlight, losses to windows and cold surfaces) • temperature gradient (cold feet, hot head) • air freshness/stuffiness • air movement – stagnation, draughts • humidity • consistency/variability of thermal conditions • any aesthetic effects of thermal design How effective overall has the design for thermal comfort been?							
Consider the thermal environmental design: • heavy/light construction • opportunities for solar penetration • thermal insulation • treatment of surfaces, inside and outside • designed method of emitting heat: suitability, efficiency • designed methods of controlling the thermal environment: air temperature, ventilation, air movement, humidity How effective overall has the thermal environmental design been?							

Stage 2 – detailed appraisal: acoustics

Using the 7-point scale provided, rank the following factors associated with the aural environment of the building (1 least favourable, 7 most favourable).

Table A5 Sheet 5 (aural environment)

	1	2	3	4	5	6	7
Consider the following aural comfort factors: • subjective impressions of noise levels generally • general quality of sound – live or dead • isolation from external (street) noise • isolation from noise created within the building e.g. in the circulation space • absence of noise in spaces needing quiet How effective overall has the design for aural comfort been?							
Consider the aural environmental design features: • volume and shape of the spaces • sound insulation of partitions (floors and other space divisions) • absorbency/reflectance of surfaces: floors, walls, ceilings, other surfaces;their effect on noise levels and quality of sound • design of movable furniture: effect on noise levels and quality of sound How effective overall has the aural environmental design been?							

Stage 2 – detailed appraisal: environmental impact

Using the 7-point scale provided, rank the following factors associated with the environmental impact of the building (1 denotes least favourable, 7 most favourable).

Table A6 Sheet 6 (environmental impact)

	1	2	3	4	5	6	7
Consider any energy-efficient design strategies present in the building: • daylighting • energy-efficient lighting • natural ventilation • 'free' cooling • heating and cooling control strategies • visible signs of energy consumption information to occupants and visitors							
Consider water usage in the building: • possible rainwater collection or grey water use • water efficient appliances and sanitary facilities • visible signs of water consumption information to occupants and visitors							
Consider the construction materials used for the building: • use of recycled construction materials • use of materials according to their environmental impact							
Consider the employees/visitors and goods transport facilities: • availability to public transport and cycling facilities • use of local suppliers for materials, equipment, etc.							
Consider contribution to pollution and ozone depletion from the building: • low NO_x emissions plant and non-ozone-depleting refrigerants • visible signs of regular monitoring of building pollution contribution							
Consider the ecological impact of the building; e.g. protection of pre-existing ecological features such as trees, hedges, water courses, etc.							
How effective do you rate the overall environment impact of the building?							

Stage 3 – re-appraisal

Summarise the assessments made so far, again using the 7-point scale (1 least favourable, 7 most favourable).

Table A7 Sheet 7 (Re-appraisal)

	1	2	3	4	5	6	7
First impressions of total environment							
Functional requirements							
Space							
Visual environment							
Thermal environment							
Aural environment							
Environmental impact							
Re-assessment rating of total environment							

Use this space to note down any factors which have changed your assessment from the first overall appraisal (sheet 1). Note any environmental factors not considered.

Notes

1 A useful general website for up-to-date information on the implementation of the EPBD is found at www.buildup.eu.
2 More details can be found at www.usablebuildings.co.uk.
3 A report was published in the *Building Services Journal* and can be found in www.usablebuildings.co.uk under PROBE.

5   Health and wellbeing

5.1   Introduction

The previous chapter introduced the post occupancy evaluation of buildings and argued its importance focusing mainly on energy performance. It included elements of occupant evaluation of their environment and referred to environmental assessment methods, with an emphasis mostly on energy efficiency, which was the main driver in regulations and guidelines until recently. These initiatives at national, continental and international levels have yielded results in the form of regulated energy-efficient design and provision of the residual required heat and electricity from renewables; the aim is to achieve nearly zero-energy buildings (NZEB) for all new designs and retrofitted buildings; this is a necessary action.

As we know, energy-efficient buildings are built to provide shelter to occupants that is *safe*, *healthy* and *comfortable*, which are the fundamental functions of buildings. Conditions inside buildings are influenced by external ambient conditions (climate); the design (facade, geometry, materials); and heating, ventilation and air conditioning (HVAC) services, as well as the way that the building is used by the occupants. To this end, environmental design describes the design process for buildings where the (a) external environment, (b) building fabric, (c) operation of the building and (d) building services (including renewable sources) work together to provide a healthy and comfortable internal environment. This is shown diagrammatically in Figure 5.1, which also indicates the four elements of occupant comfort – thermal, visual, acoustical and air quality – to correspond with how we sense the internal environment through our skin (thermal comfort), our eyes (visual comfort), our ears (acoustics) and our nose (air quality). The combination of the four elements is usually referred to as indoor environmental quality (IEQ), which impacts the health and wellbeing of occupants.

5.2   Indoor air quality (IAQ)

IAQ has increased in importance and is currently at the centre of considerations in the design of building services at equal terms with energy efficiency. This is because certain energy efficiency measures might have impacted negatively on air quality because of incorrect operation; for example, increased airtightness combined with misuse of purpose-provided ventilation might result in increased pollution levels inside the building, such as moisture in residential buildings, leading to mould growth. On the other hand, external pollutants might be brought to the building from outside if appropriate filtration for the pollutant is not provided. The importance of suitable ventilation rates for the use of the building is explained in

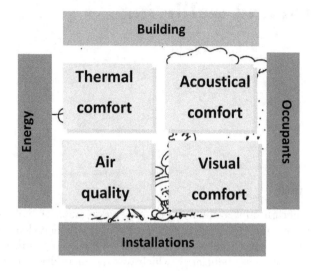

Figure 5.1 Buildings and environmental conditions

more detail in the next chapter. In this chapter, the source of pollutants and their effect on occupants (health) is outlined. So, what are the indoor pollutants, what are their sources and how can they be controlled?

5.2.1   *Indoor air pollutants, their sources and current limits*

The following pollutants are considered harmful to health when they exceed certain concentrations in the air (Kukadia and Upton 2019).

Volatile organic compounds (VOCs) are compounds that easily become vapours or gases; they are released from burning fuel, building materials and furnishings or consumer products such as aerosols and cigarettes; they should be less than 300 $\mu g/m^3$ for an 8-hr interval (CLG 2010).

Formaldehyde is also a volatile organic compound present in building materials such as resins, wood-based products and urea formaldehyde foam cavity wall insulation, and is a constituent of cigarettes and of combustion gases from burning fuel. In view of its widespread use, toxicity and volatility, formaldehyde poses a significant danger to human health. It should be less than 100 $\mu g/m^3$ for a 30-min average concentration (WHO 2010).

Particulate matter (PM) is a complex mixture of small particles and liquid droplets in the air. The main sources of PM are from combustion and transport, construction and demolition activities and certain cooking activities (e.g. frying). Exposure to particle pollution is linked to a variety of significant health problems such as asthma. The size of the particle is important in terms of its impact on human health; PM 2.5 (up to 2.5 μm in diameter) and PM 10 (up to 10 μm in diameter) are commonly quoted. PMs are attracting more attention recently, as they are connected to air pollution in urban areas. Current guidelines for PM concentration are 20 $\mu g/m^3$ for PM 10 and 10 $\mu g/m^3$ for PM 2.5 annual mean (WHO), while the 24-hr mean is 50 $\mu g/m^3$ for PM 10 and 25 $\mu g/m^3$ for PM 2.5 (WHO 2010 as quoted in EN 16798–1 2019). Figure 5.2 shows the size of different particulates and other pollutants. Particles to be removed from the outdoor air are small and usually not visible. All particles

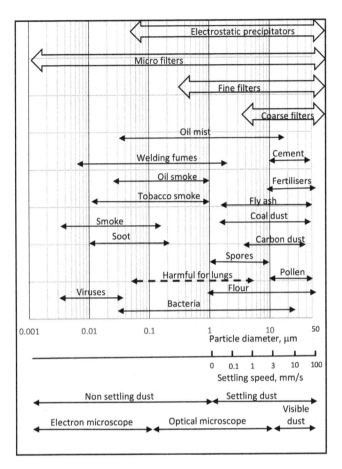

Figure 5.2 Size of different particulates and pollutants with some indication of required filters and settling speed

with a diameter less than 1 μm will behave like gases; they do not settle but are carried with air flow. Large particles settle slowly – the settling velocity increases with particle diameter.

Ozone (O_3) in this context refers to ground-level ozone (to be distinguished from upper-atmosphere ozone, which protects us from ultraviolet radiation from the sun) and arises from photochemical reactions in the ambient air and is also produced indoors by some electronic equipment such as printers and photocopiers. It is short-lived, as it reacts with surfaces and other pollutants; however, it can trigger health problems such as asthma. The current guideline for O_3 concentration is 100 μg/m³ 8-hr mean (WHO 2010 as quoted in EN 16798–1 2019).

Nitrogen dioxide (NO_2) increases the likelihood of respiratory problems. The major outdoor source of NO_2 is the burning of fossil fuels, and in the cities most comes from transportation-related sources. In buildings the major source is from unflued gas heaters and cookers. Current guidelines for NO_2 are 200 μg/m³ 1-hr mean and 20 μg/m³ annual mean (WHO 2010 as quoted in EN 16798–1 2019)

Carbon monoxide (CO) is an odourless, colourless and toxic gas. The impact on health can vary greatly from person to person depending on age, overall health and the concentration and length of exposure, and it can be fatal in high concentrations. It is produced by the incomplete combustion of most fuels, and therefore indoor sources are unvented kerosene and gas space heaters; leaking chimneys and furnaces; gas stoves, generators and other gasoline-powered equipment; automobile exhaust from attached garages; and tobacco smoke. In the UK and other countries CO detectors are recommended to be fitted in houses. Current guidelines for CO are 100 mg/m³ 15-min mean, 35 mg/m³ 1-hr mean, 10 mg/m³ 8-hr mean and 7 mg/m³ 24-hr mean (WHO 2010 as quoted in EN 16798–1 2019).

Other pollutants include biological pollutants generated from house dust mites, mould, fungal particles, bacteria and pollen, domestic animals and pests; odour generated from internal or external sources;industrial pollutants from local sources of industry;and ground pollutants such as radon or previous use of the building site.

Finally, last but not least, important compounds are considered pollutants that are also used traditionally as IAQ indicators.

Indoor carbon dioxide (CO_2) sources are human and animal respiration and combustion products and vegetation. We also know that CO_2 is a greenhouse gas, usually called atmospheric CO_2, and is present in outdoor air; the outdoor concentration depends on the location of the building, but globally the concentration is slightly more than 400 ppm in 2019.[1] Such low concentration does not have an impact on human health, but we try to keep it low because of the possible global warming impact. CO_2 starts to have an impact on human health at high concentrations, usually more than 5,000 ppm; for example, the UK Health and Safety Executive Workplace Exposure Limits are 5,000 ppm 8-hr average and 15,000 ppm 15-min average. CO_2 is also used as an indicator of IAQ, and guidelines recommend much lower limits; these depend on the IAQ we aim for in a building, as will be explained in more detail in Chapter 6 (Tables 6.1–6.3). As a general guideline, for most buildings the recommended CO_2 limits are 800–1,000 ppm (including atmospheric CO_2) to assure good IAQ. These limits refer to enough ventilation provided in the space so that all pollutants are kept below the recommended limits. It is also a parameter traditionally used in building management systems (BMS) for the control of IAQ in non-domestic buildings. Figure 5.3 shows a typical graph of CO_2 concentration in an office building over two days. You notice that CO_2 increases from 7:00 a.m. when office workers arrive and reaches a steady

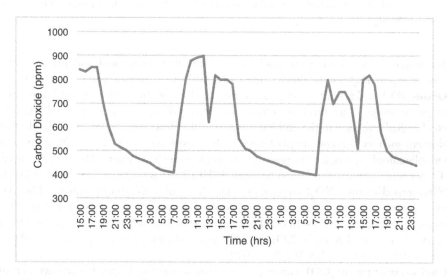

Figure 5.3 Metabolic carbon dioxide in an office

state before the lunch break, when CO_2 starts dropping until office workers return and then it starts increasing again until about 3:30 p.m. when workers start leaving. It then declines until it reaches atmospheric levels during the unoccupied period overnight.

Moisture sources in buildings are respiration and activities such as cooking and washing. Moisture is also present in the air and is a well-known parameter when designing HVAC systems, using psychometric charts to calculate the moisture content in the air in a variety of units. Excessive moisture in the indoor air can lead to dampness and mould growth, which has an impact on health and also could affect comfort both at high and low concentrations. Most building regulations include specific limits for moisture, usually in terms of relative humidity. It is also a parameter traditionally used in extract ventilation systems for the control of IAQ in domestic buildings.

5.2.2 *Control of indoor air pollutants*

In order to maintain acceptable IAQ in buildings, sensors are used to control the ventilation rates of HVAC systems or actuators in natural ventilation systems to provide additional ventilation should the sensed parameter be above recommended guidelines. As mentioned for IAQ, CO_2 and humidity used to be the sensed parameters (alongside air temperature for ensuring thermal comfort). However, increasing the air flow rate will impose a large energy penalty and, depending on the pollution levels, might not be able to provide acceptable IAQ. Therefore, filters are used in mechanical ventilation systems to capture the ingress of outdoor pollutants. Large particles such as pollen are relatively easy to remove from the air flow, but typical particles in the outdoor air are much more difficult, requiring a higher category of filters with better capture efficiency. Filters with high efficiency also have a higher pressure drop and thus higher energy consumption due to higher requirements of the fan power. Table 5.1 shows different filter categories recommended for combinations of outdoor air and supply air categories.

In recent years controlling IAQ based on CO_2 or humidity levels is not enough for many buildings; this is because (a) certain pollutants, in particular VOC and PMs, might not correlate well with CO_2 because they are generated by different sources and (b) low-cost IAQ sensors have become available. Table 5.2 gives an indication on what sensors are suitable to what spaces.

Low-cost IAQ sensors are available, but in many cases they are based on detecting one substance (such as VOC) and extrapolated to others (in a similar way that CO_2 sensors are used). Other sensors are based on measuring PMs (10 and 2.5 separately or integrated). Recently, results have been published comparing favourably low-cost PM sensors suitable for use in building controls systems, especially for detecting events.

Table 5.1 Recommended minimum filter classes (according to EN 16798–3–2017). F7, F9, M5 and M6 are types of filters.

Outdoor Air Quality	Supply air with concentration of particulat ematter and/or gases				
	Very low	Low	Medium	High	Very high
ODA1	M5 + F7	F7	F7	F7	
ODA2	F7 + F7	M5 + F7	F7	F7	M5
ODA3	F7 + F9	F7 + F7	M6 + F7	F7	F7

Table 5.2 The suitability of air quality sensors for demand-controlled ventilation for different spaces from a technical and cost point of view

Space	Occupancy	Mixed gas	CO_2	CO_2-mixed gas	CO_2/CO
Public spaces	no	satisfactory	excellent	good	good
Offices	satisfactory	satisfactory	excellent	good	good
Residential	satisfactory	satisfactory	excellent	good	good
Restaurants	satisfactory	satisfactory	poor	satisfactory	good
Garages	satisfactory	satisfactory	poor	satisfactory	satisfactory

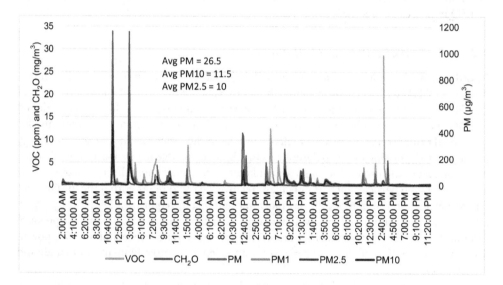

Figure 5.4 Example of VOC, formaldehyde and PMs monitored in a kitchen.

Figure 5.4 shows measured concentrations of pollutants in the kitchen of a multi-occupancy apartment. We can see that although guideline concentration limits are not exceeded, there are spikes for certain pollutants at times, indicating cooking/eating activities.

5.3   Thermal comfort

Thermal comfort provision is as complicated as IAQ provision because human thermal comfort is influenced by a number of parameters, which in turn are influenced by external climatic conditions, construction of the building and solar/internal heat gains. As we know, humans should maintain a core body temp at 37°C, generate heat (at approximately 100 W at rest to 1000 W when active) and need to dissipate this heat. We dissipate this heat by (a) sensible heat loss through skin (convection and radiation) – when the surrounding temperature is lower than the skin temperature – and (b) latent heat through sweating and respiration (evaporation). How much heat we dissipate depends on six parameters, four related to environmental conditions and two related to human behaviour as follows:

- air temperature
- relative humidity
- surface temperature

- air speed
- clothing
- metabolic rate.

In an attempt to describe thermal comfort in a simple form, several indexes have been developed combining two or more of the notes parameters. In many professional guidelines (including CIBSE and ASHRAE) operative temperature is an example of such an index; it combines air temperature and radiant temperature, and in some cases corrections for air speed, moisture and season are included. The same guidance also refers to two thermal comfort models, as described in (EN 16798–1 2019); these are the adaptive and predicted mean value/predicted percentage dissatisfied (PMV-PPD) thermal comfort models. In general, the PMV-PPD model is used for air-conditioned buildings, while the adaptive model is used for free-floating buildings (e.g. naturally ventilated or mechanically ventilated without cooling).

The PMV-PPD model is based on the heat balance between the human body and its environment; PMV is the predicted mean value of votes on the ASHRAE scale of warmth (−3 to +3) of a large group of persons exposed to the same environment and with identical clothing and activity; it can be calculated based on the heat balance equations. As PMV is a mean value, it is useful to know the people who would be dissatisfied (i.e. people who would vote >+1 or <−1 on the sensation scale. PPD is calculated from the PMV using the following equation (BS EN ISO 7730):

$$\text{PPD} = 100 - 95 \exp\left[-\left(0.03353\ \text{PMV}^4 + 0.2179\ \text{PMV}^2\right)\right] \tag{5.1}$$

The use of these two indices is widespread, and in most cases the calculation of PMV and PPD is included in thermal simulation programmes or freely available apps that can be downloaded.

The adaptive thermal comfort model takes another approach and is based on comfort surveys in operational buildings, which has shown that the human response to thermal conditions is related to ambient temperature. Therefore, equations have been developed to calculate a neutral temperature given the external conditions. The expression that describes thermal comfort temperature is:

$$\theta_c = 0.33\theta_{rm} + 18.8 \tag{5.2}$$

where θ_c is the optimal operative temperature and θ_{rm} is the exponentially weighted running mean of the daily mean outdoor air temperature, given by:

$$\theta_{rm} = \left(1 - \alpha\right)\left(\theta_{ed-1} + \alpha\theta_{ed-2} + \alpha^2\theta_{ed-3} \cdots\right) \tag{5.3}$$

where $\alpha°$ is a constant (<1) and θ_{ed-1}, $°\theta_{ed-2}$, etc., are the daily mean outdoor air temperatures from yesterday, the day before yesterday and so on.

By using data from comfort surveys around Europe, the best value to estimate the thermal comfort temperature is when $\alpha = 0.8$.

Due to the weight of α as can be seen from Eq. 5.3, the temperature influence is more pronounced when the day is closer to the day for which the thermal comfort is calculated. When a whole range of days are not available, Eq. 5.4 gives an estimated calculation by using the last seven days:

$$\theta_{rm} = \frac{\theta_{ed-1} + 0.8\theta_{ed-2} + 0.6\theta_{id-3} + 0.5\theta_{ed-4} + 0.4\theta_{ed-5} + 0.3\theta_{ed-6} + 0.2\theta_{ed-7}}{3.8} \tag{5.4}$$

Table 5.3 Thermal comfort limits (after EN 16798–1 2019)

Category	Level of expectation	PPD	PMV	Adaptive	
				upper	lower
				limit	
I	High	<6	0.2<PMV<+0.2	$\theta_0 + 2$	$\theta_0 - 3$
II	Medium	<10	0.5<PMV<+0.5	$\theta_0 + 3$	$\theta_0 - 4$
III	Moderate	<15	0.7<PMV<+0.7	$\theta_0 + 4$	$\theta_0 - 5$
IV	Low	<25	1.0<PMV<+1.0		

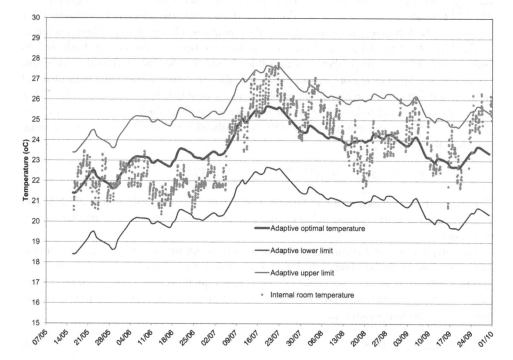

Figure 5.5 Adaptive thermal comfort optimal temperature, with both higher and lower limits, in an office building

It is also accepted that neutral temperature can have a range in a building and will depend on comfort expectations. So, buildings have been categorised and comfort temperature ranges given as shown in Table 5.3.

An example of a calculation of the adaptive thermal comfort range in an office building (category I) is shown in Figure 5.5.

5.4 Visual comfort

Similarly to IAQ and thermal comfort, visual comfort provision is complicated and has an impact on both the health and general wellbeing of occupants. Exposure to light can have both positive and negative impacts on human health – impacts that can become evident soon after exposure or only after many years.

In controlling lighting in buildings, illuminance is used as the main parameter to control. The illuminance E is defined as the luminous flux density on a small surface element, and its unit of measurement is the lux.

$$E = \frac{F}{A} \qquad [lx = lm\ /\ m^2] \qquad\qquad\qquad (5.5)$$

where the luminous flux F is a measure of the flow of energy related to the quantity of light; the unit in which it is measured is the lumen [lm], and A is the area of the surface element.

Most guidance in lighting design specifies suitable illuminance for the task performed in the space. For example, CIBSE (SSL 2012) in the UK has the following recommendations:

- 100 lux for rarely used interiors, with visual tasks confined to movement and casual seeing without perception of detail, e.g. corridors, stores
- 200 lux for interiors where the visual tasks do not require perception of detail, e.g. foyers and entrances
- 300 lux for interiors where visual tasks are moderately easy, e.g. libraries, sports and assembly halls, teaching spaces, lecture theatres, background office lighting.
- 500 lux for interiors where the visual tasks are moderately difficult and also where colour judgement may be required, e.g. offices (desks), kitchens, laboratories, retail shops.

These recommendations do not identify the source that is required to provide these illuminances, and the recommended levels may be met using either daylight or electric light. However, the use of daylight depends on the external daylight availability, and it is normal to design an electric lighting system to provide the required illuminance, even if the daylight provides sufficient light for most of the time.

But illuminance levels are not the only parameter affecting the occupants' visual comfort. Other parameters are:

- Glare – The term 'glare' covers a number of situations in which extremes of illuminance in the field of view reduce one's ability to perceive detail in less well-illuminated areas and may even cause discomfort. The impact of glare on visual comfort has been studied extensively, and a limiting glare index is usually calculated for every artificial lighting installation.
- Colour rendering – This depends on the type of lamps used, which are usually rated by the manufacturers.
- Ratio direct/indirect light – This needs to stay within a specific range, as big changes in illuminance between vertical and horizontal surfaces will create visual discomfort.

In many designs the lumen method is used, which calculates the illuminance of artificial lighting; the method is part of most software used for lighting design.

In addition, most guidance and regulations encourage the integration of daylighting in lighting design. This is because of high potential energy savings in buildings so as to reduce CO_2 emissions. Also, and most relevant to this chapter, many studies have shown that people prefer daylit buildings. Daylight appears to have particular benefits, as it can provide variability and contact with the outside, as well as high levels of light, thereby improving mood.

Some of the benefits of daylighting are:

1   flow of light and modelling of objects within the room
2   variation of the colour of natural lighting

3   ability to view the outside world
4   Variety of lighting with time of day, weather and season.

In addition, inadequate exposure to daylight in winter months can contribute to psychological illnesses, such as seasonal affective disorder (SAD).

However, there are some challenges with its design, such as:

1   limited entry to a room
2   increased heat gains and losses through glass compared with those through heavier building fabric
3   glare from windows
4   availability of daylight

Availability and variability of daylight pose particular design challenges in providing the specified illuminance in a space. By definition, daylighting design exploits the use of light from the sky as a whole. This is usually referred to as diffuse light; because of the variability of external illuminance, the daylight factor (DF) is used to measure daylighting in a space. Direct sunlight is excluded for both values of the illuminance.

$$Daylight\ Factor = \frac{Indoor\ Illuminance}{Simultaneous\ Unobstructed\ Outdoor\ Illuminance} \tag{5.6}$$

This DF may be an average value for a room or may relate to a particular position in the room, for example, the point most distant from the window. In some cases architects and engineers have a statutory duty to provide a certain minimum DF (e.g. 2% in school classrooms).

Recommended values of the average and minimum DF for different types of spaces in a few varieties of buildings are included in the lighting guidance in each country. Average recommended values range from 1% to 5%. Minimum recommended values can be as low as 0.3% (school assembly halls) and as high as 3.5% (sports halls), but are more typically in the range of 1–2%.

The DF can be subdivided into the following components:

DF1   received directly from the sky
DF2   reflected by external surfaces, including other buildings and the ground
DF3   that has undergone reflections within the room before arriving at the reference surface.

The three components of the DF are shown schematically in Figure 5.6.

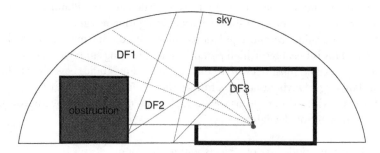

Figure 5.6 Sources of daylight at a point within a room

An estimate of the average DF on a horizontal working plane in a room is given by the following formula based on work by Lynes, Crisp and Littlefair. It is sometimes referred to as the BRE/CIBSE approximate formula and is applicable to cuboidal rooms with a height of about half the width. The resulting formula is

$$DF = \frac{WT\theta}{A(1-R^2)}$$ (5.7)

where
W the window area
T the window light transmittance (about 85% for clean single glazing)
A the total area of all room bounding surfaces
R the area-weighted average reflectance of all the room bounding surfaces
θ the angle subtended by the sky visible from the window centre; this is determined in the vertical plane perpendicular to the window.

If a multi-storey building is to be completely lit by daylight, there are limits on its overall plan depth. If a daylit room is too deep, the rear will look gloomy compared to the brightly lit area near the windows. If a daylit room is lit by windows in one wall only, the depth of the room *L* should not exceed the limiting value given by:

$$\frac{L}{W} + \frac{L}{H_W} < \frac{2}{(1-R_b)}$$ (5.8)

where *W* is the room width, H_w is the window head height above floor level and R_b is the average reflectance of surfaces in the rear half of the room (away from the window). If *L* exceeds this value, the rear half of the room will tend to look gloomy, and supplementary electric lighting may be required.

In the last few years two developments in lighting design were developed:

- Climate-based daylight modelling (CBDM) has been introduced to calculate actual daylighting illuminances and to predict them using specific weather files when daylight levels are above and below a set threshold. This approach uses the useful daylight illuminance (UDI) metric. UDI is the annual occurrence of illuminances across the work plane that are within a range considered 'useful' by occupants. CBDM delivers predictions of absolute quantities that are dependent both on the geographical location and the fenestration orientation, in addition to the space's geometry and material properties. Prediction of actual illuminances rather than a DF would give more precision in the calculations. However, the method is more computer intensive than the calculation of DF.
- In the last few years, we have also greatly increased our knowledge about the impact of light on human physiology and psychology. Scientists have found a sensor in the eye that sends signals to the brain to control the hormones that make us alert or sleepy (Ticleanu and Littlefair 2019). It has been shown that adequate exposure to wide-spectrum light during daylight hours is essential for regulation of the circadian system (or 'body clock') and helps facilitate healthy sleep duration and quality, as well as improving task performance during the waking hours (CIBSE Guide A 2015). Disruption of the circadian rhythm can also influence mood and vitality. This

discovery on occupant wellbeing has led to the development of 'human-centric circadian lighting', which emulates the natural environment and the 24-hour cycle of natural light. It includes not only intensity of lighting but also its colour; during the daytime, maximum levels of cool white light and at night reduced light levels with warm light.

5.5   Acoustical comfort

Noise can have physical and psychological health impacts if an occupant is exposed to consistently elevated sound levels. This section does not focus on workplace noise due to the nature of the work undertaken, as this is closely regulated and employers should take mitigation measures. The section focuses on noise levels which are generated within the building because of its use (e.g. offices, lecture rooms, theatres) or noise ingress from outside because of the location of the building. Noise is sometimes referred to as a pollutant and as an annoyance. Noise annoyance can be affected by many parameters, as indicated in Figure 5.7. These many factors can be divided into two groups: (a) those which are characteristics of the noise source and (b) those which are characteristics of the environment.

For the noise source, annoyance is likely to be due to the loudness, which can be assessed in dB(A) or noise rating (NR), the frequency characteristics and the intermittency. The environment or background level is affected by the type of area, the time of day and to some extent by the time of year. Hearing is a subjective response, and so personal circumstances (including age) could have an impact.

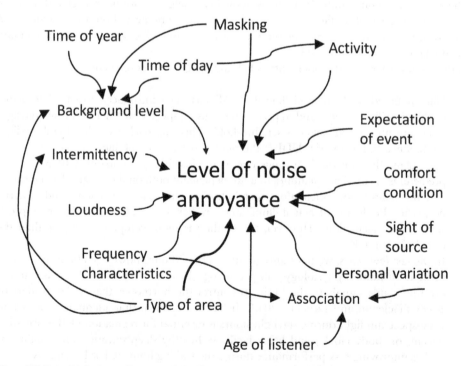

Figure 5.7 A multiple-cause diagram of the factors associated with noise annoyance

Table 5.4 Examples of design-equivalent continuous sound levels for continuous sources

| Building | Type of space | Equivalent continuous sound level dB(A) | | |
		I	II	III
	Level of expectation			
Residential	Living room	<30	<35	<40
	Bedroom	<25	<30	<35
Hospitals	Wards	<32	<36	<40
	Operating theatres	<35	<40	<45
Hotels	Rooms	<25	<30	<40
	Reception/lobbies	<30	<35	<40
Offices	Small office	<30	<35	<40
	Open plan	<35	<40	<45
	Conference room	<30	<35	<40
Schools	Classroom	<30	<34	<38
	Gymnasiums	<35	<40	<45
Commercial	Retail stores	<35	<40	<45
	Department stores Supermarkets	<40	<45	<50

Source: EN 16798–1 2019

Of all the parameters, loudness can be measured, but its impact on people depends on the frequency and the response of the human ear. Two indices are used for loudness; the first is weighted decibel (dB) and the second NR curve.

Decibels can be measured with a sound level meter, but the measurements will not be representative of the loudness that the human ear will hear. This is because perceived loudness is related to the frequency of the noise. Decibel is a logarithm scale – a 10 dB difference would mean double the loudness;a less than 3 dB difference is not noticeable. For typical noise levels inside buildings the A-weighted scale is used for low to medium amplitudes up to 55 dB. Time exposure to a noise level is also taken into consideration, and the equivalent continuous A-weighted sound pressure level (LAeq, T) index is used for guidelines inside buildings. This is the value of the 'A-weighted sound pressure level in decibels (dB) of a continuous steady sound that, within a specified time interval, T, has the same mean-squared sound pressure as the sound under consideration that varies with time' (definition by BS 8233–2014). Therefore, the following design values are recommended in Table 5.4.

The NR curves were developed to predict acceptable noise levels for offices and are widely used for rating inside buildings. NR values cannot be converted to dbA values, but the following approximate relationship applies:

$$NR = dBA - 6 \tag{5.9}$$

5.6 Indoor environmental quality assessment methods and tools

As mentioned, several environmental assessment tools exist that award credits for different environmental categories. An example in the UK is Building Research Establishment Environmental Assessment Method (BREEAM), and Figure 5.8 shows the various broad categories it addresses. As can be seen in Figure 5.8, Health and Wellbeing is one of the categories

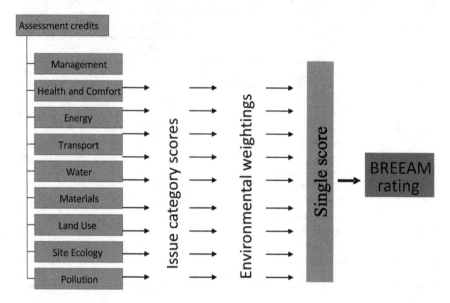

Figure 5.8 Method of environmental scores – example of BREEAM

which accounts for about 15% of the weighting. The subcategories within Health and Well-being include scores for Indoor Air Quality, Thermal Comfort, Lighting and Noise to reflect the humans' perception of their environment. Similarly, other environmental assessment tools (such as Leadership in Energy and Environmental Design [LEED], Green Star, Estimada), which were described briefly in Chapter 4 focusing on energy assessment, include health and wellbeing categories.

In 2015, the WELL Building Standard (IWBI 2016) was launched in the United States, which focuses exclusively on the health and wellness of the people in buildings and is performance-based. It heavily relies on existing standards and best-practice guidelines where available and attempts to integrate existing thresholds and requirements. WELL is a points-based system, similar to other sustainability assessment tools.

WELL includes ten concepts (categories):[2] Air, Water, Nourishment, Light, Movement, Thermal Comfort, Sound, Materials, Mind and Community. According to WELL 'each concept is comprised of features with distinct health intents each includes preconditions and optimisation options'. Most relevant for this chapter are the categories of Thermal Comfort, Light, Sounds, Air and Materials.

> *Thermal Comfort*: The precondition is that the majority of building users find the thermal environment acceptable by referring to guidance as described in Section 5.2. Points are scored for enhanced thermal performance, thermal zoning, individual thermal control, radiant thermal comfort, thermal comfort monitoring and humidity control.
>
> *Light*: The preconditions are light exposure and education and visual lighting design. These refer to (a) appropriate light exposure in indoor environments by using daylighting or electric lighting strategies and to provide users with education about the importance of light for health and (b) provide appropriate illuminances on work

planes for regular users of all age groups while taking into account light levels required for the tasks performed in the space. Points are scored for circadian lighting design, glare control, enhanced daylight access, visual balance, electric light quality and occupant control of lighting environments.

Sounds: The precondition is sound mapping, which refers to 'strategic interior planning and site zoning to create an acoustical plan that identifies internal and external noise sources that can negatively impact the acoustical environment of interior spaces'. Points are scored for maximum noise levels, sound barriers, sound absorption and sound masking.

Air and Materials: These are two categories both referring to IAQ.

Air: There are four preconditions (a) fundamental air quality (provide acceptable air quality levels); (b) smoke-free environment; (c) ventilation effectiveness (fresh air from the outside through mechanical and/or natural means in order to dilute human- and product-generated air pollutants; and (d) construction pollution management (protect IAQ during building construction and renovation). Points are scored for enhanced air quality, enhanced ventilation, operable windows, air quality monitoring and awareness, pollution infiltration management, combustion minimisation, source separation, air filtration, active VOC control and microbe and mould control.

Materials: There are three preconditions: (a) fundamental material precautions (restriction of hazardous-ingredient components in newly installed building materials); (b) hazardous material abatement (the application of protective practices during repair, renovation or maintenance), and (c) outdoor structures (restriction of wood preservatives and lead, including lead paint, in outdoor structures). Points are scored for waste management, in-place management, site remediation, pesticide use, hazardous material reduction, cleaning products and protocol, volatile compound reduction, long-term emission control, short-term emission control, enhanced material precaution and material transparency.

As mentioned, other environmental assessment methods that are based on a points system such as BREEAM and LEED have developed links with WELL to describe alignments so that certification/rating by both tools becomes easier and compatible. The health and wellbeing categories of BREEAM and LEED are presented next along with Green Star (Australia) and Estimada (United Arab Emirates).

In BREEAM the health and wellbeing category has a weighting of 17% of the total score and includes minimum requirements for safely and ventilation according to UK and international standards. The health and wellbeing of the indoor environment category gives credits for thermal comfort, visual comfort, IAQ, acoustical comfort and water quality. Health and wellbeing issues are also addressed in other categories, such as ecology and outdoor space, active/healthy lifestyle, safety and security and health and wellbeing of users in surrounding area (pollution), as well as in management/consultation/handover.

In LEED, health and wellbeing area ddressed in all six assessment categories and in particular sustainable sites, materials and resources and indoor environmental quality.

In Green Star, which is a voluntary scheme in Australia, health and wellbeing is addressed within the indoor environmental quality and materials categories.

Estimada, which is a sustainable urban planning tool developed by the Abu Dhabi Urban Planning Council, is based on the four pillars of sustainability: environmental, economic, cultural and social. Unlike other environmental assessment tools, which are voluntary, the Pearl rating system used in Estimada is incorporated into Abu Dhabi's building codes. It is a points-based system with mandatory components, but because of the climatic conditions, a different emphasis is put, for example, on water conservation. Health and wellbeing is addressed in all six assessment categories.

5.7 Summary

This chapter described the elements of indoor environmental quality which should be satisfied in all energy-efficient buildings to provide a healthy and comfortable environment for occupants. It focused on:

1 IAQ. It listed air pollutants with their corresponding safe limits so that the impact on health is minimised. The use of filters is discussed.
2 Thermal comfort and parameters affecting this for humans. Two thermal comfort models are presented; PMV/PPD is usually applied to air-conditioned buildings, and the adaptive model is suitable for buildings without cooling.
3 Visual comfort provided by daylighting and/or artificial lighting was presented, highlighting the benefits of daylighting, as well as the challenges for its design. DF and CBDM are briefly discussed, as well as circadian lighting and its impact on human health.
4 Acoustical comfort is discussed, as well as indices used to measure noise annoyance.

The final section of the chapter highlights the increasing inclusion of health and wellbeing in all indoor environmental assessment methods, which are increasingly being used to provide sustainability indicators for buildings. The WELL building method is briefly described.

Notes

1 www.esrl.noaa.gov/gmd/ccgg/trends/
2 https://v2.wellcertified.com/v/en/concepts

6 Energy-efficient ventilation

6.1 Introduction

Ventilation is necessary in buildings, mainly for the following reasons:

- to provide fresh air for occupants
- to dilute and exhaust pollutants
- to protect the buildings against moisture in certain climatic conditions
- to provide air for fuel-burning appliances
- to provide cooling in summer.

Ventilation provision is thus related either to indoor air quality (IAQ) or thermal comfort. Until recently, most regulations and guidelines on ventilation provision were based on IAQ requirements. However, the function of ventilation to improve thermal comfort in certain situations is also being addressed, mainly by guidelines and newer standards.

This chapter first outlines ventilation requirements for various types of buildings and available ventilation strategies. It then introduces some parameters used to measure ventilation efficiency within a single space. It follows with an outline of how the ventilation rate due to the natural forces of wind and buoyancy can be calculated and enhanced by the use of fans. The chapter concludes with a description of ventilation strategies useful to provide thermal comfort in a building, thus avoiding the need for artificial cooling in certain circumstances.

6.2 Ventilation requirements

Two standards are usually quoted in relation to ventilation requirements; these are the European Standard EN 16798–1 (2019) (adopted as a British Standard in the UK) and the American Society of Heating, Refrigerating and Air-conditioning Engineers' Standard 62.1 (ASHRAE 2019a). The standards include calculation methods, where the recommended ventilation rates can be found tables listing values for different types of spaces. Before detailing ventilation requirements, it is worth mentioning BS EN 16798–3 Part 7 (2017) includes classifications of outdoor and indoor air, which are useful to know, as they impact on ventilation requirements.

EN 16798–3 Part 7 (2017) divides outdoor air into three categories (Table 6.1), and EN 16798–1 (2019) divides IAQ into four categories (Table 6.2). The quality of indoor air can then be defined by its level of CO_2 above outdoor levels (Table 6.3). The outdoor CO_2 concentration varies normally between 350 and 420 ppm. The internal CO_2 level is a good indicator for the emission of human bio-effluents and is therefore suitable to use as a proxy indicator of IAQ within buildings that are occupied by people.

Table 6.1 Classification of outdoor air quality (ODA)

Category	Description
ODA 1	Pure air which may be only temporarily dusty (e.g. pollen)
ODA 2	Outdoor air with high concentrations of particulate matter and/or gaseous pollutants
ODA 3	Outdoor air with very high concentrations of gaseous pollutants and/or particulates

Source: EN 16798–3 Part 7 (2017)

Table 6.2 Classification of indoor environmental quality (IEQ)

Category	Description
IEQ I	High indoor environmental quality
IEQ II	Medium indoor environmental quality
IEQ III	Moderate indoor environmental quality
IEQ IV	Low indoor environmental quality

Source: EN 16798–1 (2019)

Table 6.3 CO_2 levels in rooms

Category	Default design CO_2 concentrations above outdoor concentration assuming a standard CO_2 emission of 20 L/(h per person)	
	Default value	*Default value for bedrooms*
IEQ I	550	380
IEQ II	800	550
IEQ III	1350	950
IEQ IV	1350	950

Source: EN 16798–1 (2019) Annex B

In addition, pollution from building materials is considered; these can be classified as follows:

- *very low polluting buildings*: buildings where an extraordinary effort has been made to select low-emitting materials, activities that emit pollutants are prohibited and no previous emitting sources (e.g. tobacco smoke) were present
- *low polluting buildings*: buildings where an effort has been made to select low-emitting materials, and activities that emit pollutants are limited or prohibited
- *not low polluting buildings*: old or new buildings where no effort has been made to select low-emitting materials and activities that emit pollutants are not prohibited.

An example of a low polluting building is the following: The building is low polluting if the majority of the materials are low polluting. Low polluting materials are natural traditional materials, such as stone and glass, which are known to be safe with respect to emissions, and materials which fulfil the following requirements:

- emission of total volatile organic compounds (TVOC) is below 1,000 $\mu g/m^3$
- emission of formaldehyde is below <100 $\mu g/m^3$
- emission of carcinogenic compounds (IARC) is below 5 $\mu g/m^3$

(EN 16798–1 (2019), Annex B)

Following these definitions and applying the calculation methods described in EN 16798–3 Part 7 (2017) based on the requirements by people and pollution level by building materials, ventilation rates can be calculated usually in litres per second per square metre of floor area (l/s/m²) or litres per second per person (l/s/p). As an example, Table 6.4 gives calculated values for non-residential buildings for four categories of pollution from the building extracted from EN 16798–1 (2019), Annex B.

In the UK, ventilation requirements are specified for various buildings by Part F of the Building Regulations.[1] They mainly cover IAQ requirements, while thermal comfort requirements are covered to some extent by Part L of the Building Regulations. Requirements are traditionally divided into domestic and non-domestic buildings. The requirements of the 2010 regulations are presented in Table 6.5.

For the air supply in office buildings, the specified whole-building ventilation rate is 10 l/s/p. This can be achieved by natural ventilation (reference is made to CIBSE 2005a), mechanical ventilation or alternative approaches (reference is made to CIBSE 2000a, 2005a, 2015, 2016). It can also be provided by other ventilation systems, provided that they meet specified moisture and air quality criteria (performance-based ventilation).

Table 6.4 Examples of recommended ventilation rates for non-residential buildings for four categories of pollution from the building itself

Category	Air flow per person l/s/pers	Air flow for building emission pollutions (l/s/m²)		
		Very low polluting building	*Low polluting building*	*Non–low polluting building*
I	10	0.5	1	2
II	7	0.35	0.7	1.4
III	4	0.2	0.4	0.8
IV	2.5	0.15	0.3	0.6

Source: EN 16798–1 (2019), Annex B

Note: Rates are given per person or per m² floor area.

Table 6.5 Ventilation requirements for domestic buildings, extract ventilation rates and whole-dwelling ventilation rates

Room	Intermittent extract	Continuous extract	
	Minimum rate	*Minimum high rate*	*Minimum low rate*
Kitchen	30 l/s adjacent to hob; or 60 l/s elsewhere	13 l/s	Total extract rate should be at least the whole-dwelling ventilation rate
Utility room	30 l/s	8 l/s	
Bathroom	15 l/s	8 l/s	
Sanitary accommodation	6 l/s	6 l/s	

Whole-dwelling ventilation rates
Number of bedrooms in dwelling

	1	2	3	4	5
Whole-dwelling ventilation rate l/s	13	17	21	25	29

Source: Building Regs, Part F.

For other building types, reference is made to professional publications and standards. CIBSE Guide A (2015) includes tables with a suggested general air supply rate at 10 l/s/p for most building types. It varies for specific areas, such as hospital operating theatres (0.65–1.0 m³/s), hotel bathrooms (12 l/s/p), ice rinks (3 ACH – air changes per hour) and sports hall changing rooms (6–10 ACH).

In addition, airtightness specifications for envelopes and ducts have been introduced in many countries; these are usually related to energy efficiency measures in buildings, but obviously have a direct effect on ventilation requirements and systems that provide them. For example, in the UK, there is a minimum standard of 10 m³/h.m² of the building envelope at 50 pascals (Pa), and all buildings must be tested to comply with this. In addition, ductwork leakage testing should be carried out for systems served by fans with a design flow rate greater than 1 m³/s. Obviously, airtightness specifications will affect the infiltration rate considered in ventilation calculations.

6.3   Ventilation strategies

Ventilation is normally provided by the following means:

- infiltration (which is determined by the external envelope airtightness)
- purpose ventilation, which can be further divided into

  ○   natural
  ○   mechanical
  ○   combination (hybrid or mixed-mode ventilation).

Ventilation is controlled in buildings because the health and comfort of the occupants are at risk if low ventilation is provided; but if too much ventilation is provided, especially during the heating season, a high energy penalty will be incurred.

Air infiltration is the process by which air enters into a space through adventitious leakage openings in the building fabric. Infiltration is considered unwanted ventilation because it cannot be controlled. As mentioned in the previous section, in many countries, stringent regulations on the restriction of infiltration exist.

Natural ventilation is the process by which outdoor air is provided to a space by the natural driving forces of wind and the stack effect. These forces are constantly fluctuating; the challenge of natural ventilation design is to use these forces so that the air flow into a space is maintained at the desired rate. Natural ventilation is usually provided by one of the following configurations:

- single-sided ventilation (Figure 6.1)
- crossflow ventilation (Figure 6.2)
- stack ventilation (Figure 6.3).

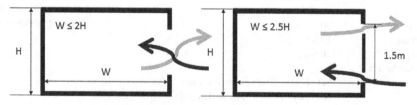

Figure 6.1 Schematic showing the principles of single-sided ventilation

Source: Redrawn from BRE (1994)

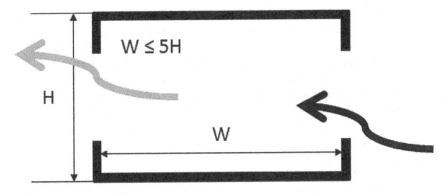

Figure 6.2 Schematic showing the principle of crossflow ventilation

Source: Redrawn from BRE (1994)

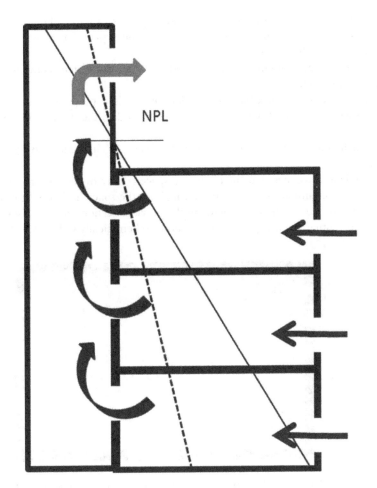

Figure 6.3 Schematic showing the principles of stack ventilation

Source: Redrawn from BRE (1994)

Mechanical ventilation uses fans to supply air to, and exhaust air from, rooms in buildings. Depending on demand, the supply air may be heated, cooled, humidified or dehumidified. The ventilation system may be equipped to recover heat from the exhaust air. The system may also recirculate extracted air. Windows may be sealed or operable. During the last decade, major developments have taken place or been further refined, such as various kinds of demand-controlled ventilation, systems with improved air flow characteristics at room level (e.g. displacement ventilation), heat recovery systems with efficiencies up to 90%, major developments in fan characteristics (e.g. direct current and inverter drive variable-speed fans) and low-pressure air distribution systems. In mechanically ventilated buildings, the ventilation air may also be conditioned before it is supplied to the rooms via duct systems (this is discussed in more detail in Chapter 8).

Hybrid ventilation (Heiselberg 2002) *or mixed-mode ventilation* (CIBSE 2000a) is normally used to improve the reliability of natural ventilation or to increase its range. A hybrid system can include a natural system combined with a full independent mechanical system. Hybrid ventilation techniques include:

- natural and mechanical: two fully autonomous systems in which the control strategy either switches between the two systems or uses one system for some tasks and the other system for other tasks
- fan-assisted natural ventilation: this is based on a natural system combined with an extract or supply fan
- stack and wind-assisted mechanical ventilation: this is a predominantly mechanical system in which natural driving forces are used to add to the driving pressure.

Within a single space, there exist two extreme modes of air circulation:*mixing and displacement ventilation*.

Mixing ventilation assumes that the air in the space is fully mixed; this means that the concentration of contaminants is the same throughout the space and equal to that in the exhaust. Ventilation requirements presented in the previous section include this assumption of a fully mixed ventilation strategy, and this is shown schematically in Figure 6.4. In reality,

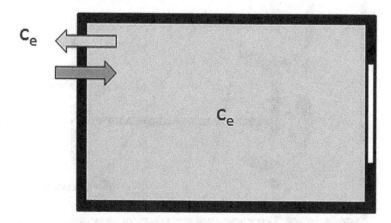

Figure 6.4 Fully mixed flow in a single space results in the mean contaminant concentration in the space and in the exhaust being identical

Source: Adapted from Mundt (2004)

however, the room air is seldom fully mixed. Figure 6.5, parts (a) and (b), illustrate an extreme case where there is a short circuit in the ventilation system. The contamination released in the room in both figures is the same, as is the concentration in the exhaust. The mean concentration in the room, however, differs because of the position of the contamination source.

Displacement ventilation is an alternative strategy to mixing ventilation. The principle is based on air density differences, where the room air separates into two layers: an upper polluted zone and a lower clean zone (Skistad 2002). This is achieved by supplying cool air at a low velocity in the lower zone and extracting the air in the upper zone.

For displacement ventilation systems, the selection of the design-supply air temperature is an important task (Schild 2004). If it is too low, the draught risk increases in the occupied zone. Figure 6.6 shows a simplified chart for temperature changes in the room with displacement

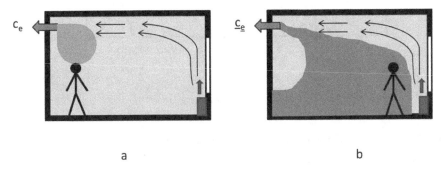

a   b

Figure 6.5 Diagram shows a different situation of ventilation effectiveness in a single space, depending on the location of contaminant sources

Source: Adapted from Mundt (2004)

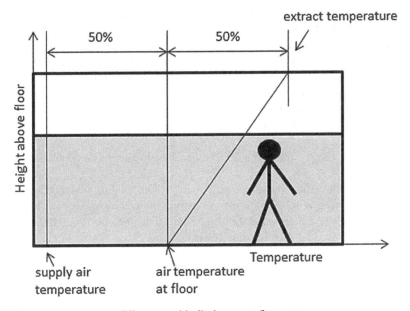

Figure 6.6 Average temperature differences with displacement flow patterns

Source: Adapted from Skistad (2002)

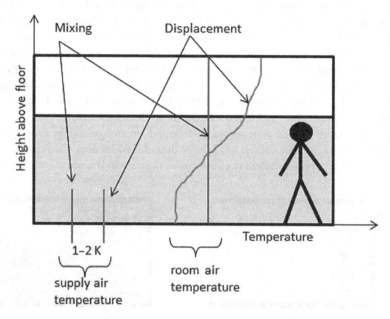

Figure 6.7 Temperature distributions in missing and displacement flow patterns

Source: Adapted from Skistad (2002)

ventilation. The temperature difference between extracted air and room air at floor level is about the same as the temperature difference between room air at floor level and supply air.

The advantages of the displacement ventilation air distribution system can be seen in Figure 6.7. Due to vertical temperature distribution, the supply air temperature in the cooling situation can be higher than that of the mixing air flow pattern. With a higher supply of air temperature, free cooling with outdoor air can be used for longer periods, and when mechanical cooling is used, the coefficient of performance of the compressor cycle is better with a higher temperature of the evaporator. The vertical temperature difference also leads to a higher extract air temperature with a displacement system, which is an advantage when heat recovery from ventilation air is applied.

In summary, the advantages of the displacement flow pattern for air distribution are:

*   For a given air quality, there are indications that displacement ventilation needs less supply air.
*   Displacement ventilation has more potential for free cooling and needs less cooling energy than mixing ventilation. This is most pronounced in rooms with high ceilings.
*   Diffusers for displacement ventilation need less of a pressure drop than diffusers for mixing ventilation, and thus less fan power.

6.4   Ventilation efficiency

In many cases, the circulation of air within a single space would be between mixing and displacement ventilation. For this reason, indices have been developed to measure the ability of the ventilation system to remove contaminants. The following definitions have appeared in the ventilation literature since the 1980s (Liddament 1993, 1996; Sutcliffe 1990), and the descriptions that follow have been adapted from Mundt (2004).

6.4.1   Air change efficiency

At the design stage when the use of the space is unknown, the ventilation should be designed to give a rapid air exchange in the room. The air change efficiency is a measure of this. In order to explain this, the concept of *age of air* is introduced first, which measures how old the air is in a space. Thus, the local mean age of air at a given point is a measure of the air quality at that point. In a fully mixed situation, the local mean age of air will be the same in the whole room. If there is a shortcut from supply to exhaust, the local mean age of air will be low in the short-circuited zone, and high in the stagnant zone. This is shown diagrammatically in Figure 6.8, where another concept, *nominal time constant*, is also introduced; this is the local mean age of air at the exhaust, defined as:

$$\tau_n = \frac{V}{q_v} \tag{6.1}$$

where
τ_n = the nominal time constant of ventilation h
V = the room volume, m^3
q_v = the supply air flow, m^3/h

From the definition, the nominal time constant depends on the air flow rate and volume of the room and is independent of the ventilation pattern.

Figure 6.8 illustrates how nominal time constant and *local mean age of air* are related.

> The local mean age of air of a small volume in the room is the average time from the time that the molecules in the room enter the room. The local mean age is thus given by the time at which the concentration of original molecules falls to zero at point P. The local mean age increases linearly from zero at the entrance to the nominal time constant to the exit. The average room mean age of air is therefore half of the nominal time constant in the case of piston flow.
>
> (Mundt 2004)

Air change efficiency can then be defined as the shortest possible air change time (which is the nominal time constant) over the actual air change time:

$$Air\ change\ efficiency = \frac{nominal\ time\ constant}{actual\ air\ change\ time} \tag{6.2}$$

The definition of the air change efficiency can also be explained as the ratio between the lowest possible mean age of air (nominal time constant over 2) and the room mean age of air.

$$Air\ change\ efficiency = \frac{nominal\ time\ constant}{2 \times mean\ age\ of\ air} \tag{6.3}$$

The range of air change efficiency for four broad classifications of ventilation patterns in one space is summarised in Table 6.6.

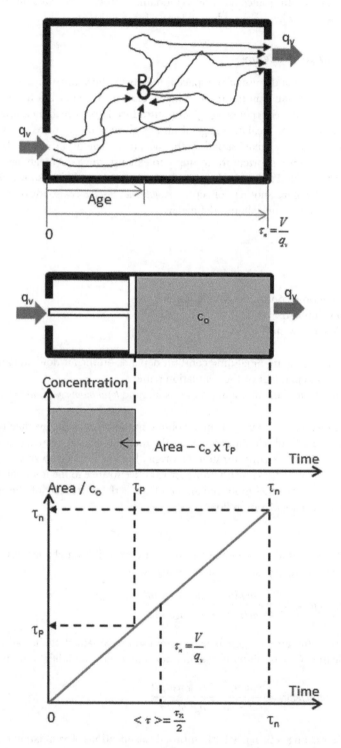

Figure 6.8 Relationship of age of air and nominal time constant for an ideal piston flow in a space

Source: Adapted from Mundt (2004)

Table 6.6 Air change efficiency for different flow conditions in one space

Flow pattern	Air change efficiency
Ideal piston flow	100%
Displacement flow	50–100%
Fully mixed flow	50%
Short-circuit flow	<50%

6.4.2 Contaminant removal effectiveness

In many cases we are interested in knowing the ability of a ventilation system to remove contaminants, and this can be done by comparing the concentration in the exhaust with the mean concentration in the room. This measure is called *contaminant removal effectiveness*:

$$Contaminant\ removal\ effectiveness = \frac{concentration\ in\ the\ exhaust}{mean\ concentration\ in\ the\ room} \tag{6.4}$$

A satisfactory result is achieved when the concentration of contaminants in the space is low compared with the concentration in the exhaust air. The concentration in the exhaust is dependent on the release rate of the contaminant and the ventilation flow rate, and is independent of the ventilation arrangements. The exhaust concentration can also be compared to local values of the concentration of contaminants; these indices are called *local air quality indices*. This index is used in large spaces with local pollution sources, with a focus on the specific location of breathing zones of occupants or workers. These can be calculated using computational fluid dynamics (CFD) models or tracer gas measurements.

6.5 Calculating ventilation rate due to natural driving forces

When the ventilation strategy for a building is for natural or mixed mode, it is necessary to calculate the air flow provided by natural forces. This is difficult to carry out accurately, considering the number of possible air paths into the building and the variability of external driving forces of wind and temperature. Guidance exists on how to carry out quick calculations at the feasibility stage and use detailed computer simulation tools, either stand-alone (for ventilation calculations only) or integrated within thermal simulation models.

In the following section, the process of feasibility-stage calculation methods is described, based on publications such as CIBSE Guide B2 (2016), BS EN5925 (1991) and CIBSE AM10 (2005a).

6.5.1 Flow through purpose-designed openings

For a given applied pressure, the nature of air flow is dependent on the dimensions and geometry of the opening itself. For well-defined, purpose-provided openings such as vents, air flow is usually assumed to be turbulent and is often approximated by the orifice equation given by:

$$Q = C_d A \left[\frac{2}{\rho} \Delta p \right]^{\frac{1}{2}} (\mathrm{m^3/s}) \tag{6.5}$$

Where:
Q = air flow rate $(\mathrm{m^3/s})$
C_d = discharge coefficient

ρ = air density (kg/m³)
ΔP = pressure difference across opening (Pa)
A = area of opening (m²)

In this instance the area, A, is the total net physical openable area. This is usually much less than the dimension of the vent itself, because a significant part of the vent area is taken up with an insect screen or loose filling. In this equation, A represents the area of a single opening, which is equivalent in resistance to all the resistances between the inlet and the outlet (CIBSE 2005a). If there are multiple resistances, the resistance needs to be summed in series, using:

$$\frac{1}{A^2} = \sum_{i=1}^{i=j} \frac{1}{A_i^2}$$

(6.6)

where A_i is the area of the ith element in the path

In cases of openings on opposite walls, using the orifice equation (Eq. 6.5), the total opening area A on each facade can be calculated from the area (A_E) between the two facades, as follows (BS EN5925 1991):

$$\frac{1}{A_E^2} = \frac{1}{A_1} + \frac{1}{A_2}$$

(6.7)

The discharge coefficient

The discharge coefficient is dependent on the opening geometry and on the direction of the approaching flow (e.g. wind direction). For a flat-plate orifice in which the air stream is directed at right angles to the opening, it typically has a value of approximately 0.61–0.65. Such a range is widely used in preliminary design calculations. However, for practical components, the actual value will be dependent on the component itself, as well as the wind direction. For more detailed analysis, therefore, the component should be tested to obtain its actual flow characteristics.

6.5.2   Estimating wind-induced pressure

In general it is observed that relative to the static pressure of the free wind, the time averaged pressure acting at any point on the surface of a building may be represented by the equation:

$$P_w = \frac{\rho}{2} C_p v^2$$

(6.8)

Where:
P_w = wind-induced pressure (Pa)
ρ = air density (kg/m³)
C_p = wind pressure coefficient
v = wind velocity at a datum level (usually building height) (m/s)

Terrain and shielding

Since the strength of the wind close to the Earth's surface is influenced by the roughness of the underlying terrain and the height above ground, a reference level for wind velocity must be specified for use in the wind pressure calculation. When calculating the wind impact on ventilation, the wind velocity is commonly expressed as the measured speed at building height. As a general rule, 'on-site' wind data are rarely available, and therefore data taken from the nearest meteorological station must usually be applied. Before such data can be used, however, it is essential that such measurements are corrected to account for any difference between measurement height and building height, and intervening terrain roughness must also be considered. By nature of the square term in Eq.6.8, wind pressure is very sensitive to the wind velocity and, as a consequence, the arbitrary use of raw wind data will invariably give rise to misleading results. This is perhaps one of the most common causes of error in the calculation of air infiltration rates and wind-induced air flow rates.

Suitable correction for the effects of these parameters may be achieved by using a power law wind profile equation of the form:

$$\frac{v}{v_m} = \alpha z^\gamma \tag{6.9}$$

Where:
z = datum height (m)
v = mean wind speed at datum height (i.e. height of building) (m/s)
v_m = mean wind speed at meteorological station (m/s)
α and γ are coefficients according to terrain roughness (see Table 6.7)

This equation and the associated coefficients are taken from BS EN5925 (1991). Example coefficients are given in Table 6.7.

Such an approach is generally acceptable for winds measured between roof height and a recording height of 10 m. It is inappropriate for the reduction of wind speeds measured in the upper atmosphere. Alternative methods of wind correction are also used based on a log law profile.

Pressure coefficient

The pressure coefficient, C_p, is an empirically derived parameter which is a function of the pattern of flow around the building. It is normally assumed to be independent of wind speed, but varies according to wind direction and position on the building surface. It is also significantly affected by neighbouring obstructions, with the result that similar buildings subjected to different surroundings may be expected to exhibit markedly different pressure coefficient patterns.

Accurate evaluation of this parameter is one of the most difficult aspects of natural ventilation and air infiltration modelling and, as yet, is not possible by theoretical means alone. Although pressure coefficients can be determined by direct measurements of buildings, most information comes from the results of wind loading tests made on scale models of isolated buildings in wind tunnels.

Purpose-designed tests for specific buildings and shielding conditions may be performed, but this is an expensive exercise and is therefore rarely possible.

Table 6.7 Terrain coefficients for use with Eq. 6.9

Terrain coefficient	a	γ
Open, flat	0.68	0.17
Country with scattered windbreaks	0.52	0.20
Urban	0.35	0.25
City	0.21	0.33

For low buildings of up to typically three storeys, pressure coefficients may be expressed as an average value for each face of the building and for each 45° sector, or even 30° sector, in wind direction.

For taller buildings, the spatial distribution of wind pressure takes on much greater significance, since the strength of the wind can vary considerably over the height range. In these instances spatial dependent data are essential. Solutions include scale wind tunnel modelling and the use of CFD to predict the surrounding pressure field.

Representative pressure coefficients for open and urban environments are shown in Table 6.8.

Table 6.8a Example of pressure coefficients for low-rise buildings in open terrain

		OPEN TERRAIN/SHIELDING							
		Wind angle							
Location		0	45	90	135	180	225	270	315
Face 1		0.7	0.35	−0.5	−0.4	−0.2	−0.4	−0.5	0.35
Face 2		−0.2	−0.4	−0.5	0.35	0.7	0.35	−0.5	−0.4
Face 3		−0.5	0.35	0.7	0.35	−0.5	−0.4	−0.2	−0.4
Face 4		−0.5	−0.4	−0.2	−0.4	−0.5	0.35	0.7	0.35
Roof	Front	−0.8	−0.7	−0.6	−0.5	−0.4	−0.5	−0.6	−0.7
(<10° pitch)	Rear	−0.4	−0.5	−0.6	−0.7	−0.8	−0.7	−0.6	−0.5
Roof	Front	−0.4	−0.5	−0.6	−0.5	−0.4	−0.5	−0.6	−0.5
(11–30° pitch)	Rear	−0.4	−0.5	−0.6	−0.5	−0.4	−0.5	−0.6	−0.5
Roof	Front	0.3	−0.5	−0.6	−0.4	−0.5	−0.4	−0.6	−0.4
(>30° pitch)	Rear	−0.5	−0.4	−0.6	−0.4	0.3	−0.4	−0.6	−0.4

Table 6.8b Example of pressure coefficients for low-rise urban buildings

		URBAN							
		Wind angle							
Location		0	45	90	135	180	225	270	315
Face 1		0.2	0.05	−0.25	−0.2	−0.25	−0.3	−0.25	0.05
Face 2		−0.25	−0.3	−0.25	0.05	0.2	0.05	−0.25	−0.3
Face 3		−0.25	0.05	0.2	0.05	−0.25	−0.3	−0.25	−0.3
Face 4		−0.25	−0.3	−0.25	−0.3	−0.25	0.05	0.2	0.05
Roof	Front	−0.5	−0.5	−0.4	−0.5	−0.5	−0.5	−0.4	−0.5
(<10° pitch)	Rear	−0.5	−0.5	−0.4	−0.5	−0.5	−0.5	−0.4	−0.5
Roof	Front	−0.3	−0.4	−0.5	−0.4	−0.3	−0.4	−0.5	−0.4
(11–30° pitch)	Rear	−0.3	−0.4	−0.5	−0.4	−0.3	−0.4	−0.5	−0.4
Roof	Front	0.25	−0.3	−0.5	−0.3	−0.4	−0.3	−0.5	−0.3
(>30° pitch)	Rear	−0.4	−0.3	−0.5	−0.3	0.25	−0.3	−0.5	−0.3

6.5.3 The stack effect

Assuming a uniform air temperature, the pressure of an air mass at any height z above a convenient datum level z_o (for example, ground or floor level) is given by:

$$p_z = p_o - \rho gz \ \text{(Pa)} \tag{6.10}$$

Where:
p_z = pressure at required height (Pa)
P_o = pressure at datum level z_o (Pa)
g = acceleration due to gravity (m/s^2)
z = height above datum (m)

The resultant pressure gradient is therefore:

$$\frac{dp}{dz} = -\rho g \tag{6.11}$$

which becomes, by consideration of the ideal gas law, the following equation:

$$\frac{dp}{dz} = -\rho_o g \frac{273}{\theta} \tag{6.12}$$

Where:
p_o = air density at 273K (kg/m^3) = 1.293 (kg/m^3)
θ = absolute temperature of the air mass (K)

Thus, the pressure gradient is inversely proportional to the absolute temperature of the air mass. For two openings on opposite walls, as illustrated in CIBSE AM10 (2005a), the stack pressure difference can be calculated by:

$$p_s = -\rho_o g 273 (h_2 - h_1) \left[\frac{1}{\theta_e} - \frac{1}{\theta_i} \right] \text{(Pa)} \tag{6.13}$$

Where:
θ_e = absolute temperature of the outdoor air (K)
θ_i = absolute temperature of the indoor air (K)

Stack pressures are often comparable with wind-induced pressures, and many ventilation designs concentrate on developing the stack pressure to drive ventilation air flow.

6.5.4 Combining wind with stack pressure

The total pressure, p_{t_i}, acting at an opening, i, due to the combined impact of wind and stack effect is given by:

$$p_{t_i} = p_{w_i} + p_{s_i} \tag{6.14}$$

It is important to understand that summing the pressures due to the stack and wind effect at each opening is not the same as summing the flow rates determined by calculating the flow rates due to wind and stack pressure separately.

6.5.5   *Calculating the natural ventilation rate*

The calculation of naturally induced air flow through the building requires the following steps:

- calculate the wind pressure for each path
- calculate the stack pressure for each path
- determine the total pressure (wind + stack) for each path
- apply the general flow equation to each path
- determine an internal pressure for the space such that the total air flow into a space is balanced by the total air flow out of the space.

This final step is the central component of ventilation calculations. Except for very simple networks, the calculation is not direct and the process of 'iteration' is required. It is this element that makes ventilation calculation approaches so difficult to follow.

Taking all the flow paths, the conservation of mass requires a flow balance between the ingoing and outgoing air flow. This is expressed by:

$$\sum_{i=1}^{j} \rho_i Q_i = 0 \qquad \text{(kg/s)} \tag{6.15}$$

Where:
ρ_i = density of air flowing through flow path i(kg/m³)
Q_i = volume air flow rate through flow path i (m³/s)

This method of calculating air flow rate for single zones is relatively easy to incorporate in thermal models, and the algorithm on how it can be done is included in CIBSE Guide A (2015).

For multi-zone network models, achieving a solution becomes more complex. Unlike the 'single-zone' approach, where there was only one internal pressure to determine, there are now many values. This adds considerably to the complexity of the numerical solution method.

As mentioned, multi-zone ventilation models are incorporated into most thermal and energy simulation models to consider the energy impact of natural ventilation paths. In addition, there exist multi-zone models focusing on ventilation and contaminant distribution in buildings such as CONTAMN (2019).

6.6   Fans

Fans are used in hybrid ventilation designs to increase air flow rate when natural driving forces are not enough to achieve the required ventilation rate. Their properties depend on the design, the size and the rotating speed; this will affect their energy consumption.

When a fan is connected to a duct system, the fan type and size are selected according to the air flow and the resistance of the duct system and components such as filters, dampers and heating/cooling coils. The fan must work against this resistance to move the air at the desired flow rate and to the desired position in the building.

When a fan is connected to a duct system, the flow in the system stabilises at the flow rate with which the pressure generated by the fan is equal to the pressure drop in the complete system. The fan must work against the resistance of losses in the duct system to move the air at the desired flow rate and to the desired position.

6.6.1 Power demand of fans

Power input to the air flow is the product of air flow and pressure rise:

$$P = \Delta p q_v \tag{6.16}$$

Where:
P = the input power of fan into the air flow, W
Δp = the total pressure difference across the fan, Pa
q_v = the air flow through the fan, m³/s

The power demand to run the fan, however, is much greater due to losses in the fan impellor, the fan drive and the motor. Often, all these efficiencies are lumped into the total efficiency of the fan. The electrical power required to run the fan can be calculated from the equation:

$$P = \Delta p q_v / \eta_{tot} \tag{6.17}$$

Where:
η_{tot} is the total efficiency of fan, including fan itself, motor (including speed control, etc.), drives and losses in the built-in situation

The power demand is thus affected by the air flow, the pressure difference and the efficiency. The fans are designed for one velocity air flow, where the flow has lowest losses and the efficiency has its maximum value. If the flow is smaller or larger, the efficiency is decreased.

If the rotation speed of a fan increases, the air flow changes in relation to the rotating speed, pressure proportional to the second power of the speed and power demand to the third power of the fan speed.

$$\frac{q_{v1}}{q_{v2}} = \frac{n_1}{n_2} \tag{6.18}$$

$$\frac{\Delta p_1}{\Delta p_2} = \left(\frac{n_1}{n_2}\right)^2 \tag{6.19}$$

$$\frac{P_1}{P_2} = \left(\frac{n_1}{n_2}\right)^3 \tag{6.20}$$

Where:
n = fan speed, 1/min

Therefore, variable-speed fans have become popular for energy efficiency reasons.

6.6.2 Specific power of fans

The term 'specific power' of each fan is used to define the overall efficiency of the air-moving system.

$$P_{SFP} = \frac{P}{q_v} = \frac{\Delta p}{\eta_{tot}} \tag{6.21}$$

Where:

P_{SFP} = the specific fan power in W/m³/s
P = the input power of the motor for the fan, W
q_v = the nominal air flow through the fan in m³/s
Δp = the total pressure difference across the fan, Pa
η_{tot} = the total efficiency of fan, motor and drive in the built-in situation

In the air handling system the coefficient is valid for the nominal air flow with clean filter conditions and all bypasses closed. It is related to an air density of 1.2 kg/m³.

In many countries, the allowed specific power of the fan is stipulated in the regulations and guidelines to help with energy efficiency measures. For example in the UK, for non-residential buildings, the notional building (according to Part L Building Regulation) specific fan power ranges from 0.3 W/l/s for a local terminal unit to 1.8 W/l/s for central ventilation units.

6.7   Ventilation for cooling

Increasingly, ventilation is used to provide internal thermal comfort in buildings, i.e. controlling temperature inside buildings and thus avoiding overheating. In this case, the ventilation rates required are quite different (usually much higher) than ventilation rates required for IAQ purposes. In many cases, ventilation is provided by natural means, although hybrid strategies and coupling with internal thermal mass are increasingly popular. In some cases, mechanical ventilation is used for providing thermal comfort, although this option is usually energy consuming because of the energy required by the fans. Ventilation for controlling internal temperatures is location and building operation dependent. In very broad terms the following climatic classifications might apply:

Climatic regions with a high cooling load

In such climates, ventilation strategies are usually designed to provide some cooling to reduce reliance on active air conditioning systems. Such ventilation strategies would most likely be implemented in climates with a hot cooling season and winters requiring no or very little heating. Ventilation strategies for thermal comfort (cooling) would usually be combined with other passive and/or active cooling methods in addition to passive and/or active heating methods.

Climatic regions with a high heating load

These ventilation strategies are mainly designed to provide IAQ. They would most likely be implemented in climates with a cold heating season and a summer requiring none or very little cooling. Ventilation strategies for IAQ (heating) would, in most cases, be efficient mechanical ventilation strategies, perhaps combined with passive cooling strategies for the summer.

Climatic regions with moderate heating and cooling loads

In these climates, ventilation strategies are designed to provide thermal comfort in the summer. Such ventilation strategies might also be implemented in climates where extra high moisture levels might impose an additional load. Natural ventilation strategies might be able to satisfy cooling load requirements for a range of buildings with moderate internal heat gains.

Therefore, outside air, if colder than thermal comfort temperature, can be used to cool an indoor space. Levermore (2002) gives the heat transfer equation as:

$$\Phi_v = \tfrac{1}{3} NV (\theta_f - \theta_o)(1 - e^{-x}) \tag{6.22}$$

Where:
Φ_v = heat transfer by ventilation (W)
N = air change rate (h^{-1})
V = room volume (m^3)
θ_f = surface temperature of the internal surfaces of the building fabric (°C)
θ_o = outdoor air temperature (°C)

And exponent x is given by:

$$x = \frac{4.8A}{\tfrac{1}{3} NV} \tag{6.23}$$

where A is the area of opening (m^2).

If free cooling through external air is provided with mechanical ventilation, the electrical energy consumption of fans should be considered when calculating energy efficiency improvements.

An international project on ventilative cooling was completed in 2018.[2] Its outputs include a design guide for ventilation cooling, detailed case studies of operational buildings using ventilation cooling from various climates and a tool to assess the potential of ventilative cooling in various climates and building types.

In addition to free cooling as described earlier, ventilative cooling can be coupled with the thermal mass of the ground and/or thermal mass in the building. Two such strategies are described next.

Ground-coupled ventilation

Ground-coupled air systems (sometimes called earth-to-air heat exchangers [ETAHE]) are mainly used for preconditioning outdoor air during the summer but can also be used during winter to preheat the air. An ETAHE draws ventilation air through ducts buried underground, as shown in the simplified diagram in Figure 6.9. A typical ideal operation of the system is shown in Figure 6.10. Figure 6.11 shows an operational building from Portugal where an ETAHE is used as part of the low-energy design of the building.

Most existing ETAHEs are installed in mechanically ventilated buildings, in which electrical fans provide the air flow-driving forces. In such systems, an ETAHE can be a single duct or multiple parallel ducts made of pre-fabricated metal, PVC, or concrete pipes with diameters at a magnitude of 10 cm. In case of the parallel pipe systems, the distance between the pipes should be kept approximately 1 m from each other in order to minimise the thermal interaction. The size of an ETAHE depends on the designed air flow rate and the available space. A maximum air velocity of 2 m/s is normally recommended for smaller systems, and larger systems can be designed for air velocity up to 5 m/s. Due to the high velocity and small duct size, a large amount of energy has to be spent on the mechanical ventilation systems to deliver air through the ETAHEs (Perino 2008).

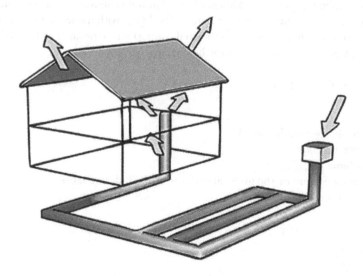

Figure 6.9 Indicative diagram of an ETAHE system

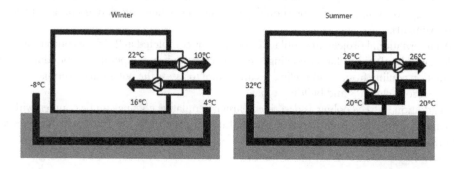

Figure 6.10 ETAHE system schematic for winter and summer incorporating a heat exchanger for winter

Source: Adapted from Hollmuller (2011)

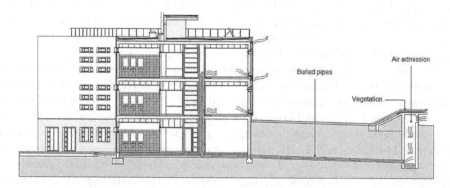

Figure 6.11 The SOLAR XII building in Portugal in which ETAHE is used as part of the low-energy
        cooling strategy

Source: Courtesy of the AdVent Project

The cooling potential of the systems depends on the ground temperature distribution. A depth of 2–4 m for the pipes is generally recommended, but this depends on the location and the condition of the soil (Zimmerman and Remund 2001).

The energy-saving potential of an ETAHE has attracted many simulation studies since the 1980s. The main efforts have been on the development of simulation methods. However, some simplified tools for the prediction of outlet temperature and sizing of ETAHE systems have also appeared. These methods are usually based on statistical analysis of simulation results and not independent of external conditions (Grosso and Raimondo 2008; Santamouris 2006; Santamouris and Asimakopoulos 1996; Warwick *et al.* 2009). Many thermal simulation models include modules to consider ETAHE systems during the design phase.

Night cooling

Night cooling is a low-energy cooling strategy where the building is ventilated during the night. It works by using natural or mechanical ventilation to cool the surfaces of the building fabric at night so that it can absorb heat during the day. Night-time ventilation is suitable for areas with a high diurnal temperature range and where night-time temperature is not so cold as to create discomfort. Figure 6.12 shows temperature measurements to demonstrate the effect of natural night ventilation in practice (Kolokotroni 1998). It shows that night ventilation reduces air and surface temperatures and delays the peak internal temperature until later in the day.

Night ventilation is effective where a building includes an exposed internal thermal mass so that heat can be absorbed during the day. Night ventilation can affect internal conditions during the day in four ways:

- reduces peak air temperatures
- reduces air temperatures throughout the day, and in particular during the morning hours
- reduces slab temperature
- creates a time lag between external and internal temperatures.

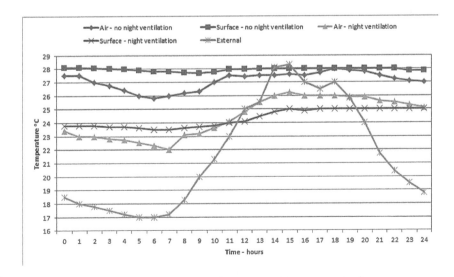

Figure 6.12 Measured hourly temperatures in an exposed thermal mass office with and without night ventilation

Source: Adapted from Kolokotroni (1998)

Parameters directly affecting the cooling potential of night ventilation are:

- day ventilation rate
- night ventilation rate
- exposed internal thermal mass
- internal heat gains
- weather (temperature and solar gains).

Figure 6.13 presents the effect of day and night ventilation and the exposed thermal mass on the maximum internal temperatures for a library in southeast England. It can be seen that, for example, a 2.5°C reduction could be achieved for the maximum day temperature in a construction with exposed thermal mass to which night ventilation is provided at a rate of 5 ACH (Kolokotroni 2001).

Night ventilation systems are classified as direct or indirect as a function of the procedure by which heat is transferred between the thermal storage mass and the conditioned space (Santamouris 2004). In direct systems, the cool air is circulated inside the building zones and heat is transferred in the exposed opaque elements of the building. The reduced temperature mass of the building contributes to reduce the next day's indoor temperature through convective and radiative processes. Circulation of the air can be achieved by natural or mechanical ventilation. In direct systems, the mass of the building has to be exposed and the use of coverings or false floors or ceilings has to be avoided.

In indirect systems, the cool air is circulated during the night through a thermal storage medium where heat is stored and then recovered during the day period. In general, the storage medium is a slab covered by a false ceiling or floor while the circulation of the air

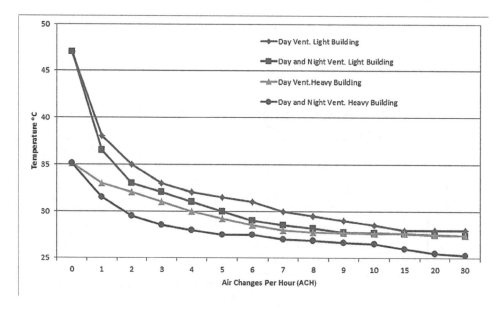

Figure 6.13 Predicted maximum temperatures for library buildings as a function of ventilation rate and exposed thermal mass

Source: Adapted from Kolokotroni (2001)

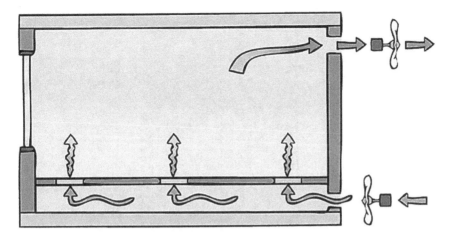

Figure 6.14 Indicative diagram of an indirect night cooling system

Source: Author's compilation with Warwick

is always forced (Figure 6.14). ETAHE systems, as described in Section 6.7.1, can be classified as indirect night cooling systems. Direct and indirect systems are used many times in a combined way. Their combination has been successfully demonstrated in the past, and they are now mature technologies, for example, in the Schwerzenbacherhof Office and Industrial Building in Schwerzenbach, Switzerland (Zimmerman and Anderson 1998).

Although it is a powerful strategy for cooling, it presents important limitations. Moisture and condensation control is necessary, particularly in humid areas. Pollution and acoustic problems (especially in urban areas), as well as problems of privacy, are associated with the use of natural ventilation techniques. Another important limitation is associated with climatic conditions within cities. The increase of the ambient temperature, especially during the night because of the urban heat island phenomenon, as well as the marked decrease of wind speed in urban canyons, might considerably reduce the potential of night cooling strategies.

6.7.4 Developments based on traditional systems and strategies

Finally it should be mentioned that in the drive for energy efficiency and low-energy heating/cooling methods for buildings, traditional ventilation components and strategies are explored, developed for modern buildings and reach the market successfully. As an example, the windcatcher ventilation component and the passive downdraught evaporative cooling systems are described briefly.

Windcatcher

The windcatcher system is a passive ventilation system which uses both buoyancy and wind-driven forces to provide ventilation in a space. A diagram of the component is shown in Figure 6.15. Windcatcher systems have been employed in buildings in the Middle East for many centuries, and they are known by different names in different parts of the region.

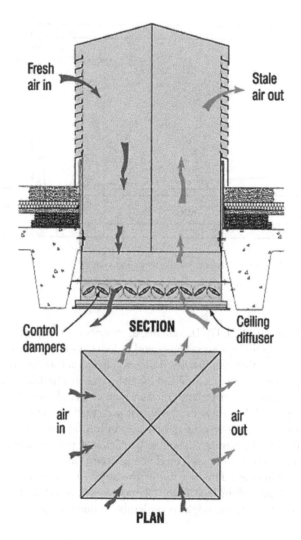

Figure 6.15 Modern windcatcher® design by Monodraught Ltd

Source: Courtesy of Monodraught, UK

Modern-day variants have been used increasingly in locations where a strong wind-driving force is available.

The modern wind tower of a windcatcher consists of a vertical stack which, in cross-section is divided into quadrants. This normally penetrates and terminates at the ceiling level in a space. Wind directed at the windward-facing quadrant (or quadrants) is driven down the stack to ventilate the space below. Simultaneously, stale air is sucked out of the space driven by the negative pressures generated by the wind on the leeward-facing quadrant or quadrants. Dampers located in ceiling-mounted diffusers control the rate of air flow.

The advantages of the components are that air is drawn in at a high level where pollutant concentration is usually lower than at street level; they can be integrated with a hybrid fan to ensure reliable operation under low wind speed conditions; and they have the possibility to supply air into deep plan spaces.

PDEC

The passive downdraught evaporative cooling (PDEC) system has again been traditionally developed in hot dry climates. In the past, evaporative cooling was achieved through water-filled porous pots within the supply air stream or the use of a pool of water at the base of the supply stack. In modern systems, water sprayed high into the supply air stream cools the air stream and increases the supply air density, thereby augmenting the buoyancy-induced pressure differences that drive air flow.

Modern applications can be found in moderate climates, for example, the School of Slavonic and East European Studies (SSEES) building of University College London (Cook and Short 2005).

6.8 **Summary**

In this chapter the principles of energy-efficient ventilation were outlined; ventilation can be provided through natural or mechanical means or a combination of the two (hybrid or mixed-mode ventilation). It was stated that ventilation provision has a dual purpose: first to provide satisfactory IAQ and second (depending on the external climate and type/use of the building) thermal comfort. These two functions of ventilation must be considered separately to provide ventilation in an energy-efficient manner.

This chapter first outlined ventilation requirements for various types of buildings and available ventilation strategies. It then introduced some parameters used to measure ventilation efficiency within a single space. It followed with an outline of how ventilation rates due to the natural forces of wind and buoyancy can be calculated and enhanced by the use of fans. The chapter concluded with a description of ventilation strategies useful to provide thermal comfort in a building, thus avoiding the need for artificial cooling in certain circumstances.

- Ventilation requirements for various types of buildings are presented according to regulations and guidelines of professional institutions.
- Ventilation strategies are outlined and parameters used to measure ventilation efficiency within a single space are outlined.
- Calculations of ventilation rates due to natural forces of wind and buoyancy are presented. These calculations are useful for the initial design stage in a building when the ventilation strategy has not been decided yet.
- Fan characteristics related to energy efficiency are presented.
- Ventilation strategies useful to provide thermal comfort in a building are presented. These include ventilative cooling, ground-coupled ventilation, night ventilation, wind-catchers and natural evaporative cooling.

Notes

1 Building Regulations are available online from www.planningportal.gov.uk/buildingregulations/approveddocuments/ (accessed 19 June 2019).
2 The outputs of the IEA Ventilative cooling project can be found at https://venticool.eu/annex-62-home/ (accessed 20 June 2019)

7   Air conditioning systems

7.1   Introduction

For low-energy consumption and greenhouse gas emissions, the emphasis in the design of thermal environment control systems for buildings should be on using passive or very low-energy systems such as natural or mechanical ventilation, discussed in Chapter 6. This, however, may not be always possible in office or other commercial and institutional buildings due to high internal and external heat gains in the summer months. To overcome heat loads and provide thermal comfort throughout the year, designers of building services systems are frequently called upon to select the most appropriate air conditioning system for a specific application. To carry out the selection process correctly, they must have a good knowledge of the various air conditioning systems available on the market, their characteristics and applications. They must also be aware of the implications of selecting a particular system on the capital, running and life cycle costs. Other important criteria in the selection process include spatial requirements, maintainability, reliability, flexibility and environmental impacts. All these factors are interrelated, and so the designer should understand their relative importance for each design job. Other important selection criteria include the specific requirements of the client in terms of aesthetics, maximisation of rental income and the saleability of the property.

   In this chapter we consider the different types of air conditioning systems and their applications.

7.2   Classification of air conditioning systems

Air conditioning systems can be classified into a number of broad categories, as follows:

   Central systems or unitary systems. Central systems employ one or more air handling units which are served by heating and cooling equipment located outside the conditioned area. In unitary systems, all the components required for conditioning the supply air are housed in a single package. Cooling is provided by a direct expansion coil housed in the package.

   Single-zone or multi-zone systems. A single-zone system serves only one zone in the building. The spaces in this zone have similar load characteristics, and the supply conditions are controlled from a single sensing point within the zone. In a multi-zone system terminal devices are used to control the supply conditions to individual zones. Each zone is controlled from its one sensing and control unit.

   Constant air volume (CAV) and variable air volume (VAV) systems. In CAV systems, the supply air volume to each zone is kept constant. Variations in the load at part-load conditions are met by varying the temperature of the supply air. In VAV systems the supply air temperature is kept constant. Variations in load are catered for by varying the supply air volume flow rate.

7.3   Unitary systems

Unitary systems are the simplest air conditioning equipment. They are factory-assembled units with all the components contained in a small number of enclosures. The units can provide cooling only or can be configured to provide both cooling and heating by using the vapour compression cycle in reverse, i.e. heat pump cycle. A simpler method of providing heating is to use electric resistance heaters with a non-reversible vapour compression cooling system.

The simplest types of unitary systems are through the wall units in which all components are contained in one enclosure, and split systems. Split air conditioning systems consist of an outdoor and an indoor unit. The outdoor unit contains the compressor, the condenser/fan assembly and the expansion valve. The indoor unit contains the evaporator and the supply fan. The outdoor and indoor units are connected by refrigerant piping. The supply fan recirculates air from the conditioned space. A fixed quantity of fresh air may also be provided through an opening on the wall. Split systems are available in larger capacities than room units.

Advantages

*   simple to install
*   lower cost than central systems in providing individual space control.

Disadvantages

*   no humidification capability
*   higher energy usage than central systems for the air conditioning of large buildings
*   limited air distribution capability within the conditioned space.

Applications

*   small buildings
*   commercial premises such as banks
*   hotel apartments.

7.3.1   Multi-split and variable refrigerant volume systems

In multi-split systems one external unit serves a number of indoor units. The systems can be single-zone, where all indoor units are controlled by a single thermostat, or multi-zone, where each unit is controlled by its own thermostat and can provide individual space temperature control. A limitation, however, is that if cooling is required in one area, it is not possible to provide heating in a different area served by the same system because the compressors of the outdoor unit will function in only the cooling or the heating mode. This limitation can be overcome by the variable refrigerant volume system (VRV). A schematic diagram of a typical VRV system is shown in Figure 7.1.

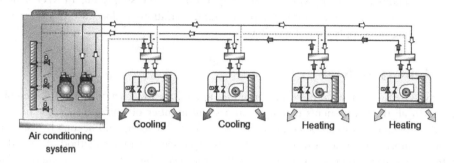

Figure 7.1 Schematic of a VRV system

With VRV systems each indoor unit may provide cooling or heating independently of the other units. Where one area of the building requires heating and another cooling, heat removed from the area that requires cooling can be upgraded and used in the area that requires heating, thus improving the overall efficiency of the system.

Ventilation air to the spaces, if desired, can be provided by separate heat recovery units. With these systems, an equal quantity of outside air and exhaust air are drawn through a heat recovery heat exchanger so that the exhaust air is used to heat or cool the ventilation air.

Advantages

• lower cost than central systems for single-zone air conditioning
• VRV units provide individual zone control.

Disadvantages

• no humidification capability
• units sited in indoor spaces may create noise problems
• multiple units sited indoors may occupy valuable rentable space.

Applications

• office and other commercial buildings
• multi-tenanted buildings – each tenant served from a separate multi-split or VRV system
• renovation work – retrofit applications.

7.3.2 Packaged systems

These are larger-capacity single-zone systems, up to 100 kW. They can be used to serve a single space or a number of spaces within a zone by siting the system centrally and distributing the supply air to the individual spaces through a duct distribution system. Each zone is served by its own packaged unit.

Packaged systems can be mounted indoors in basements, utility rooms, attic or crawl spaces, outdoors on the ground or on rooftop. Unlike through the wall and split systems, packaged units are supplied with fans capable of operating with ductwork.

Advantages

- lower cost than central systems for single-zone air conditioning
- can provide individual air distribution in the conditioned spaces
- can provide ventilation air at all times
- units are available with complete and self-contained control systems
- can provide individual zone control.

Disadvantages

- no humidification capability
- air-cooled units should have access to outdoor air
- units sited in indoor spaces may create noise problems
- multiple units sited indoors may occupy valuable rentable space.

Applications

- office and other commercial buildings to provide floor-by-floor air conditioning
- multi-tenanted buildings – each tenant served from a separate packaged system
- open-plan spaces to provide central air distribution
- renovation work – retrofit applications.

7.4   Central air conditioning systems

Central systems employ one or more air handling units which are served by heating and cooling equipment located outside the conditioned area.

Applications

Central air conditioning systems can be used:

- For the conditioning of spaces which have a uniform load. These are usually large, open spaces with small external loads and high internal loads such as theatres, department stores and the public spaces of commercial buildings.
- For precise control of the conditions within a small space. Such spaces can be precision laboratories or manufacturing units and operating theatres requiring cleanliness and accurate control of temperature, humidity and air distribution.
- As a source of conditioned air for other systems. Some multi-zone systems utilise the plant in individual zones to offset part of the sensible load. The latent load and the remainder of the sensible load are handled by the central system, which conditions either fresh air or a mixture of fresh and recirculated air. This arrangement reduces the amount of air handled by the central plant, and as a consequence, the size of the ductwork is reduced. Such systems can be used in multi-storey buildings where space is at a premium.

Figure 7.2 shows the main components of a central single-zone air conditioning system. The majority of these components are described in detail in Chapter 8.

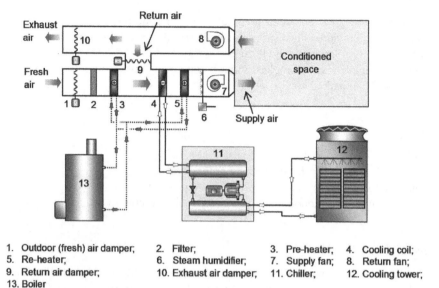

1. Outdoor (fresh) air damper;   2. Filter;   3. Pre-heater;   4. Cooling coil;
5. Re-heater;   6. Steam humidifier;   7. Supply fan;   8. Return fan;
9. Return air damper;   10. Exhaust air damper; 11. Chiller;   12. Cooling tower;
13. Boiler

Figure 7.2 Main components of a central air conditioning system

7.5   All-air central air conditioning systems

7.5.1   *Introduction*

In all-air central air conditioning systems, the only medium providing both sensible and latent cooling in the conditioned space is air. All cooling and humidification/dehumidification are provided in the central plant, and the cold air is distributed to the various zones within the building. Heating may be provided either at the central plant or within the air stream in individual zones.

All-air systems may be classified into two major categories: a) constant-volume variable-temperature systems and b) variable-volume constant-temperature systems.

These systems may also be classified as *single-duct systems* and *dual-duct systems*. Single-duct systems employ a common duct distribution system, and all the heating and cooling coils are arranged in a series flow path within the air handling unit (AHU). Dual-duct systems employ two duct distribution systems with one duct conveying the hot air stream and the other duct the cold air stream.

Advantages

* Quiet operation and centralised maintenance – all mechanical equipment is located remotely from the conditioned spaces. This facilitates noise isolation and allows for easy maintenance.
* Design simplicity – wide choice of zoneability and humidity control and simultaneous availability of heating and cooling if required. Wide flexibility in the design of air distribution within the conditioned spaces and minimum interference from furniture, windows, curtains etc.

- Economy of operation – outdoor air can be used directly in marginal weather conditions to provide 'free cooling', reducing the need for refrigeration.
- Adaptability to heat recovery – they can easily be adapted to accommodate heat recovery equipment.

Disadvantages

- duct clearance – the requirement for clearance either at the ceiling or floor level and clearance for duct risers reduces the usable space, which may be critical, especially in high-rise buildings
- complicated balancing – multi-zone systems with a large number of air outlets are very difficult to balance to achieve the design flow rates
- coordination at the design stage – they require greater coordination between the architect and the mechanical and structural designers at the design stage to provide easy access to terminal devices.

Applications

All-air systems are applied to buildings with a large number of zones that require individual control of space conditions. Such buildings include department stores, supermarkets, common areas in hotels, office buildings, theatres, cinemas and hospitals. They can also be used in specialised applications where there is a need for close control of temperature and humidity; these include clean rooms, hospital operating theatres, computer rooms and textile factories.

7.5.2 *Constant air volume variable-temperature system*

In the constant-volume variable-temperature system, the supply volume to the conditioned spaces remains constant and the heating and cooling loads are satisfied by varying the supply temperature of the air. The supply temperature can be varied by a number of methods, the most common being, a) control of the cooling capacity of the cooling coil, b) air reheat control and c) terminal reheat system.

Cooling coil capacity control

In direct expansion coils, the control of cooling capacity can be achieved by a number of ways, such as on-off control of single or multiple reciprocating compressors arranged in parallel, cylinder unloading of reciprocating compressors, hot gas bypass, control of the inlet guide vanes in the case of centrifugal compressors and variable-speed control. Variable-speed control is the most efficient and currently the most commonly used method in medium-to-large capacity systems.

Control of the capacity of chilled water coils can be achieved through modulation of the water flow through the coil. The flow control can be achieved either by a two-way valve or by a three-way diverting valve and a space thermostat, as shown in Figure 7.3. Two-way valves are used in variable flow pumping systems and three-way diverting valves in constant flow systems.

When a diverting valve is used, a regulating valve is placed in the bypass line and set at the same pressure drop as the coil and the isolating valves to balance the flow.

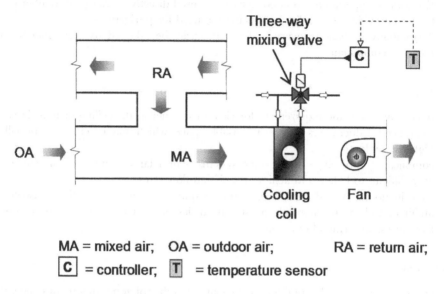

MA = mixed air;   OA = outdoor air;   RA = return air;
[C] = controller;   [T] = temperature sensor

Figure 7.3 Chilled water flow control

At part-load conditions, as the load on the space falls, the space thermostat, T, sends a signal to the controller, which progressively reduces the mass flow of chilled water through the coil, thus maintaining a constant space temperature. With this method of control, the off-coil temperature and the relative humidity of the air in the space will vary with the load, whereas the space and return air dry-bulb temperatures will remain constant.

Reheat control

This is the best method of control in terms of maintaining the design space temperature and relative humidity at part-load conditions. The system is illustrated in Figure 7.4. It employs a humidistat, a thermostat, a controller and heating and cooling coils. The controller compares the signals from the humidistat and the thermostat, and if the humidistat signal is higher, the coil control valve is modulated accordingly to control the space humidity. If in the process the temperature falls below the set point, the controller energises the heating coil control valve to raise the space temperature to the design value. If the temperature rises above the set point, the signal from the thermostat is used to control the cooling coil valve in sequence with heating to reduce the space temperature. The term 'sequence control' is used to indicate that either the heating coil or the cooling coil is used at a particular time, but not both.

The space humidity may also be controlled by a dry-bulb temperature sensor placed in the duct immediately after the cooling coil. This sensor controls the flow in the cooling coil to maintain the off-coil temperature constant. Although the apparatus dew point and the moisture content of the air leaving the coil will vary, the variation is very small. This method of control is therefore considered to provide a constant off-coil air dew point temperature and moisture content, which is known as dew point control.

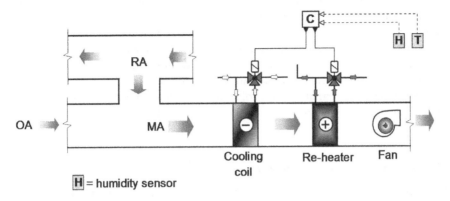

Figure 7.4 Cooling and dehumidification with reheat

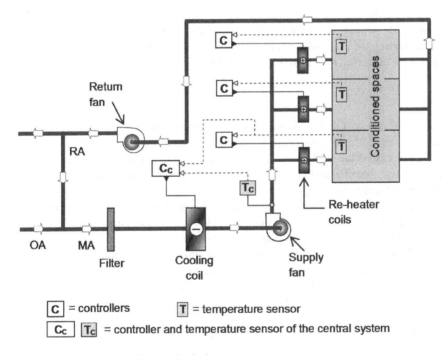

Figure 7.5 Schematic diagram of a terminal reheat system

Terminal reheat system

The terminal reheat system can be used to provide temperature and humidity control to zones or spaces that have unequal loadings and require high ventilation rates. Applications include hospitals, office buildings, schools and laboratories. A schematic diagram of the system is shown in Figure 7.5.

The cooling coil, with a pre-heater coil if required, is placed in the central plant, and heating coils are inserted in the duct system in individual zones. Heating may be provided by steam, hot water or electricity. The central plant provides cooling to satisfy the design cooling and dehumidification loads of the building. The temperatures in the various zones or spaces are controlled by reheating the air stream of the particular zone to the required temperature. The reheaters can also be used to provide winter heating as required.

The main disadvantage of terminal reheat systems is their high operating costs because they provide simultaneous cooling and heating of the supply air.

To save energy, a load analyser control system can be employed, which allows the off-coil temperature to rise in response to a reduction in demand from the greatest cooling load.

7.5.3   Variable air volume system

VAV systems satisfy the cooling and heating loads of a building by varying the supply air volume to the conditioned spaces. In their simplest form, VAV systems consist of a central plant which conditions the air and a supply fan which delivers the air to zone terminal devices. The terminal devices regulate the air flow to the conditioned spaces. Terminal devices are available in various configurations, the simplest being a thermostatically controlled damper. The ratio of the minimum air volume flow to the design flow at the central plant is termed the system turn-down ratio, whereas the ratio of the minimum flow to the design flow at the terminal device is termed the terminal turn-down ratio.

Advantages

- *Low operating cost.* The fan power and refrigeration power follow the reduction in air flow at reduced loads, producing savings on the operating cost of the system. In inter-mediate seasons all outdoor air can be used for 'free cooling', saving in refrigeration power. Also, heating and cooling cannot occur simultaneously, resulting in savings in chiller and boiler running costs.
- *Low capital cost.* The system has lower capital cost than other systems providing indi-vidual space control because it employs single runs of ducts and simple control at the terminal devices. The system can also exploit diversity in heat gains, and thus it is possible to size the equipment on the simultaneous heat gain rather than the individual maximum heat gains, resulting in lower-capacity equipment.
- *Simple individual space control and simple operation.* The system provides individual space temperature control through a thermostat and a flow control terminal device. Changeover of operation from summer cooling to winter heating simply involves stopping the refrigeration equipment and starting the heating equipment at the central plant.
- *Centralised equipment.* All major equipment is centralised, allowing for easy maintenance.

Disadvantages

The traditional VAV system has a number of disadvantages, which can be overcome to a greater or lesser extent by employing more sophisticated equipment and control systems. These include:

- *Reduced space air movement at reduced loads.* This problem may be overcome by using special terminal devices and introducing design modifications to the basic VAV system.

- *Reduced dehumidification capacity at reduced loads.* At low sensible loads, if the latent gains remain relatively constant, the relative humidity in the space may rise to unacceptable levels.
- *Reduced ventilation capacity at low loads*, if a fixed percentage of outdoor air is used. This problem may be overcome by using a flow sensor in the outdoor air duct to control the mixing dampers.

Applications

Constant-temperature variable-volume systems are suitable for applications where there is a relatively constant load throughout the year, such as the interior spaces of office buildings and department stores. These systems can also be used to provide summer cooling in perimeter zones of buildings which are served from a separate wet distribution heating system. Such buildings include hotels, hospitals, department stores and office buildings.

7.6 Air and water central air conditioning systems

7.6.1 Introduction

In air and water air conditioning systems, both air and water are distributed to the conditioned spaces to provide heating and cooling. Because of the higher heat-carrying capacity of water compared to air (higher specific heat and density), lower quantities of water are needed to transfer or remove the same amount of heat energy from the conditioned spaces than air. This reduces the space requirement for pipe distribution compared to ductwork distribution.

The combination of air and water – with water normally satisfying the sensible load and the air normally satisfying ventilation and humidity control requirements – enables considerable reduction of the space required for fluid flow distribution in the building, while maintaining some of the performance capabilities of all-air systems. The reduced air requirement combined with high-velocity distribution can minimise the space requirement of air and water systems. The pumping power required to circulate water in the conditioned spaces is less than the fan power requirement in all-air systems, resulting in lower running costs.

The most commonly used air and water systems are a) *induction systems* and b) *primary air fan coil systems*.

In all-water systems, the load is satisfied entirely through the distribution of water to the conditioned spaces. Air is not conditioned centrally but may be drawn directly into the space from the outside through an opening in the wall for ventilation purposes. All-water systems do not provide humidity control, and they cannot be classified as 'full' air conditioning systems. The most common types of all-water systems are a) space ventilation fan coil system and b) the unitary heat pump system.

The following sections of this chapter consider the characteristics, operation, control and psychrometrics of air and water and all-water air conditioning systems.

7.6.2 Air and water system characteristics and applications

Air and water systems consist of a central air conditioning plant, a duct distribution system and a room unit. The central plant provides constant-volume air to the room unit. This air stream is usually referred to as primary air. The room unit draws air from the room and passes it through a coil, where it is either cooled or heated, as required, before it is discharged back to the room. The room air recirculated by the room unit is usually termed the secondary air. The primary air quantity is very small and is designed to satisfy:

- the ventilation requirements of the space or spaces in the building
- part of the sensible cooling requirements to supplement the cooling provided by the coil of the room units
- the maximum cooling load at changeover from summer to winter cycle
- the latent load of the building.

The primary air is cooled and dehumidified in the summer and heated and humidified in the winter at the central plant. Return air from the building can be recirculated, but the recirculated quantity is very small and thus it may not be economic if the capital cost of the return ducts is considered.

When 100% outdoor air is used and the outdoor temperature is likely to fall below freezing, a pre-heater is necessary. The use of a re-heater at the central plant depends on the design of the system.

The water side consists of piping and a pump, which circulates water through the coil in the room units. The water is either cooled by the chiller serving the cooling coil of the central plant or heated by a boiler or an alternative water heating system such as a heat pump.

Advantages

- *Low space requirements.* The use of water as a thermal energy distribution medium reduces the space required for fluid flow distribution in the building.
- *Reduced size of the central air handling system.* The lower air requirement reduces the size of the air handling equipment at the central plant.
- *Individual space temperature control.* Individual room temperature control can be provided by adjusting the thermostat, which controls the flow of water in the room unit.
- *Energy savings.* Using water as the main source of transmission of cooling and heating to the various spaces in the building reduces the power requirement for fluid flow distribution.
- *No cross-contamination for 100% outdoor air systems.* If the outdoor air is used only for ventilation and humidity control purposes, then recirculation of air from the building is not required. Recirculation is achieved within the conditioned room by the fan of the room unit and so cross-contamination is avoided.

Disadvantages

- *Difficult control during intermediate season operation.* Because of the low quantities of air supplied by the central plant, the control of space conditions during intermediate weather conditions is difficult if space demand alternates between heating and cooling. Correct timing of changeover of the plant from cooling to heating and vice versa requires careful consideration and considerable experience.
- *Limited control of humidity.* Humidity control is provided at the central plant or at individual zones serving a number of spaces. Individual space humidity control is not provided, and so humidity in individual spaces is allowed to vary within limits acceptable for thermal comfort.

- *Risk of condensation requiring careful design.* Because of the low quantities of air supplied by the central plant, to satisfy dehumidification requirements, the air must be supplied at a low dew point temperature. To avoid condensation of the coils of the room units, the surface temperature of the coil should be maintained above the dew point temperature of the supply air.
- *Low ventilation capability.* Because of the low quantity of supply air, the system is unsuitable for use in applications requiring high ventilation rates such as laboratories. In these applications, all-air systems or air and water systems with supplementary ventilation are used.

Applications

Air and water systems are mainly used in applications with highly variable sensible loads and a fairly constant latent load and where accurate control of humidity is not very important. Such applications include the perimeter zones of buildings, such as hospitals, schools, office buildings and hotels. The low space requirements for fluid flow distribution make air and water systems particularly suited to high-rise building applications where usable space is a very important design criterion.

7.6.3 *Air and water induction system*

In air and water induction systems, primary air (usually 100% outdoor air) is conditioned at the central plant and is distributed at high velocity to the room induction units (Figure 7.6). The primary air flows through nozzles in the induction unit, and the reduction of pressure, which results from the increase in velocity, induces secondary air from the room that flows through the coil(s) in the unit. The secondary air is either heated or cooled depending on the temperature of the water flowing through the coil(s) in the unit. The primary air is then

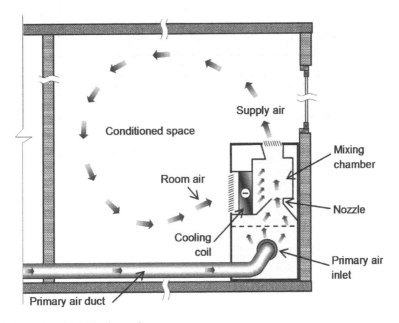

Figure 7.6 Air and water induction unit

mixed with the secondary air in the mixing chamber of the unit, and the mixture is accordingly discharged to the conditioned space.

Design considerations

Air and water induction unit systems are normally designed for 100% outdoor air. The cooling coil of the central plant, usually referred to as the primary coil, is designed to provide all the dehumidification requirements of the system. The coil in the induction unit, usually referred to as the secondary coil, is designed to provide only sensible cooling or sensible cooling and heating, depending on the design of the system. Ordinarily, no latent cooling is accomplished at the secondary coil. A drain pan, however, is usually provided in case condensation occurs at abnormal operating conditions.

The volume of secondary air induced depends on the induction ratio of the unit. The induction ratio is defined is the ratio of the secondary air induced to the primary air volume flow rate.

A wide variety of air and water induction system configurations exist. They can be classified into two broad categories: a) non-changeover systems and b) changeover systems.

Both non-changeover and changeover systems may employ two-pipe, three-pipe or four-pipe induction unit configurations.

Non-changeover induction systems

Non-changeover systems are used in mild winter climates or for building zones with large winter solar loads. A schematic diagram of the system is shown in Figure 7.7.

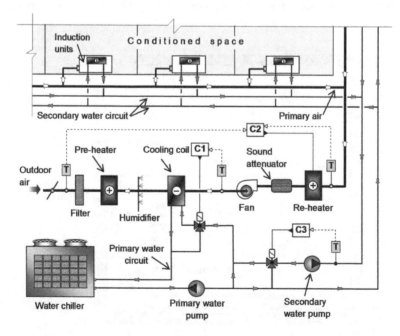

Figure 7.7 Schematic diagram of a two-pipe air and water induction system

Notes: C = controller; T = temperature sensors

The system provides chilled water to the secondary coils all year round. During summer operation, the system provides cold and dehumidified primary air to the induction units. As the outdoor temperature drops in winter, the primary air is reheated accordingly to meet the increased heating load of the building.

Changeover induction systems

The changeover induction system is suitable for severe climates with sharply defined seasons. During summer operation, the system supplies cold primary air and chilled secondary water to the induction units. As the outdoor temperature falls, the primary air is progressively reheated to offset transmission losses and prevent rooms with low cooling loads from becoming too cold. In intermediate seasons, the water in the secondary coils remains cold.

As the outdoor temperature drops further, the changeover temperature is reached and the secondary water is now supplied hot to overcome the higher transmission losses. Because changeover in intermediate seasons may be troublesome due to wide oscillations in outdoor temperature above and below the changeover temperature, changeover to hot water is limited to times of protracted cold weather.

The changeover induction system may be of a two-pipe, three-pipe or four-pipe configuration:

- *Two-pipe system.* In this configuration, the induction unit employs one coil with one water supply and one water return pipe. In summer and intermediate weather conditions, the coil is supplied with cold water, whereas in winter weather conditions, the coil is supplied with hot water.
- *Three-pipe system.* In a three-pipe configuration, the terminal unit employs one coil with one hot water supply, one cold water supply and a common return pipe. With this system, terminal units can be supplied with either hot or cold water simultaneously. This configuration is rarely used because of its high energy consumption due to mixing of the hot and cold return water streams.
- *Four-pipe system.* This configuration normally has two secondary coils – one cold and one hot, with one hot water supply, one hot water return, one cold water supply and one cold water return pipe. The primary air is supplied cold to the induction units all year round. In intermediate seasons, both cold and hot water is available to the induction units, which can be operated independently at any level from maximum cooling to maximum heating. The four-pipe configuration has several advantages over the two-pipe configuration, which include:

  - simplicity of operation and control
  - flexibility and quick response to changes in load
  - lower operating costs
  - no need for summer–winter changeover.

The main disadvantage of the four-pipe system is its higher capital cost.

7.6.4 *Primary air fan coil system*

The primary air fan coil system is very similar to the induction system. The only difference is in the terminal unit, where the induction unit is replaced by the fan coil unit. The fan coil unit basically consists of a filter, a finned tube coil and a centrifugal fan. Sometimes two coils

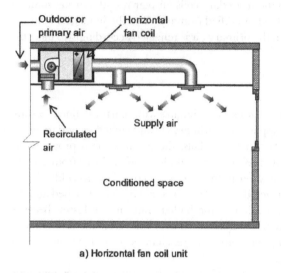

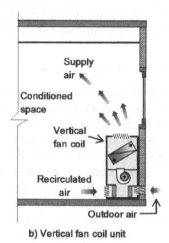

a) Horizontal fan coil unit   b) Vertical fan coil unit

Figure 7.8 Schematic diagram of vertical and horizontal fan coil units

are used: a hot water coil and a cold water coil. The fan runs continuously and recirculates air from the space through the coil, which is supplied with either hot or cold water. The primary air can be supplied directly to the space or through the fan coil unit. When the air is supplied directly to the space, low-velocity distribution can be used, resulting in savings in fan power. The supply temperature, however, should be maintained high enough to avoid cold drafts in the conditioned space.

When the primary air is supplied through the fan coil unit, two configurations are possible. The primary air may mix with the secondary air at the supply outlet of the unit after the secondary air has passed through the coil. Alternatively, the primary air may mix with the secondary air before the coil with the mixed stream, passing through the coil before it is discharged to the space. With the second configuration, the primary air supply may vary during capacity control of multiple units because the primary air flow to each unit is dependent to a certain extent on the operation of the fan in the unit.

Fan coil units may be of the horizontal or vertical type, as shown in Figure 7.8. Horizontal fan coil units are usually ceiling mounted, while the vertical fan coil units are normally installed under the window sill. Vertical units installed under the window sill prevent cold drafts from the cold window surfaces by discharging warm air in the opposite direction, which raises the inner surface temperature of the window. For places that have mild winter climates, where cold window drafts are not a problem, horizontal ceiling-mounted units are preferred because they occupy less floor space.

System control

For a two-pipe fan coil unit, control is achieved by a thermostat, which actuates a two-way valve to control the water flow to the coil. The thermostat may be of the summer–winter mode of operation type. For a four-pipe system in which two coils are used, each coil is controlled by a two-way valve. In the summer, as the space temperature drops, the thermostat progressively reduces the chilled water flow to the cooling coil until the valve is completely shut off. A further drop in temperature will cause the thermostat to actuate the two-way valve of the hot water coil. The hot water flow through the valve will increase in response to the decrease in the space temperature. The capacity of the fan coil unit can also be controlled by air bypass over the coil and by modulating the fan speed.

Advantages

The primary air fan coil system has the general advantages of air and water systems over all-air systems. Its greatest advantage over other air and water systems is its flexibility in design and layout.

Disadvantages

The system suffers from the general disadvantages of air and water air conditioning systems. An additional disadvantage over the induction system is the fan power consumption of the induction unit and fan-generated noise in the unit.

Applications

The primary air fan coil system is generally used for the perimeter areas of multi-room and multi-storey buildings such as office buildings, hotels, schools and hospitals. The system can also be designed to cater for interior spaces.

7.7 All-water air conditioning systems

The most common types of all-water air conditioning systems are a) *the space ventilation fan coil system* and b) *the unitary heat pump*. Neither of these systems is considered to be a 'full' air conditioning system, since they do not provide humidity control to the conditioned space.

7.7.1 *Space ventilation fan coil system*

The space ventilation fan coil system is supplied by chilled or hot water, which satisfies the total load of the space. Ventilation is provided by outdoor air, which is induced into the fan coil unit through an opening in the external wall.

Advantages

- *Lower capital cost*. The system does not require a central air handling unit to condition the air and no ductwork to distribute the air to the spaces.
- *Lower space requirement*. The absence of air distribution ductwork reduces the space required for fluid flow distribution in the building.
- *Lower running cost*. The absence of a central air distribution fan reduces the total fan power consumption.
- *Greater flexibility*. The absence of primary air distribution ductwork increases flexibility in the positioning of the fan coil units.

Disadvantages

- *Sufficient outdoor air ventilation is not guaranteed.* The amount of outdoor ventilation air induced into the fan coil unit depends to a large extent on outdoor weather conditions and is influenced by wind pressure and direction.
- *Condensation on the coil of the fan coil unit.* The coil of the fan coil unit will provide a degree of dehumidification in the summer. The moisture content of the outdoor air will also vary, and with certain conditions, condensation will take place on the coil of the unit. This increases design provisions for condensate collection and disposal, and the danger of bacterial growth in the unit exists.
- *Increased noise and dirt from the outdoor air.* The opening in the wall will increase the transmission of outdoor noise and dirt into the space. Better filtration and more frequent cleaning and replacing of the filters will be required than in the case of the primary air fan coil unit.

Because of the difficulties with ventilation and maintenance, the primary air fan coil system is usually preferred over the space ventilation fan coil system.

7.7.2 *Reverse-cycle unitary heat pump system*

This system utilises water-to-air, reversible, unitary heat pump units, which can be arranged in a modular configuration around the building (further information on heat pump systems is given in Chapters 8 and 9). The heat pumps operate on the reversible refrigeration cycle and consist of a compressor, a finned refrigerant-air coil, an expansion device, a refrigerant-water coil and a reversing valve. The refrigerant-water coil is connected to a two-pipe water distribution system, which is maintained at approximately a constant temperature. The water provides the heat source and the heat sink for the heat pumps. The water circuit includes a closed heat rejection system and a boiler or heat pump. A schematic diagram of the system is shown in Figure 7.9.

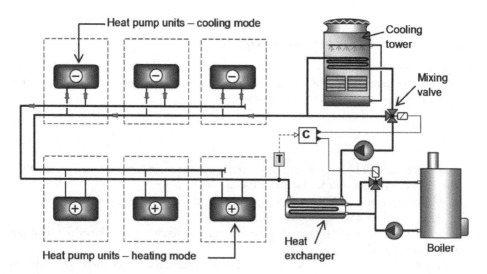

Figure 7.9 Reverse-cycle heat pump system

Each individual heat pump unit is controlled by a space thermostat. If the thermostat calls for heat, the heap pump operates in the heating mode of operation with the refrigerant-water coil acting as the evaporator, extracting heat from the water, and the refrigerant-air heat exchanger rejecting heat to the space. When the thermostat calls for cooling, the cycle is reversed and the refrigerant-air coil acts as the evaporator, removing heat from the space. The refrigerant-water coil acts as the condenser, rejecting the heat to the circulating water.

If half of the heat pumps in the system operate in the heating mode and half in the cooling mode, the system is in thermal equilibrium. The water is maintained at a temperature of about 27°C and no heat is added to or removed from the system. If more units operate in the cooling mode than in the heating mode, excess heat will be generated by the system, and this heat is rejected to another system or to the ambient air through the closed-loop cooling tower. If more units operate in the heating cycle than in the cooling cycle, additional heat will be required by the system, and this is provided by a boiler.

Outside air for ventilation may be provided by a central plant, or alternatively, it may be drawn directly from the outside into each heat pump unit.

Advantages

- *Economy of operation in applications with simultaneous heating and cooling requirements.* Heat rejected from some units is used as a heat source for other units, improving the overall coefficient of performance of the heat pumps.
- *Flexibility in design and operation.* The system does not require zoning due to the simultaneous availability of heating and cooling. Each unit can be controlled independently to provide cooling, heating or air circulation.

Disadvantages

- *Poor control of humidity.* The heat pump units do not provide active humidity control, although a degree of humidification and dehumidification can be provided if primary air is conditioned in an air handling unit.
- *Noise problems.* The compressors of the heat pump units are located in the conditioned space, and compressor and fan-generated noise may be a problem.
- *High maintenance costs.* The heat pump units within the conditioned space will require more frequent maintenance than other terminal units.

Applications

The unitary heat pump system is suitable for perimeter areas of buildings with diverse heat gains, so that for the major part of the year some of the areas will require cooling while an almost equal number will require heating, resulting in an overall load balance. Such buildings include office buildings, hotels and hospitals.

7.8 Chilled ceilings and beams

Chilled ceilings and beams are, in their basic form, static cooling and heating devices that, in recent years, emerged as alternatives to traditional air conditioning systems. Chilled ceilings are primarily radiant devices, which consist of copper pipes attached on the back of ceiling panels (tiles). When combined with ventilation, chilled ceilings can create a comfortable environment in which heat is dissipated by radiation without causing draughts or noise. The cooling capability of chilled ceilings is relatively low, between 45 and 65 W/m². Where larger cooling capacities are required, for example, to cater for solar loads in perimeter glazed areas of the building, chilled ceilings can be used in combination with chilled beams.

Chilled beams employ water that flows in ceiling-mounted radiators. The cold surface of the radiator cools the surrounding air, creating a continuous cycle in which the cooler air drops into the occupied space and is replaced by rising warmer room air. Two main categories of chilled beam systems are available: a) *passive chilled beams* and b) *active chilled beams*.

7.8.1 Chilled ceilings

In their simplest form, chilled ceilings can be formed by embedding pipes carrying chilled water into the underside of concrete floor slabs. Although this approach can provide relatively high cooling capacity, it also introduces a number of complexities, such as difficult control and risk of condensation on the underside of the slab. In order to prevent the ceiling from running wet, a suspended variation is usually employed. This normally consists of chilled water pipes attached on the back of metal ceiling panels suspended from the ceiling. The option is also available to supply primary ventilation air through the suspended ceiling, as shown in Figure 7.10.

Due to the circulation of water, the surface temperature of the chilled ceiling is reduced a few degrees below room temperature. This together with the ventilation air increases the convective heat transfer effect of the system. Chilled ceilings can also be used to provide heating through the radiant effect.

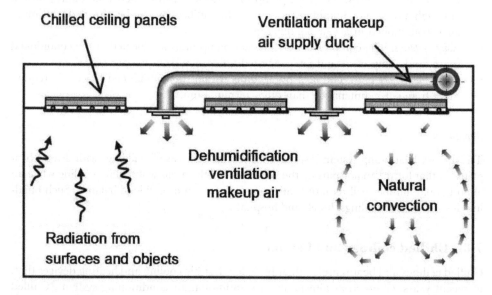

Figure 7.10 Example of a chilled ceiling system with ventilation

7.8.2   *Chilled beams*

A chilled beam system consists of a central air conditioning plant, a duct distribution system and terminal devices in the conditioned space (room units). The central plant provides constant-volume air to the room unit, which as in the case of induction and fan coil systems, is referred to as primary air. In the case of the active chilled beam, the room air is recirculated through the pipe or coil of the chilled beam. This room air is usually termed secondary air.

The primary air quantity is very small and is designed to satisfy some or all of the following requirements:

*   the ventilation requirements of the space
*   part of the sensible cooling requirements to supplement the cooling provided by the chilled beam
*   the latent load of the space.

The main concern with chilled beams is condensation. Since the air surrounding the chilled beam is sensibly cooled, its ability to hold moisture is reduced. Therefore, if the surrounding air is cooled to dew point temperature, the moisture will condense and will drop into the occupied space, as there is no possibility of draining the condensate from the terminal device. Consequently, for a chilled beam design to be successful, the following should be considered:

*   the latent load of the occupied space
*   the operating temperature of the chilled beam
*   the required ventilation rate.

A way of avoiding this problem is to always maintain the temperature of the chilled water supply at a value above the room dew point temperature. This can be done by floating the chilled water supply temperature in response to the room moisture content, as well as automatically controlling openable windows to limit infiltration of outdoor air during periods of high external absolute humidity.

Advantages

*   smaller air handling unit and less ductwork compared to all-air air conditioning systems
*   lower fan energy consumption than all-air and primary air fan coil systems
*   low noise levels
*   lower maintenance costs compared to heating, ventilation and air conditioning (HVAC) terminal units that have moving air side parts and dampers.

Disadvantages

*   lower cooling capability than conventional air conditioning systems (all-air or fan coil systems)
*   cost of active beams higher than conventional terminal devices
*   limited ability to handle high latent loads and risk of condensation that requires careful humidity control
*   slow response to varying cooling loads.

Passive chilled beams

In the passive chilled beam, chilled water circulates through pipework or the heat exchanger of the beam. This cools the air close to the beam, which becomes denser and descends, permeating through openings (perforations) on the face panel of the beam. Warmer room air rises to replace the cooler air that has moved downwards, and this creates air circulation around the beam, as shown in Figure 7.11. The face panel that is cooled by the air stream also acts as a radiant panel exchanging thermal energy with the building structure. The combination of natural convection and radiation increases the cooling capacity of the system compared to chilled ceilings. Heating to the space is normally provided by a separate perimeter heating system.

Active chilled beams

An active chilled beam system is an air–water system that can satisfy higher cooling loads than a passive chilled beam. The system utilises primary air from the central plant to induce and recirculate room air through the heat exchanger of the unit in a similar way to the induction system. As a result, the system operates with forced convection compared to free convection for the passive beam, increasing the heat transfer rate and cooling capability of the system. Due to this induction effect, the active chilled beam is also able to provide relatively high heating capacities, which are capable of meeting heating load requirements in temperate climates.

The operation of the active chilled beam can be explained by referring to Figure 7.12. Primary air is delivered from the central plant into the primary air plenum. The air then flows through nozzles into the mixing chamber, and this creates a low-pressure area at the nozzle outlet. This low pressure induces air from the room to flow through the coil and mix with the primary air in the mixing chamber. The mixed air then is accelerated as it passes through the beam diffuser and is discharged to the room.

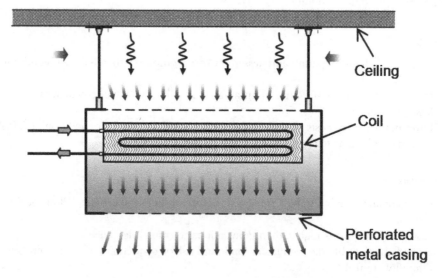

Figure 7.11 Passive chilled beam

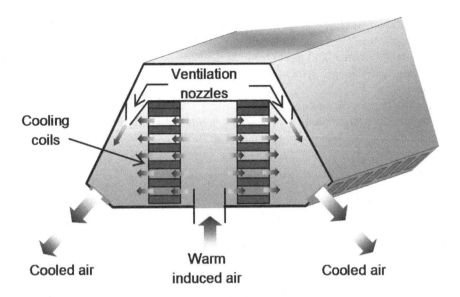

Figure 7.12 Typical construction of active chilled beam system

7.9   Air conditioning system selection and evaluation

The main air conditioning system selection criteria are:

*   capital cost
*   comfort criteria
*   energy use and running costs
*   maintenance, reliability and equipment life
*   appearance and room noise level
*   environmental issues
*   flexibility
*   space used for central plant and distribution.

This list is not ordered in terms of importance, and system selection should be based on a carefully designed evaluation methodology. This is a method of ranking priorities and using weighting factors to compare alternative systems against a set of criteria.

7.9.1   Capital cost

The capital cost of air conditioning installations varies considerably depending on the level of complexity of the design and the thermal performance of the building.

One method that can be used to compare the capital cost of various system options is to use cost indices rather than absolute costs. These can be generally derived from job records of installations carried out over recent years. Approximate costs of air conditioning systems can also be obtained from several price books, which are updated regularly. Cost data can vary from one source to another, but for reasonable estimates in the UK, it is recommended to use the latest edition of the *Spon's Mechanical and Electrical Services Price Book*.

7.9.2 *Energy and running costs*

The accurate calculation of energy demand for an air conditioning system is complex due to the requirement to model the transient response of the building and the dynamic operation of the HVAC system and its controls. At the early design stage, this is even more difficult because the building and the systems under consideration would only be at the outline design stage.

Many consultancy practices use computer programmes at the outline design stage to carry out energy analyses. These analyses make many assumptions in the input data, but the results can be useful when comparing the performance of the different systems under consideration.

7.9.3 *Maintenance, reliability and equipment life*

The requirements for maintenance of building services systems should be considered at the outline design stage of a project. When considering the system options, the maintenance requirements of each of the systems should be included in the analysis.

The maintenance requirements of a system are important because they not only affect the running cost of a building but also the reliability of the system. The cause of many problems with building services systems, which result in complaints and dissatisfaction of the building occupants, are due to lack of correct maintenance rather than faults in design or installation. The maintenance requirements of any system will depend fundamentally on the quality, and therefore the cost of the plant and equipment, and on the particular design and specification.

7.9.4 *Appearance and room noise level*

The appearance of the system within the occupied areas is one aspect to be considered in system analysis. Generally, it is possible to design all systems to supply and extract air through the ceiling with the terminal units located within the ceiling void, in which case room appearance is not normally an important factor. If under-sill perimeter units and enclosures are considered, then the loss of floor space at the perimeter is a very important factor.

Room noise levels for various systems cannot be generalised and depend on the particular equipment selections. In induction unit systems, the high-velocity primary air passes through an attenuator, and the pressure at the nozzles required to induce secondary air causes some noise generation. This necessitates careful selection of the units to meet the required room noise levels. Reverse-cycle heat pumps can be a cause of complaint in private offices with double glazing, as the on/off operation of the unit makes the noise more intrusive.

VAV system terminal units include attenuators to reduce the noise generated by the volume control damper and the air velocity–generated noise within the high-velocity supply ductwork; thus, they do not inherently have any design problems with regard to room noise levels. The design of the primary supply ductwork must avoid high-pressure differences between terminal units, as the noise level of the unit increases as the upstream duct static pressure is increased.

Fan coil units do not normally pose a design problem in terms of selection to achieve the required room noise levels. They are usually selected for low fan speed operation. Noise problems only normally occur if there is a fault on the fan, its bearings or mountings.

Chilled beam/ceiling systems do affect the room appearance more than the other systems, as it is necessary that the design of the ceiling is closely integrated with that of the cooling system. In the case of a chilled beam system, this requires that the ceiling has at least 30% open area.

7.9.5   *Environmental issues*

There is growing awareness of environmental issues related to air-conditioned buildings. These are primarily concerned with energy consumption and the indirect emissions of CO_2 at power stations, direct emissions from the leakage of hydrofluorocarbon (HFC) refrigerants and the problem of sick building syndrome. Wet cooling towers and the risk of legionnaires' disease is also a factor.

The concern regarding energy consumption of air conditioning systems is due to the perceived view that an air-conditioned building is inefficient in terms of energy. However, this may not necessarily be entirely correct, as in many cases high comfort standards and good energy efficiency can be achieved with properly designed and operated air-conditioned buildings.

When considering the energy used by a system, it is useful to know the relative CO_2 emission factors of the final forms of energy, normally electricity and natural gas. In the UK the emission factor for electricity has been reducing rapidly in recent years due to the increasing share of electricity generation by renewable energy sources, particularly, wind and solar. In 2019 average emission factors for the UK were 0.31 $kgCO_2$/kWh for electricity and 0.184 $kgCO_2$/kWh for gas (BEIS 2019a).

Sick building syndrome is a complex problem that does not have easily identifiable causes, but of the many studies which have been undertaken, it can be concluded that the risk is greater within air-conditioned buildings. One of the main contributing factors in this has been found to be the inadequate provision of outside air. The indoor air quality within an air-conditioned building is improved if the rate of outside air ventilation is increased. Generally, air–water systems are designed with ductwork sized to provide the minimum recommended outdoor air quantities, which give about 1.5–2.0 air changes per hour within the occupied space. All-air systems with full fresh air capacity operate on average throughout the year with typically twice this quantity and therefore offer a higher standard of outdoor air ventilation.

Many of the environmental considerations associated with the built environment are taken into account by the Building Research Establishment Environmental Assessment Method (BREEAM) environmental assessment method for new office building designs devised by the Building Research Establishment *(BREEAM 2019)*. This assessment includes many factors relating to the building services systems and their energy consumption, amongst other parameters. The methodology involves the allocation of credits and weighting factors for many aspects that influence the environmental impact of the building during its lifetime. Manipulation of the credits and weightings leads to an overall percentage score, which can be used to benchmark the performance of the building against other BREEAM-rated buildings.

7.9.6   *Flexibility*

Flexibility requires that systems be designed for change and can be adapted to suit changes that occur. The flexibility of an air conditioning system and the electrical services distribution system has an influence on the longer-term usefulness of a building. If the difficulty and expense of altering the existing systems are too great, it often means that the systems are not adapted correctly to suit new office layouts. This can result in operational problems and complaints of discomfort from the occupants.

Two types of flexibility are required for an air conditioning system. One is to allow for modifications to partitioned office locations that can occur quite regularly in some buildings. The other is the flexibility to provide additional cooling to areas, which may have a higher-than-designed cooling load.

To provide flexibility for partitioning rearrangements, air inlets are usually associated with the planning grid and light fittings and should not be located on possible partitioning lines.

Connections between the air inlets and terminal units can be made in flexible ductwork so that relocation of inlets is relatively simple.

Air–water systems can be designed to provide additional cooling to areas of high cooling demand by means of oversizing the chilled water distribution pipework within the ceiling void. This allows the tenant to replace a terminal unit with a larger model and increase the chilled water flow to meet the load required for a specific area. These systems can therefore be designed to incorporate good flexibility to provide additional cooling.

With an all-air system it is not as easy to provide additional cooling capacity if required by a tenant to a specific area. Some additional capacity can be provided by oversizing the air handling plant and distribution ductwork, but this is an expensive measure and requires an increase in services plant and riser space.

A very difficult design decision regarding flexibility is the question of providing individual room control to individual offices that have de-mountable partitioning and will be subject to modification. The cost of the air conditioning system can be greatly increased if a large number of terminal units are provided to give individual temperature control to each partitioned office. In designing for an owner/occupier or designing the fit-out where it is possible to determine the partitioning arrangement, it would be appropriate to design the system with sufficient terminal units to provide individual temperature control. In most designs where the partitioning layout is not known, a compromise is to use one terminal unit to serve two offices and controlled from averaging thermostats. Many systems are designed so that terminal units can be added if required or modified to provide individual temperature control.

7.9.7 *Space used for plant and distribution*

The space required for the main plant and distribution of an air conditioning system can have a large influence on the selection of the system. This includes the effect the system selection has on the horizontal service zone and hence the overall height of the building.

In general, all-air systems require greater plant, distribution and horizontal service zones than air–water systems. This is because the space required for the transportation of energy by ducted air is larger than that required when piped water is used. Figure 7.13 illustrates the difference in space required to provide equal amounts of cooling using a circular duct and a chilled water pipe; in both cases the space shown includes the thermal insulation required.

In the case of the all-air system, all the cooling required is provided by the supply of air. In the case of an air–water system, the majority of the cooling load is provided by the provision of chilled water, and only a small part of the total cooling load is dealt with by the provision of the fresh air supply. The all-air system therefore has considerably larger air handling units and distribution ducts than an air–water system. The all-air system typically has four times the total air capacity of the air–water system, but the distribution space is not

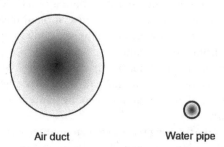

Air duct   Water pipe

Figure 7.13 Comparison of air duct and water pipe for the same rate of energy transfer

four times the area. This is because the ductwork for an all-air system such as a VAV system is sized on medium-velocity design criteria, whereas for an air–water system such as a fan coil system, the ductwork is sized on conventional low design velocities.

7.9.8 Evaluation methodology

A standard evaluation methodology consists of establishing criteria by which each of the systems under consideration will be evaluated. A weighting factor is assigned to each of the criteria, which establishes its priority and importance relative to the other criteria. The weighting factors are selected and distributed such that their summation equals 100. An example of an evaluation matrix for three different systems is shown in Table 7.1. The evaluation of each alternative involves three steps. First, for each of the three systems a rating is given for each criterion between 1 and 10 (10 being the best). Second, for each of the criteria, the rating is multiplied by the weighting factor. Third, the summation of these products is tabulated to yield a total value. The highest total value provides an indication of the most appropriate system for the project.

An example of an evaluation matrix of a single system, in this case, a chilled beam with displacement ventilation, is shown in Table 7.2.

Table 7.1 Example of an evaluation matrix

Weighting Factor (WF)	Capital Cost (20)	Space Required (20)	Maintenance Required (15)	Energy Cost (15)	Flexibility (10)	Acoustics and Comfort (10)	Environmental Issues (10)	Total Value (100)
	Rating Values (R)							
Scheme 1	7	8	5	6	8	3	4	615
Scheme 2	8	10	4	8	7	8	7	760
Scheme 3	10	10	10	9	8	9	8	935

Notes:
Weighting factors (WF) are distributed by priority; their summation equals 100.
(R) indicates the rating value of each alternative between 1 and 10, with 10 being the highest.
Total value = (WF × R) e.g. for scheme 1, = (20 × 7) + (20 × 8) + ... + (10 × 4) = 615.

Table 7.2 Example of a chilled beam system with displacement ventilation

Item	Weight	Poor 2	Mediocre 4	Good 6	Very Good 8	Excellent 10	Score
Maximum office space	18	o	o	x	o	o	108
Flexible office layout	16	o	o	x	o	o	96
Capital cost	15	o	x	o	o	o	60
Maintenance	12	o	o	o	o	x	120
Control	10	o	x	o	o	o	40
Energy consumption	9	o	o	o	x	o	72
Construction time	6	o	x	o	o	o	24
Acoustics	5	o	o	o	o	x	50
Indoor air quality	4	o	o	o	o	x	40
Building height	3	o	o	x	o	o	18
Minimum plant space	2	o	x	o	o	o	8
Total score	100						636

7.10   Summary

In this chapter the main types of air conditioning systems and their principal classification have been considered. The principles of operation of these systems have been described, as well as their advantages, disadvantages and applications. The chapter also presents the main criteria for system selection and methodologies that can be used for the evaluation of system alternatives for a particular application.

Unitary and packaged air conditioning systems are normally used in small buildings to provide heating and cooling and in certain cases limited ventilation but do not normally provide 'full' hygrothermal control of the indoor environment and effective ventilation.

Central air conditioning systems, which are normally classified as all-air and air and water systems, are used for large buildings and where good control of the indoor environment is required with relatively low energy consumption. All-air systems are normally employed where very good control of indoor conditions is required and where spaces experience large and varying latent loads. Such buildings include restaurants, theatres, dance halls, etc. Air and water systems are used where accurate humidity control is not very important;where space is at a premium, as is the case for multi-storey buildings; and where spaces experience fast and wide variation of sensible loads, such as the perimeter areas of buildings.

The most important criteria for air conditioning system selection are capital cost; energy use and running costs; maintenance, reliability and equipment life; appearance and room noise level; environmental issues; flexibility; space used for central plant; and thermal energy distribution.

For selecting the most suitable system for a particular application, it is appropriate to use an evaluation methodology to rate each system against the most important criteria for the project.

8 Energy-efficient thermal energy generation and distribution in buildings

8.1 Introduction

Air conditioning, as well as other thermal environment control systems for buildings, relies on a number of major technologies and equipment for the generation of the required cooling and heating energy, as well as the distribution of this energy to the various air handling plant and conditioned spaces. This chapter describes the various major cooling and heating technologies and equipment employed for this purpose, their important characteristics and selection criteria.

The most commonly used equipment for cooling includes vapour compression and sorption refrigeration systems. While producing cooling at the evaporator, both these systems reject heat at the condenser. This heat can either be upgraded to a higher temperature for water heating using heat pump systems or is rejected to the ambient air.

Depending on the type of thermal control systems employed in a building, the cooling or heating energy is distributed to the conditioned spaces through air or water in ducting and piping systems, respectively. This chapter outlines the main design principles employed in ducting and piping system design.

8.2 Refrigeration equipment

The cooling and dehumidification in air conditioning systems is provided in most cases either by chilled water or direct expansion refrigerant (DX) coils. Both chilled water and DX coils are served by the refrigeration plant. The most common types of refrigeration plants in air conditioning applications are:

* vapour compression
* sorption systems (absorption and adsorption).

8.2.1 Vapour compression systems

Thermodynamic systems operating on the vapour compression cycle consist of four basic components: a compressor, a condenser, an evaporator and an expansion device. The components form an elementary system, as shown in Figure 8.1. The working fluid, the refrigerant, flows round the system and transfers heat from the evaporator side to the condenser side. The compressor is the key component, its main function being to pump the refrigerant vapour from a relatively low suction pressure to a higher 'head' pressure. The high-pressure, high-temperature vapour then flows through the condenser where, becoming liquid, it

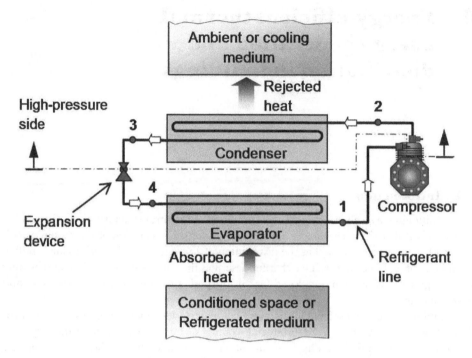

Figure 8.1 Schematic diagram of a simple vapour compression cycle

releases heat to the heat-receiving medium, for example, water or air. The refrigerant exits the condenser as a high-pressure, high-temperature liquid and passes through the expansion device, which lowers the pressure from the high-pressure condenser side to the low-pressure evaporator side of the system.

The drop in pressure is accompanied by a drop in temperature so that the refrigerant enters the evaporator as a low-pressure, low-temperature mixture of liquid and vapour. Finally, the refrigerant passes through the evaporator, where it draws heat from the heat source area (the supply air in the case of air conditioning systems), it changes state to vapour and enters the compressor to repeat the cycle.

Refrigerants

These are the working fluids in refrigeration systems. They absorb heat by evaporating at a low temperature and pressure and reject heat by condensing at a higher temperature and pressure. Refrigerant selection for refrigeration systems is important because it affects the performance and environmental impacts of the system. The suitability of a fluid for use as a refrigerant in a vapour compression refrigeration system depends primarily on its thermodynamic properties alongside cost, safety and environmental issues. The properties of refrigerants are usually displayed on a pressure/enthalpy diagram known as the Mollier diagram.

Figure 8.2 shows the pressure/enthalpy diagram for a simple saturated vapour compression cycle.

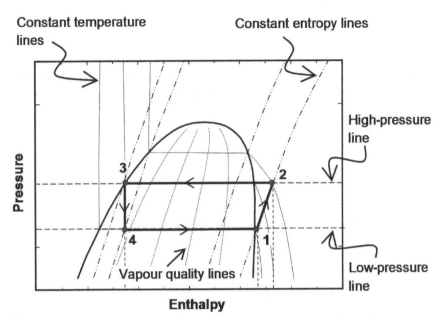

Figure 8.2 Pressure/enthalpy diagram

The thermodynamic processes involved are as follows:
Process 1–2 is entropic compression. The compressor work is given by:

$$w = h_2 - h_1 \tag{8.1}$$

Process 2–3 constant pressure heat rejection. The condenser duty can be calculated from:

$$q_c = h_3 - h_2 \tag{8.2}$$

Process 3–4 is isenthalpic expansion $h_3 = h_4$.
   Process 4–1 constant pressure heat addition or refrigeration effect. The refrigeration effect is given by:

$$q_e = h_1 - h_4 \tag{8.3}$$

It can be seen from this that a vapour compression system transfers thermal energy from a region of low temperature to a region of higher temperature. The term 'refrigerator' is used for a machine whose main function is to extract heat from a space or a process requiring cooling, whereas the term 'heat pump' is used for a machine whose main function is to supply heat at an elevated temperature to a space or process requiring heating.
   The performance of refrigeration and heat pump systems is expressed as the ratio of useful heat transferred to work input and is usually called the coefficient of performance (COP).
   For a refrigeration system,

$$COP_{ref} = \frac{h_1 - h_4}{h_2 - h_1} \tag{8.4}$$

and for a heat pump system,

$$COP_{hp} = \frac{h_2 - h_3}{h_2 - h_1}$$ (8.5)

The cooling capacity of the refrigeration system can be calculated by multiplying the refrigeration effect by the refrigerant flow rate in the system.

$$Q_e = \dot{m}_r (h_1 - h_4)$$ (8.6)

The heat rejection in the condenser is given by:

$$Q_c = \dot{m}_r (h_2 - h_3)$$ (8.7)

The work input to the compressor can be calculated from:

$$W = \dot{m}_r (h_2 - h_1)$$ (8.8)

Both the cooling and heating COPs of the vapour compression cycle vary considerably with variations in the evaporating and condensing temperatures. An increase in the condensing temperature with the evaporating temperature constant causes a reduction in both the cooling and heating COPs. On the other hand, an increase in the evaporating temperature with the condensing temperature constant causes an increase in both the cooling and heating COPs.

In practical refrigeration cycles, the refrigerant vapour entering the compressor should be superheated to ensure that no liquid refrigerant is carried to the compressor. The cycle is also designed for a small degree of subcooling of the refrigerant liquid at the condenser outlet to reduce flashing of the liquid during expansion. The degree of superheating of the refrigerant at the evaporator outlet is usually controlled by the expansion device. The most widely used expansion device is the thermostatic valve. Electronic refrigerant flow control devices, which have been introduced in the last few years, are slowly gaining wider acceptance by the refrigeration industry.

Refrigerant superheating at the compressor inlet and subcooling at the expansion valve inlet can be achieved outside the evaporator and condenser, respectively, by introducing a heat exchanger to transfer heat from the refrigerant liquid, leaving the condenser to the vapour leaving the evaporator. Liquid/suction heat exchangers enable maximum utilisation of the heat transfer surface of the condenser and evaporator.

Compressors

The compressor is the heart of the refrigeration system. Its function is to pump the refrigerant round the circuit and provide the required pressure differential between the low-pressure evaporator side of the system and the high-pressure condenser side of the system. The types of compressors normally used in refrigeration systems are:

- reciprocating
- rotary
- centrifugal.

The reciprocating and rotary compressors are positive displacement machines because the increase in pressure of the refrigerant vapour is provided by a mechanical mechanism through a reduction of the volume of the compression chamber. In the centrifugal compressor, compression is achieved by a force action created by a rotating impeller.

Depending on the method of sealing against the atmosphere, refrigeration compressors can be classified as:

*   hermetic
*   semi-hermetic
*   open.

The hermetic compressor is a completely sealed unit with the motor and the other mechanical parts enclosed in a welded steel casing. The semi-hermetic compressor represents an integral unit, with the motor and compressor housed in a bolted rather than welded casing to permit servicing. The open compressor is driven by an external prime mover coupled to the crankshaft, which protrudes from the crankcase through shaft seals. This allows the flexibility of selecting the most suitable drive for a particular application.

Compressor performance is usually given in the form of curves obtained from compressor calorimeter measurements. These curves or maps, which are usually available from the manufacturers, provide compressor power input and refrigeration capacity as a function of evaporating temperature for a range of condenser temperatures. Each set of maps is usually generated for fixed values of condenser subcooling and evaporator superheat. A typical compressor performance map is shown in Figure 8.3.

The reciprocating compressor is the most popular type of compressor and is widely used in the smaller capacity range, up to 180 kW, for domestic, commercial and industrial refrigeration applications.

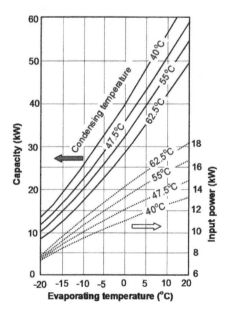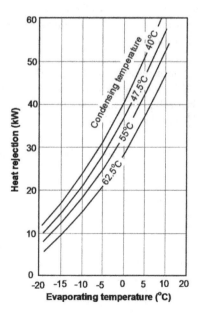

Figure 8.3 Typical compressor performance map

In small-capacity refrigeration and heat pump systems, rotary compressors such as the rotating piston and scroll have become popular in recent years due to their high efficiency, low noise and vibration and wider range of capacity control compared to reciprocating compressors. For higher capacity, up to 2.0 MW, and high-pressure ratio applications, the helical screw rotary compressor is normally employed. Screw compressors most commonly consist of two helical grooved rotors enclosed in a stationary housing fitted with suction and discharge ports. Compression is achieved by internal volume reduction through the rotary motion of grooved rotors. Helical screw compressors are popular in medium- to high-capacity refrigeration applications.

Centrifugal compressors consist basically of an impeller with radial vanes rotated by a shaft and housed in a cast-iron casing. The pressure of the refrigerant is increased through a continuous exchange of angular momentum between a steadily flowing refrigerant gas and the impeller. Centrifugal compressors are high-speed machines and are available in a capacity range between 180 kW and 3.5 MW.

8.2.2   Refrigerants

Refrigerants are the working fluids in refrigeration and heat pump systems. Currently the refrigerants employed in most commercial systems are hydrofluorocarbons (HFCs). HFCs have high global warming potential (GWP), and to control emissions from fluorinated greenhouse gases (F-gases), the European Union introduced the F-gas regulations in 2016 to limit emissions and eventually eliminate the use of F-gases. The original F-gas regulations were revised limiting the total amount that can be sold in the EU to one-fifth of 2014 sales by 2030 and banning their use in many new equipment where environmentally friendly alternatives are available.

Due to the different thermodynamic and safety properties of the alternatives, there is no 'one size fits all' solution. The suitability of a certain alternative must be considered separately for each category of product and equipment and in some cases also take into account the geographical location where the product and equipment are being used. Alternatives for HFCs include:

- Natural refrigerants (hydrocarbons, ammonia and carbon dioxide)
- HFCs with lower GWP, such as R32
- Hydrofluoroolefins (HFOs)
- HFC-HFO blends.

Each substance is assigned to a safety group, as shown in Table 8.1.

Table 8.1 Refrigerant safety groups

	Lower toxicity	Higher toxicity
No flame propagation	A1	B1
Lower flammability	A2	B2
	A2L*	B2L*
Higher flammability	A3	B3
Note: Table adapted from: *https://ec.europa.eu/clima/policies/f-gas/alternatives_en* * A2L and B2L are lower flammability refrigerants with a maximum burning velocity of ≤10 cm/s		

Table 8.2 Low-GWP refrigerants for centralised air conditioning system (chiller) applications

	Substance	GWP	Composition	Safety group	Replacement for
Natural refrigerants	R290 (propane)	3	–	A3	R134a, R407A, R410A
	R717 (ammonia)	–	–	2BL	R134a, R407A, R410A
	R718 (H₂0)	–	–	A1	R134a, R407A, R410A
	R744 (CO₂)	1	–	A1	R134a, R407A, R410A
	R1270 (propene)	2	–	A3	R134a, R404A, R407A
HFC-HFO blends	R452B	698	R32/125/1234yf	A2L	R410A
	R454B	466	R32/1234yf	A2L	R410A
	R455A	148	R32/1234yf/CO₂	A2L	R404A
	R513A	631	R1234yf/134a	A1	R134a
HFOs	R1233zd	4.5	–	A1	R134a, R410A
	R1234ze	7	–	A2L	R134a, R407A, R410A
HFCs	R32	675	–	A2L	R134a, R407A, R410A

Note: Information adapted from: *https://ec.europa.eu/clima/policies/f-gas/alternatives_en*

Table 8.3 Refrigerants for centralised heat pump system applications

	Substance	GWP	Composition	Safety group	Replacement for
Natural refrigerants	R290 (propane)	3	–	A3	R134a, R407A, R410A
	R718 (H₂O)	–	–	A1	R134a, R407A, R410A
	R744 (CO₂)	1	–	A1	R134a, R407A, R410A
HFC-HFO blends	R454C	148	R32/1234yf	A2L	R410A
	R513A	631	R1234yf/134a	A1	R134a
HFCs	R32	675	–	A2L	R134a, R407A, R410A

Table adapted from: https://ec.europa.eu/clima/policies/f-gas/alternatives_en (European Commission)

Environmentally friendly alternative refrigerants to HFCs for centralised air conditioning applications are given in Table 8.2, and refrigerants for centralised heat pump system applications are provided in Table 8.3. It can be seen that natural refrigerants such as propane, ammonia and CO_2 feature strongly in both application areas.

8.2.3 Sorption refrigeration equipment

Sorption refrigeration technologies such as absorption and/or adsorption are thermally driven systems in which the conventional mechanical compressor of the common vapour compression cycle is replaced by a 'thermal compressor' and a sorbent. The sorbent can be either solid in the case of adsorption systems or liquid for absorption systems. When the sorbent is heated, it desorbs the refrigerant vapour at the condenser pressure. The vapour is then liquefied in the condenser, flows through an expansion valve and enters the evaporator. When the sorbent is cooled, it reabsorbs vapour and thus maintains low pressure in the evaporator. The liquefied refrigerant in the evaporator absorbs heat from the refrigerated space and vaporises, producing the cooling effect.

Absorption refrigeration systems

Figure 8.4 shows a schematic diagram of an absorption refrigeration system. High-pressure liquid refrigerant from the condenser passes into the evaporator through an expansion device, which reduces the pressure of the refrigerant to the low evaporator pressure. The liquid refrigerant vaporises in the evaporator by absorbing heat from the fluid to be cooled. The refrigerant vapour then passes to the absorber, where it is absorbed by a refrigerant–absorbent solution. The solution, rich in refrigerant, is then pumped to the generator where it receives heat and separates into pure, or almost pure, refrigerant and an absorbent solution weak in refrigerant. The refrigerant then enters the condenser where it rejects its latent heat to the heat rejection fluid, whereas the weak solution returns to the absorber through a heat exchanger and an expansion device to absorb more refrigerant and carry it to the generator. The heat exchanger transfers heat from the weak solution to the strong solution and thus reduces the energy requirements of the generator. This improves the COP of the system.

The operating range and COP of absorption systems are, to a large extent, dependent on the refrigerant–absorbent fluid pair. Of the many combinations of fluid pairs that have been tried, only the lithium bromide–water and the ammonia-water pairs are commercially used in air conditioning applications.

Absorption refrigeration systems have much lower COPs than vapour compression systems, 0.7 for single-effect systems, rising to 1.2 for double-effect and 1.7 for triple-effect systems. Their main advantage, however, is that they can utilise rejected thermal energy from other plants or processes to drive the generator, reducing considerably the energy consumption of the system in situations where high-temperature reject heat is available.

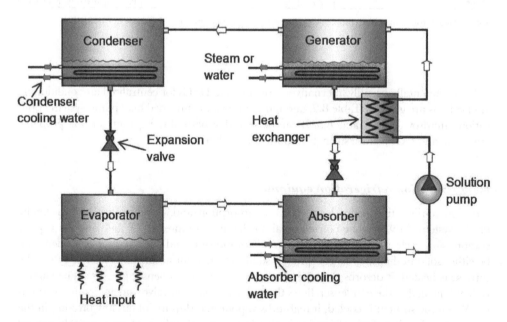

Figure 8.4 Schematic diagram of an absorption system

Adsorption refrigeration systems

Unlike absorption and vapour compression systems, adsorption refrigeration is an inherently cyclical process and multiple adsorbent beds are necessary to provide approximately continuous cooling delivery (Tassou *et al.* 2010). Adsorption systems inherently require large heat transfer surfaces to transfer heat to and from the adsorbent materials, which automatically makes cost an issue. High-efficiency systems require that heat of adsorption be recovered to provide part of the heat needed to regenerate the adsorbent. These regenerative cycles consequently need multiples of two-bed heat exchangers and complex heat transfer loops and controls to recover and use waste heat as the heat exchangers cycle between adsorbing and desorbing refrigerant.

Adsoprtion systems for air conditioning applications are commercially available from a number of manufacturers with capacities between 70 kW and 1300 kW. These systems are capable of being driven by low-grade heat, 50–90 °C, and able to give COPs of over 0.7. The main advantage of adsorption over absorption refrigeration systems is their ability to be driven by lower-temperature heat sources. Their main disadvantages are a larger footprint and higher capital cost.

8.3   Heat rejection equipment

The heat generated by the refrigeration plant can be used to satisfy some of the heat requirements of the building. If excess heat is present, then this heat is rejected to the atmosphere or to alternative sink media such as a lake or sea water. The main types of heat rejection equipment are:

- air-cooled condensers
- evaporative condensers
- water-cooled condensers
- cooling towers.

8.3.1   *Air-cooled condensers*

Air-cooled condensers consist of a casing which houses a finned refrigerant-to-air coil and a fan-motor assembly. The refrigerant enters the condenser as a superheated gas and is cooled by the air, which is drawn through the coil by the fan. The air being at a lower temperature than the refrigerant first de-superheats and then condenses the refrigerant. The refrigerant exits the condenser as a high-temperature, high-pressure liquid. A schematic diagram of an air-cooled condenser is shown in Figure 8.5(a).

The heat rejection rate from the condenser is a function of the heat transfer area, the air and refrigerant flow rates and the temperature difference between the air and the condensing refrigerant. For a given condenser size and fluid flow rates, higher ambient temperatures lead to higher condensing temperatures, which result in increased compressor power consumption.

Air-cooled condensers are used mainly in relatively small refrigeration systems where the compressor can be sited close to the condenser to avoid long runs of piping containing high-pressure refrigerant. For capacities of approximately 100 kW and above, the installed cost of air-cooled condensers is higher than the installed cost of the other heat rejection methods. Air-cooled condensers, however, have lower maintenance costs than the other heat rejection

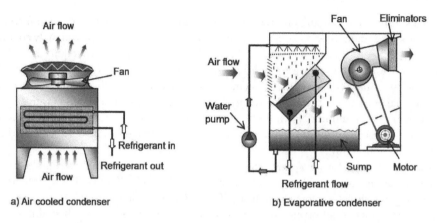

a) Air cooled condenser   b) Evaporative condenser

Figure 8.5 Air-cooled and evaporative condensers

methods. Other advantages, such as the absence of make-up water or drainage facilities, make air-cooled condensers popular in the capacity range up to 300 kW.

8.3.2 *Evaporative condensers*

Evaporative condensers consist of a refrigerant-to-air coil, a fan-motor assembly, a sump containing water and a circulating pump with a water distribution system. These components are contained in a housing, as shown in Figure 8.5(b). The water is distributed over the coil, effecting both sensible and latent heat transfer on the air side. This permits a smaller size unit for a given heat rejection rate than the air-cooled condenser, which effects only sensible heat transfer on the air side. The evaporative condenser has a common disadvantage with the air-cooled condenser in that it has to be sited close to the compressor to avoid long runs of high-pressure refrigerant piping.

Evaporative condensers generally have lower installed and operating costs than air-cooled condensers in the capacity range between 150 and 500 kW. The use of eliminators is necessary with evaporative condensers to prevent carryover of water droplets which may be contaminated.

8.3.3 *Water-cooled condensers*

Water-cooled condensers are usually shell-and-tube type coils with the refrigerant flowing through the shell and the condenser water through the tubes. The condenser water can be drawn from a river, lake or even the sea. Before such a solution is adopted, however, the cost of filtration and the maintenance requirements must be considered very carefully. The use of river and lake water may also be subject to the local water authority's approval.

Water-cooled condensers are most frequently used in combination with cooling towers. The use of a cooling tower enables the condenser to be sited close to the other refrigeration components. The cooling tower can be sited away from the condenser and the water transferred to the tower through water piping. A single cooling tower can serve a number of condensers located in different parts of the building. Water-cooled condensers in conjunction with cooling towers become economically competitive in capacities above 400 kW.

8.3.4   *Cooling towers*

Cooling towers reject heat to the atmosphere by a combination of heat and mass transfer. They cool the water from the water-cooled condenser or condensers of the refrigeration plant by exposing it to the atmosphere. The water is distributed in the cooling tower by a distribution system, which can be a combination of spray nozzles or splash bars and fill packing. Fill packing is a structure designed to provide a large surface area for water to evaporate. PVC is increasingly used as the fill material. Some of the water evaporates as it comes into contact with air flowing through the cooling tower. The latent heat of evaporation is removed from the remainder of the water and transferred to the air stream. As a result, the wet-bulb temperature of the air increases and the water temperature reduces as the two fluid streams pass through the cooling tower.

The temperature distribution within the cooling tower is shown in Figure 8.6. The temperature difference between the water entering and leaving the tower is called the range. The difference between the leaving water temperature and the entering air wet-bulb temperature is called the approach. The approach is a function of the performance capability of the cooling tower, and the larger the tower, the closer the approach. The performance of cooling towers is usually specified in terms of flow rate for a specified set of conditions such as entering and leaving water temperatures and entering air wet-bulb temperature. The rating

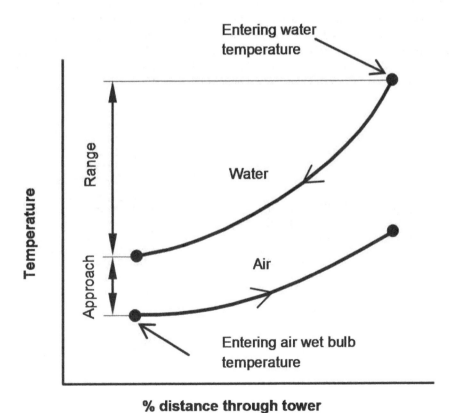

Figure 8.6 Temperature distribution within the cooling tower

or thermal capability of cooling towers is stated in terms of refrigeration capacity, which is calculated on the basis of heat rejection of 1.2 kW per kW of cooling capacity of the refrigeration plant, a range of 5 K and a water flow rate of 0.058 litres per second per kW of refrigeration capacity. This assumes that 1.0 kW of refrigeration capacity will produce approximately 1.2 kW of heat to be rejected at the condenser.

The air flow in cooling towers may be provided by mechanical means, convection currents or natural wind currents. The air may flow in a cross or counterflow direction with respect to the direction of the water flow. In air conditioning applications, the mechanical draft cooling tower is most commonly used. For this reason we will concentrate the discussion on this type of tower.

Mechanical draft cooling towers may be classified as direct contact and indirect contact towers. In direct contact towers (open towers), the fluid to be cooled is directly exposed to the atmosphere, whereas in indirect contact towers (closed towers), the fluid to be cooled does not come into direct contact with the atmosphere.

Direct contact (open towers)

In direct contact mechanical draft towers, the fan may be placed at the air inlet side (forced draft) or at the air outlet side (induced draft). The fan selection (centrifugal or axial) depends to a large extent on pressure, noise and energy consumption requirements. Cooling towers may also be classified as factory-assembled or field-assembled towers. Factory-assembled counterflow towers often use centrifugal fans in forced-draft configuration, as shown in Figure 8.7(a). Field-assembled units often use axial flow fans in induced draft configuration, as shown in Figure 8.7(b). Crossflow towers have low air side pressure drop per unit heat transfer area and produce more uniform flow on both the water and air sides compared with counterflow towers. A crossflow tower is shown in Figure 8.8.

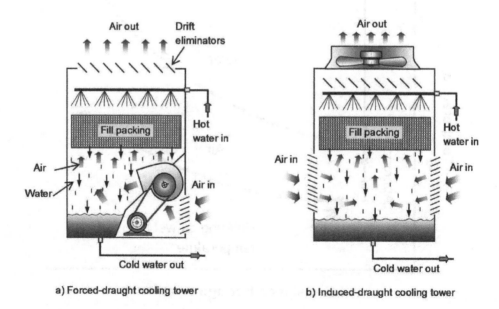

a) Forced-draught cooling tower   b) Induced-draught cooling tower

Figure 8.7 Mechanical draft counterflow cooling towers

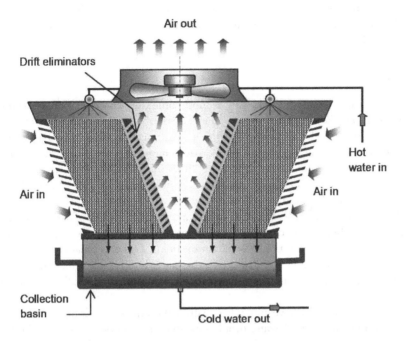

Figure 8.8 Mechanical draft crossflow cooling tower

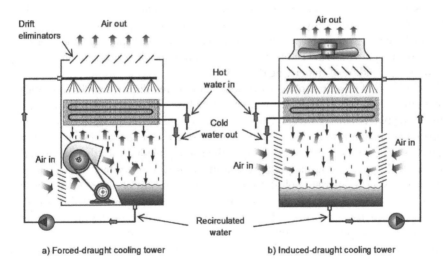

Figure 8.9 Indirect contact cooling tower

Indirect contact cooling towers

In indirect contact cooling towers, the primary fluid to be cooled circulates in the tubes of a coil in the tower. The secondary fluid (water) is sprayed on to the external heat transfer surface of the coil and removes heat from the primary fluid through evaporative cooling, as shown in Figure 8.9. Since the primary fluid does not come into direct contact with the air, this type of cooling tower can be used to cool fluids other than water. Indirect contact towers can also be used in atmospheres where there is a danger of contaminating the primary water with dirt or airborne contaminants.

Legionella in cooling towers

Legionellosis, or legionnaires' disease as it is most commonly known, is caused by members of the genus *Legionella*, which are fresh water bacteria that can be spread through the air. Water associated with cooling towers and other heat rejection equipment has been shown to be a habitat of this organism. This disease got its name from an outbreak in 1976 which caused the death from pneumonic disease of 29 members of the Pennsylvania American Legion attending a convention in Philadelphia. A number of other outbreaks have taken place around the world since then.

The *Legionella* species are widespread in aquatic habitats, but in most cases no disease is associated with their presence. Only a small number of species have been implicated as causing legionnaires' disease, the most common being *Legionella pneumophilia*. It has been found that at low temperatures, below 20°C, the bacterium remains dormant. The bacterium multiplies rapidly in temperatures between 25° and 40°C and is killed at temperatures above 55°C.

Once the *Legionella* cells have multiplied within the aquatic environment, transmission takes place through aerosols of water droplets escaping from cooling towers and other misting devices. The survival of *Legionellae* in the air depends on the relative humidity and the size of the cells. Studies have shown that relative humidities above 60% are conducive to the transmission of *Legionellae* through the air. Other factors that influence the transmission of airborne *Legionellae* are wind velocity and direction and vapour pressure.

Legionella cells of 0.5 μm in size or less can reach the inner spaces of the lungs and are associated with pneumonia. Factors influencing contraction of the disease are smoking, age, steroid therapy and other underlying diseases.

The risks associated with cooling towers and legionnaires' disease have led over the last few years to increased use of air-cooled condensers as heat rejection equipment for central air conditioning plants. Air-cooled condensers, however, result in increased energy consumption compared to cooling towers due to higher condensing temperatures. The condensing temperature is a function of the temperature of the cooling medium. In the case of air-cooled condensers, the condensing temperature is a function of the dry-bulb temperature of the ambient air, whereas in wet cooling towers, the temperature of the condenser water is a function of the wet-bulb temperature of the ambient air, which is lower than the dry-bulb temperature. The use of an air-cooled condenser in place of a cooling tower may lead to a 30% increase in the power consumption of the refrigeration system.

Prevention or control of the risk of legionnaires' disease from cooling systems can be achieved by:

- careful design and construction of cooling towers to minimise the release of aerosol
- constructing towers from durable materials with surfaces that can be easily cleaned and do not provide nutrients for bacteria to multiply
- positioning the towers away from the air intakes of air conditioning and ventilation systems
- providing easy access to the tower and components such as the eliminators and fill so that the tower can be periodically cleaned

- careful cleaning and disinfection
- water treatment.

Cooling systems should be maintained carefully. They should be cleaned and disinfected at regular intervals to prevent conditions conducive to the growth and multiplication of *Legionella* and to allow water treatment chemicals to work more effectively. Water treatment is necessary to prevent corrosion and organic growth in the system. Because *Legionellae* benefit from the presence of some host organisms, minimising organic growth indirectly affects these bacteria. A water treatment programme that controls biologic activity is therefore essential in controlling *Legionellae* in the same environment.

8.4 Heating equipment

Thermal energy for space heating and domestic hot water for commercial buildings is normally provided by boilers. As discussed in Section 3.7, however, decarbonisation of heating is driving the urgency to replace boilers that rely on the combustion of fossil fuels with lower or zero emission heating equipment. Heat pumps are such equipment, provided they are driven by low-emissions electricity and employ environmentally friendly refrigerants, discussed in Section 8.2.2. The use of heat pumps for heating and hot water production is discussed in Section 9.3.

8.4.1 Boilers

Boiler systems can be classified in a variety of ways based on type of application, operating pressure, construction materials, heat source and method of heat transfer from the combustion gases to water. Boilers can also be distinguished by their method of fabrication, for example, packaged or field erected.

Based on the heat source, boilers are often referred to as oil fired, gas fired, coal or solid fuel fired, or biomass. Gas-fired boilers are by far the most widely used for domestic and commercial space heating in Europe and the United States. In terms of method of heat transfer between the combustion gases and water, most boilers fall into two major categories: water tube and fire tube.

In a water-tube boiler, water flows through tubes inside the furnace and absorbs heat directly from the combustion gases. The heat transfer is fast, and the system can tolerate variable load requirements. Fire-tube boilers consist of a series of straight tubes that are housed inside a water-filled outer shell. The tubes are arranged so that hot combustion gases flow through the tubes. As the hot gases flow through the tubes, they heat the water surrounding the tubes and are confined by the outer shell of the boiler. Most modern fire-tube boilers have cylindrical outer shells with a small round combustion chamber located inside the bottom of the shell. Depending on the construction details, these boilers have tubes configured in one-, two-, three-, or four-pass arrangements.

Boilers are also distinguished by their method of fabrication. Smaller commercial boilers are manufactured and assembled in a factory and transported to the site as a finished product. These units are referred to as packaged boilers. Packaged boilers are generally of the shell fire-tube design. Large boilers that are not easily transportable as a single assembly due to size and weight limitations are assembled on-site from individual components or sub-assemblies. These boilers are referred to as field-erected boilers.

Boilers can be manufactured with a variety of materials. Most conventional (non-condensing) boilers are made with cast-iron sections or steel. Cast-iron boilers are composed of precast sections and thus can be more readily field assembled than steel boilers. They have a long life and are normally used in relatively small-capacity installations and where long service life is important. However, they tend to be heavier and more expensive than steel boilers. Small boilers can also be made of copper tubes, whereas condensing boilers can be made of stainless steel or aluminium.

Even though gas boilers are the most common type of boiler for space heating applications, biomass boilers are increasing in popularity due to their environmental credentials. They can burn a variety of biomass types such as wood chips, wood pellets, wood shavings, *Miscanthus* and many other materials. These fuels have different properties which determine the type of boilers and fuel handling system required.

Conventional boilers are designed such that the flue gas does not condense in the boiler. This precaution, which requires the boiler to operate with minimum water temperature of 60°C, is necessary to prevent corrosion of the construction materials of the boiler such as steel, cast iron and copper. New boilers, known as high-efficiency or condensing boilers, are specifically designed to condense the water vapour present in the flue gases using lower water return temperatures. This improves boiler efficiency to over 90% but requires the use of non-corrosive materials such as aluminium and stainless steel for the combustion chamber and heat exchanger. For this reason, condensing boilers are more expensive than conventional boilers, but the additional cost can be recovered very quickly from the savings in fuel costs.

Boiler efficiency

The efficiency of a boiler is normally defined as the net energy output measured on the water side divided by the energy input from the combustion of the fuel. The variation of typical efficiencies with load for conventional and condensing boilers is shown in Figure 8.10. It can be seen that the efficiency of conventional boilers increases as the load increases, whereas the efficiency of condensing boilers remains fairly high, in the range between 88% and 95%,

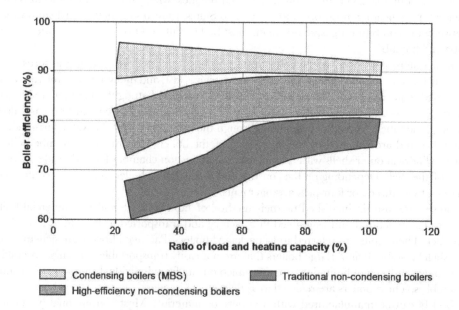

Figure 8.10 Typical boiler efficiency curves

Table 8.4 Boiler efficiency standards

Energy Related Product Directive (ErP) – Minimum seasonal boiler efficiency for all building types and fuels from April 2015

Boiler Size	Boiler seasonal efficiency
≤70 kW	86%
70 kW–400 kW	Minimum efficiency at 30% part-load 86%
	Minimum efficiency at full load 94%
Part L 2013 single gas boilers ≤2.0 MW	Minimum gross seasonal efficiency 91%

Source: Adapted from Hamworthy Heating. www.hamworthy-heating.com/Knowledge/Articles/ErP-compliant-boilers

irrespective of the load. The small increase in steady-state efficiency shown for condensing boilers at reduced loads is due to the higher potential to recover more energy from the exhaust gases at part-load.

The seasonal efficiency of boilers will be lower than the overall efficiency, which is determined at steady-state conditions, because of the heat losses from the boiler casing when the boiler is off. These losses can be reduced by continuous modulation of the firing rate of the burner as opposed to on-off firing control.

Table 8.4 shows the minimum efficiency standards for commercial boilers in new buildings specified by the Energy Related Product Directive (ErP). The Boiler Efficiency Directive applies to standard boilers, low-temperature boilers and gas-condensing boilers with an output of between 4 kW and 400 kW *(www.legislation.gov.uk/uksi/1993/3083/made)*.

8.5 Air distribution systems

The air for heating, ventilation and air conditioning of buildings is distributed from the air handling units (AHUs) to the conditioned spaces by ductwork (Figure 8.11).

The design of the ductwork and other air distribution components has a large influence on both the capital and operating cost of the system and will also be the determining factor in deciding the horizontal service zone depth.

8.5.1 Ductwork sizing methods

There are three basic methods for manual duct sizing:

- velocity reduction method
- equal friction method
- static regain method.

Each of the three methods is briefly outlined in the following sub-sections.

Velocity reduction method

With this method the ducts are sized so that the duct velocities are progressively reduced from the fan throughout the system to the final terminals. The design velocities are selected based on the limitation of noise criteria. The noise generated by the turbulent air within the

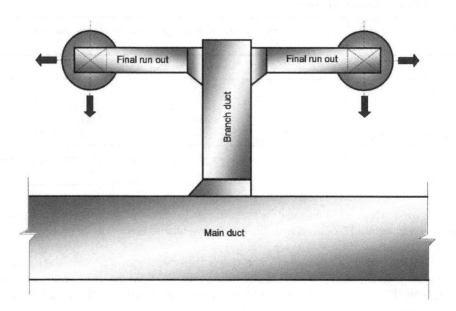

Figure 8.11 Schematic diagram of main duct, branch duct and final run outs

Table 8.5 Recommended duct velocities for low-pressure systems in relation to noise levels

Typical applications	Typical noise rating (NR)	Velocity (m/s)		
		Main ducts	Branch	Runouts
Domestic buildings (bedrooms)	25	3.0	2.5	<2.0
Theatres, concert halls	20–25	4.0	2.5	<2.0
Auditoria, lecture halls, cinemas	25–30	4.0	3.5	<2.0
Bedrooms (non-domestic buildings)	20–30	5.0	4.5	2.5
Private offices, libraries	30–35	6.0	5.5	3.0
General offices, restaurants, banks	35–40	7.5	6.0	3.5
Department stores, supermarkets, shops, cafeterias	40–45	9.0	7.0	4.5
Industrial buildings	45–55	10.0	8.0	5.0

Source: *CIBSE GUIDE B2*

ductwork increases as the duct air velocities are increased. Therefore, for a given noise criteria, there are recommended maximum duct velocities. Refer to Table 8.5 for recommended duct velocities for low-pressure systems.

Equal friction method

Using this method a design pressure drop per metre of ductwork is selected and then ducts can be sized using duct sizing charts, such as the one given in CIBSE Guide B2 (2016).

Equivalent diameters for rectangular duct sections can be determined from Eq. 8.9, where *a* and *b* are the width and depth of the duct, respectively:

$$d = 1.265 \left[\frac{(ab)^3}{a+b} \right]^{0.2}$$

(8.9)

The normal design pressure drop values selected for the equal friction method, which is commonly used for sizing low-pressure extract and supply systems, are in the range 0.8–1.2 Pa/m.

Static regain method

This method uses the principle that if the velocity within a duct is reduced, for example, at an expansion or a branch, then the reduction in velocity pressure is converted to an increase in static pressure minus any friction losses that occur. The sizing of ductwork based on this principle therefore seeks to equalise the static pressure at all the branch take-offs in a system and thus provide an inherently self-balanced system. There is a reduction in velocity pressure throughout the ductwork system where the air quantities are reduced at branches within the system. The decrease in velocity pressure results in an increase in the static pressure, and the aim of sizing the ductwork is to achieve a counterbalance of the increase in static pressure to the static pressure loss due to friction.

8.5.2   Recommendations for practical duct sizing

In practice, the most commonly used method of duct sizing is a combination of the velocity reduction method and the equal friction method. The selection of the design velocities and pressure drop criteria are generally based on experience and require an engineering judgement. This seeks to balance the requirement for smaller ducts to minimise the space required and at the same time avoid large pressure drops and too high velocities, which will increase fan pressures and energy consumption. Several criteria must be considered when sizing ducts:

- The air velocities must be limited to those required to achieve the design noise criteria within the space.
- The pressure drop between air diffusers/grilles or terminal units must not be excessive, as it will not be possible to balance the system without creating noise problems from nearly closed volume control dampers. In ductwork systems with varying lengths between the outlets, it will be necessary to provide dampers within the system. These dampers will provide additional resistance to the outlets nearest to the fan, to balance the pressure loss due to friction for the longer branches which are farthest from the fan. It is recommended that this pressure imbalance should be limited to about 50 Pa when sizing the ductwork for conventional low-velocity systems with volume control dampers within the branch ducts serving the grilles of spaces with noise criteria of NR 35. For variable air volume (VAV) terminals which have secondary attenuation, this should be limited to about 150 Pa.
- The space required for distribution ducts within the building, particularly the horizontal distribution within the ceiling voids, must be carefully considered and a balance must

be found between reducing the building costs and increasing the ductwork costs. Obviously, if the horizontal distribution space within the ceiling voids can be minimised, there will be capital cost savings on the building due to a reduced area of external cladding, structural column lengths, etc.

• The fan energy must be considered, and the duct sizing has a direct bearing on this. Unfortunately, it is difficult to determine the cost benefit in terms of reduced fan energy when sizing a ductwork system without undertaking a detailed life cycle costing exercise, which would include the building costs. Therefore, it is only possible to recommend that low fan pressures be aimed for by careful design of the ductwork systems. This is particularly important, as the fan energy can account for typically 15% of the total energy consumption of an air-conditioned office building. (For VAV systems, about 550 Pa supply and 100 Pa extract should be aimed for.)

How each of these factors is considered will vary from project to project depending on the particular emphasis given by the design team and client. However, noise will always be an overriding criterion, and therefore, the maximum duct velocities will always be one of the limiting factors.

The pressure imbalance within the system needs to be limited as described earlier, and therefore, generally these two criteria are used to determine the circular duct sizes initially. Then the space requirement criteria come into the design process, and the circular duct sizes selected using the duct sizing chart are converted into the equivalent rectangular or flat oval duct sizes using tables given by the Chartered Institution of Building Services Engineers (CIBSE).

The table used should be that for equal volume flow rate, pressure loss and surface roughness. In this case the actual mean velocity within the equivalent duct will not be the same as in the circular duct, but in practice the differences are relatively small and usually within 10%.

For a VAV system, the maximum duct velocities for an NR design level of 35 are in the range of 10 m/s for the main ducts and 8 m/s for branch ducts. The maximum recommended design velocities for supply ductwork for a VAV system are much higher than conventional systems. This is because the VAV system has terminal attenuators built into each VAV terminal box downstream of the control damper. This attenuator deals with both the noise generated by the control damper and the velocity-generated noise of the air within the upstream ductwork system. All ductwork downstream of the VAV terminal box should be sized on the conventional low velocities given in the preceding table for the final run outs.

For flexible ducts that are used in many cases for final connections to air diffusers, the maximum velocity should be limited to 0.5–1.0 m/s below the tabulated values for the final run outs.

When selecting the equivalent rectangular duct for a circular duct, it is recommended that an aspect ratio of not greater than 5:1 be used. The aspect ratio of a duct is the ratio of the longest side to the shortest side, which is normally the width to the height. This is because ducts sized with higher aspect ratios are difficult to adequately stiffen, and there is the possibility of increased noise breakout and drumming caused by movement of the long sides of the duct.

The performance and characteristics of a circular and an equivalent rectangular duct are compared in Table 8.6. Both ducts are sized to handle the same quantity of air.

Table 8.6 Comparison between circular and rectangular ducts for the same air flow

	Circular	*Rectangular*
Size	600 mm Ø	1200 mm × 300 mm
Surface area per metre length	1.88 m²	3.00 m²
Minimum sheet thickness	0.8 mm	1.0 mm
Air leakage per metre length	2.87 l/s	4.58 l/s
Horizontal service zone required for insulated duct	760 mm	430 mm
Installed cost per metre length, including 40-mm insulation	50% of rectangular duct	Reference duct
Noise breakout loss at 125 Hz	50 dB	22 dB

From the engineering and cost viewpoint, circular ducts have a distinct advantage over rectangular ducts, but unfortunately, it is not normally possible to design ductwork systems with only circular ducts because of the need to optimise the overall design of the building and space for services.

Extract duct systems are not normally sized on medium or high velocities because the ductwork, particularly if rectangular, will become unstable under large suction pressures. Therefore, extract ductwork systems are normally sized on low-velocity criteria even when the supply ductwork is sized on medium-velocity criteria.

Extract ductwork can be minimised if the ceiling void is used as an air plenum returning air into the void either through specially designed openings within the body of recessed light fittings or through grilles. It is possible to draw the air about 20–25 metres within a ceiling void, provided the air velocity within the void does not exceed 1.5 m/s.

Flexible ducts are often used to make the final connections to ceiling-mounted grilles and diffusers to ease installation and future changes. Care should be taken in limiting the lengths of flexible ducts used, as they cannot be easily supported and can become sharply bent. Flexible ducts should never be used for the connection from the main duct to VAV terminal units, as the high-velocity air generates noise, and the noise breakout through the flexible duct can be a problem.

Rectangular ducts are most commonly used for low-pressure and low-velocity (0–10 m/s) conventional ventilation and air conditioning systems. At higher velocities rectangular ducts of larger sizes can tend to drum and the sides can be difficult to adequately stiffen.

Circular ducts are particularly suitable for medium-velocity (10–20 m/s) systems. The normal method of manufacture, which involves forming the circular duct from a continuous strip with an interlocking joining seam, provides good rigidity and airtightness. This type of duct is referred to as spirally wound circular. Because of their inherent high rigidity, the noise transfer from within the duct to the surrounding area is low when compared to the rectangular duct. Standard ranges of circular duct fittings and bends are manufactured, making the ductwork system costs economical.

Flat oval ducts are formed from spirally wound circular ducts that are mechanically transformed to the flat oval shape. This duct section has a higher rigidity than the rectangular duct and is more suitable for medium-velocity systems. The flat oval duct is a compromise between the characteristics desirable from a circular duct and the space-saving advantage of a rectangular duct. The widespread use of flat oval ductwork has reduced because the bends and fittings required can become relatively complex and the ductwork system costs are higher.

8.5.3   Ductwork materials

For normal air conditioning and ventilation systems, ductwork is manufactured in galvanised sheet steel. The sheet thickness depends on the size of the duct's longer side and the classification of the ductwork, i.e. low, medium or high pressure. Typically sheet thicknesses are 0.8 mm and 1.0 mm for ducts of between 600 mm and 1600 mm on the longest side.

Other duct materials used for special applications include PVC, generally used for laboratory fume extract systems, where a high resistance to corrosion is required.

8.5.4   Standard ductwork sizes

Standard duct dimensions are used in the UK to provide uniformity in the range of sizes of ductwork and fittings (i.e. bends, tees, branches, etc.) used for systems. The standard sizes are defined by the Heating and Ventilating Contractors Association specification DW/142 (HVCA 1988; CIBSE Guide B2 2016). The standard sizes do not include very large ducts, which are manufactured to the particular sizes required. The advantages of using standard sizes are that the cost of the design, manufacture and installation of the ductwork, which can typically account for 25% of the total air conditioning system cost, can be minimised.

The standard sizes for circular and rectangular ducts are given in Tables 8.7 and 8.8, respectively.

Table 8.7 Standard circular duct size

Duct diameter (mm)													
63	80	100	125	160	200	250	315	400	500	630	800	1000	1250

Table 8.8 Standard rectangular duct sizes

Long Side (mm)	Short Side (mm)										
	100	150	200	250	300	400	500	600	800	1000	1200
150											
200											
250											
300											
400											
500											
600											
800											
1000											
1200											
1400											
1600											
1800											
2000											

8.6   Hot and chilled water systems

The design of the hydraulic system is fundamental to the correct operation and control of heating and cooling systems. This section covers the design of recirculating closed-loop heating and cooling water systems for heating, cooling and air conditioning. It includes the design aspects of system pipework circuits and primary-secondary system pumping.

8.6.1   Hot water system design

Constant- and variable-temperature circuits

Air-conditioned buildings are normally required to have two separately controlled heated water pipework systems supplied from a common boiler plant. These systems are a) constant-temperature circuit and b) variable-temperature circuit.

Constant-temperature heated water is required to serve the heating coils within AHUs. This circuit of heated water is supplied from the heating plant at a constant temperature throughout the year. The circuit can also be used to supply the primary coil of hot water storage calorifiers.

Variable-temperature heated water is required for heating circuits that include radiators, convector heaters, under-floor heating pipes and other such heat emitters that require the temperature of the heated water to vary depending on the required heating demand. The variable-temperature circuit incorporates a three-port mixing valve to allow the flow temperature to be varied by mixing a portion of the lower-temperature return water back into the flow without passing it through the boiler plant. The flow temperature is normally varied based on the outside air temperature, but can also be varied by solar or wind sensors. This type of system is referred to as a compensated heating system. The variable-temperature circuit has the benefit of reduced heat losses from the distribution pipework at times when the flow temperature is reduced, thereby improving the energy efficiency of the system.

Heating system pipework circuits

The following section outlines the basic forms of pipework circuits for a radiator heating system. The principles also apply to other heat emitters such as convectors, coils, heat exchangers and radiant panel heaters. There are three basic heating pipework circuit configurations, namely: a) single-pipe circuit, b) two-pipe system and c) reversed return system.

Single-pipe circuit

In this system, illustrated in Figure 8.12, the flow pipe from the heat source is routed in one continuous pipe loop around the system, and radiator flow and return pipes are connected to the single pipe. The water circulation through each radiator is mostly by gravity, and therefore the pressure available for flow through the radiators is very low. If thermostatic radiator valves are used to locally control the output of each radiator, they have to be of a very low resistance type to allow flow through the radiator by gravity circulation. Because the circulation through the heat emitter is mostly due to gravity circulation, as the pressure drop through the main pipe is relatively low, this system is only really suitable for low-pressure-drop heat emitters such as radiators.

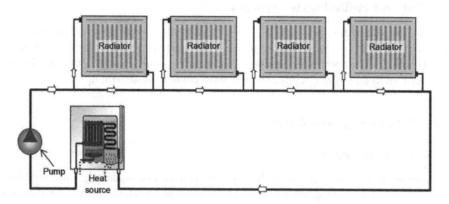

Figure 8.12 Schematic diagram of a single-pipe system

Each radiator in the circuit progressively receives a lower flow temperature from the mixing of return water into the main from the preceding radiators. The radiators at the end of the circuit receive water close to the boiler design return water temperature. This leads to the radiators needing a larger surface area compared with those at the start of the circuit. This requirement for increased radiator sizes does not always prove aesthetically acceptable.

The main advantage of the single-pipe system design is its simplicity and low cost. The single pipe is of one constant size throughout the system. The disadvantage is that the last radiators on the circuit are large in comparison with those at the start of the circuit. In practice, if this pipework arrangement is used within, for example, an open-plan office, the radiators are all sized for the same area, resulting in the radiators at the start of the circuit being oversized for aesthetic reasons.

Two-pipe system

The two-pipe system is the most commonly used pipework arrangement for heating systems. With this system all radiators receive the same flow temperature, and radiators of the same output are of equal size. The flow through each radiator is determined by the system pump pressure and can therefore be of a relatively high pressure drop.

The pipework system is more complicated than a single-pipe system, and the flow and return pipes reduce in size throughout the circuit. The inherent problem in this distribution is the pressure imbalance throughout the system. This means that the installation of balancing valves is important to adjust the resistance through each parallel path to match the required flow to the available head in the mains at each junction. A schematic of this system is shown in Figure 8.13.

Reverse return system

The reverse return system is a variation of the two-pipe system and generally has similar characteristics. The reverse return pipework distribution system, however, overcomes the problem of unequal pressures throughout the circuit and provides an approximately equal pressure at each radiator. It is not generally possible to design a system so that the entire pipework network is a reverse return circuit other than in relatively small and simple systems.

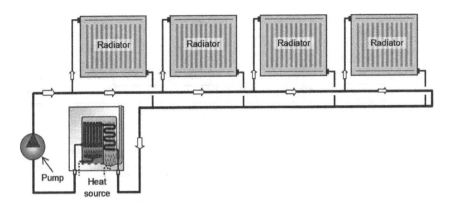

Figure 8.13 Schematic diagram of a two-pipe system

The system has the advantage of making the balancing of the system easier. The disadvantage is that an additional pipe to the flow and return pipe is needed, making the pipework cost greater and requiring more space for the additional return pipe.

Selection of heating system design temperatures

Conventional low-temperature hot water (LTHW) systems are designed for a flow temperature of 82°C and a return water temperature of 71°C ($\Delta T = 11$ K). In larger systems it may be advantageous to design the heating system for a greater temperature difference between the flow and return temperatures with, say, 80°C flow and 60°C return ($\Delta T = 20$ K). For a given heating load this allows the flow rates to be about one-half of those of a conventionally designed system. The higher temperature difference reduces the pipe sizes and the pump energy requirement. The disadvantage is that the mean surface temperature of the heat emitters is lower, resulting in the need for larger surface areas.

The wider temperature difference systems require more careful design, since they are much less forgiving with respect to design errors. Heat transfer surfaces must be selected with care, and piping must be sized correctly with low resistance.

Primary and secondary pumping

In large systems, a suitable way to avoid problems of system and control valve interaction and flow variation through boilers or heat pumps and chillers is to use primary and secondary pumping, as illustrated in Figure 8.14. This ensures that the flow rate on the boiler or chiller side remains constant and that there is no interaction between the control valves.

The primary pump is sized to pump through the primary side only of the system, and each secondary pump is sized for its own secondary circuit. The balance pipe is required to avoid the primary pump pressure, causing water to flow the wrong way through the three-port mixing valve. It is necessary to install a regulating valve to avoid short-circuiting the secondary system.

A practical adaptation of this system is to pump into a main primary header that eliminates the requirement for balance pipes and regulating valves. The header is sized for a low pressure drop so that any change in flow in the header does not result in a system pressure change. This is shown in Figure 8.15.

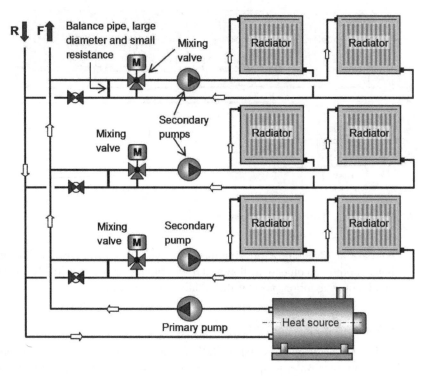

Figure 8.14 Primary and secondary pumping arrangement

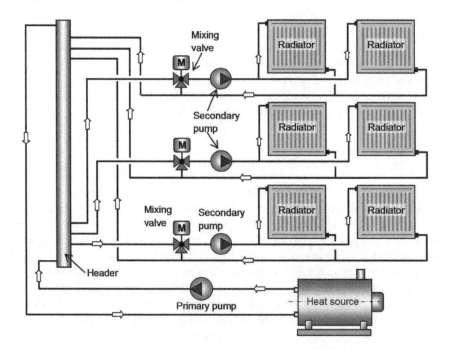

Figure 8.15 Pumping arrangement employing a header

8.6.2 *Chilled water system design*

Chilled water systems are generally designed as constant-temperature systems for 'all-air' air conditioning systems. The chilled water is supplied to the cooling coils within the AHUs at a constant supply temperature to ensure that the coil apparatus dew point provides the required dehumidification of the supply air.

In 'air–water' air conditioning systems, there is a requirement to provide two separate temperature-controlled chilled water circuits. One of the circuits supplies chilled water at a constant temperature to the AHU coils to cool and dehumidify the primary air. A second chilled water circuit is required to serve the terminal units. In the case of a fan coil unit, the chilled water may be supplied to the units at a temperature above the dew point temperature of the air so that the fan coils operate dry. In the case of a chilled beam/ceiling system, it is essential that the chilled water temperature is always above the room dew point temperature; otherwise, condensate will form on the ceiling. This type of chilled water circuit arrangement can be accomplished using an injection circuit.

An injection circuit is similar to the standard mixing circuit but is used where the secondary flow temperature is designed not to be the same as the primary flow temperature and where the secondary temperature must be limited to a predetermined level. The circuit is shown in Figure 8.16, which illustrates both the primary/secondary pumping and constant-temperature circuits for a chilled water system.

It can be seen that the injection circuit includes a bypass line between the secondary pump and the mixing valve. The regulating valve within this pipe is set to provide a constant bypass of water, which ensures that when the three-port valve opens fully to the primary circuit, the secondary flow temperature will be limited to above the room dew point by the mixing of the

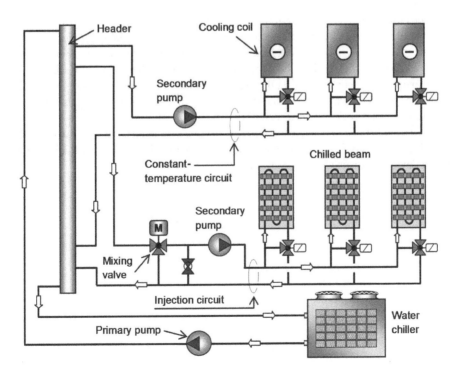

Figure 8.16 Primary/secondary pumping and injection circuit

return water within the bypass line. This circuit also allows a smaller valve for the secondary circuit and one that operates across its full range.

Chilled water systems employ primary/secondary pumping, as described in the preceding section for heating systems, as it is normally essential to maintain constant water flow through an operating chiller. A reduced flow can lead to a freeze-up of the evaporator.

Chilled water pipework circuits are normally of the two-pipe design. They can also benefit from reversed return pipework design to ease the balancing of the hydraulic system.

The standard temperatures used for chilled water supply to the cooling coils within AHUs are 6°C flow and 12°C return. On large systems where the pipework is extensive, the pipework sizes and pump energy can be reduced if the cooling system is designed with a higher temperature difference between the flow and return.

8.7   **Summary**

In this chapter, we have considered the major technologies and equipment employed to generate the cooling and heating energy for the control of the thermal environment in buildings. We have also considered aspects of design of ductwork and pipework systems for the distribution of thermal energy to the terminal devices in the conditioned spaces. The selection of heating and cooling technologies for a given application will have an influence on the energy consumption and environmental impacts of the building and should be given careful consideration at the design stage.

Vapour compression refrigeration systems have higher COPs than sorption systems, but the latter can offer a viable alternative where waste heat is available to drive the generator. Solar thermal energy can also be used for this purpose, particularly in locations with high solar insolation.

The selection of heat rejection equipment will have an influence on the power consumption of the refrigeration plant. Air-cooled condensers and 'dry coolers' (water-to-air heat exchangers) are the most commonly employed heat rejection equipment in temperate and cold climates for small to medium heat rejection duties. For higher duties than about 400 kW and in warm climates, heat rejection through cooling towers becomes economically viable.

In this chapter we have also considered the various ductwork sizing methods and the practical aspects of ductwork system design. The advantages of using standard sizes have been considered, and listings of standard sizes for circular, rectangular and oval ducts have been provided for reference. Common piping arrangements for cold and hot water distribution to terminal devices have also been discussed.

9 Low-energy approaches for the thermal control of buildings

9.1 Introduction

Concerns over global warming and the escalating costs of fossil fuels over the last few years have put pressure on governments, professional bodies and engineers to re-examine the whole approach to building design and control. There are considerable pressures on architects to design less energy-dependent buildings while maintaining or even improving thermal comfort. Low-energy thermal control technologies that have been used for centuries in various parts of the world are now being re-examined and re-engineered to fit within modern building forms.

Low-energy technologies can be considered an alternative to the traditional approaches for the environmental control of buildings described in Chapters 7 and 8. They may employ natural sources such as outdoor air, water or ground to provide cooling and heating to buildings or other facilities without lowering the desired level of indoor air quality and thermal comfort. Low-energy technologies have the potential to offer reductions in energy consumption, peak electrical demand and energy costs, provided they are properly designed and implemented.

Various low-energy design concepts can be applied to non-residential buildings, as well as to new and retrofit structures in a wide range of weather conditions. These technologies are also often used in combination with traditional cooling and heating systems. In this chapter we consider the characteristics and applications of the most promising low-energy approaches for control of the indoor thermal environment of buildings.

9.2 Thermal energy recovery

In modern commercial buildings it is estimated that 20–30% of energy is lost due to the requirement to provide and condition fresh air for the building's occupants. Fresh air is necessary for the health and productivity of the occupants and the control of relative humidity. A way of reducing the energy consumption arising from mechanical ventilation, apart from demand control ventilation (control of ventilation rate based on actual occupancy and CO_2 generated by the occupants in the building) is through thermal energy recovery.

The main methods of air-to-air thermal energy recovery are a) fixed plate heat exchangers, b) rotary air-to-air exchangers (rotating wheels), c) run-around coil systems and d) heat pipes. For all types of thermal energy recovery equipment, the efficiency of the process is defined by the total effectiveness of the process, given by:

$$\varepsilon_t = \frac{actual\ transfer\ of\ thermal\ energy\ across\ the\ device}{maximum\ possible\ transfer\ of\ thermal\ energy\ across\ the\ device} \tag{9.1}$$

9.2.1   *Fixed plate thermal energy exchangers*

Plate heat exchanger heat recovery systems (Figure 9.1) are available in many different configurations, materials, sizes and flow patterns. In their simplest form they are composed of a
cubical sandwich of thin metal or plastic plates. These plates allow the exhaust and supply
air streams to exchange thermal energy through heat transfer across the separating plate
walls. The space between the plates can very between 2.5 and 12.5 mm, depending on the
design and application. The plates are arranged into modules, which can range in capacity
from 0.01 to 5 m³/s. Multiple modules can be used in different sizes and configurations to
satisfy individual application requirements (ASHRAE 2016a). Plate heat exchangers have a
number of advantageous features, including:

*   no moving parts, reducing maintenance requirements
*   no cross-contamination, as the two air streams are kept apart and do not mix
*   very low-energy requirements, so only small additional fan power energy is required
    to overcome the pressure drop in the plates.

9.2.2   *Rotary air-to-air thermal energy exchangers*

Rotary energy exchangers or thermal wheels are composed of a revolving circular matrix
of air-permeable material which provides a large internal surface area for heat transfer (Figure 9.2). For commercial building applications, the matrix material can be aluminium foil,
paper, plastic or other synthetic material in a variety of configurations that provide small air
passages (1.5–2 mm) in the direction of air flow. The wheel is positioned across parallel ducts,
which carry the supply and exhaust air streams in a counterflow arrangement. The exhaust
air stream flows through half of the wheel, where energy is transferred from the air stream
to the cooler wheel matrix material. As the wheel rotates slowly, the warm part of the wheel
enters the air supply duct and transfers the stored heat to the cooler supply air to the building.

Thermal wheels can be used for both sensible and latent heat transfer. Latent heat transfer
can be achieved if the matrix of the wheel is coated with desiccant material. The desiccant will absorb moisture from the higher-humidity (higher vapour pressure) air stream and
desorb it (release it) to the lower-humidity (lower vapour pressure) air stream. The absorption
and desorption processes are driven by vapour pressure differences between the surface of
the desiccant and the respective air streams.

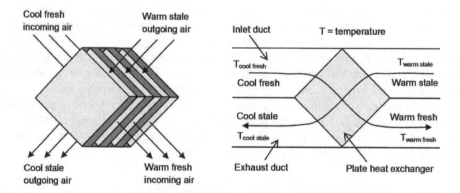

Figure 9.1 Fixed plate thermal energy exchanger

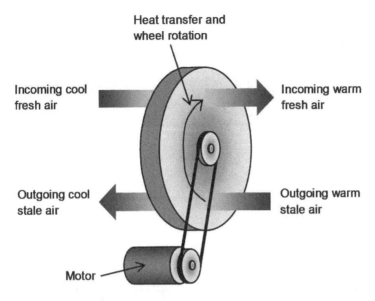

Figure 9.2 Rotary air-to-air thermal energy exchanger

The thermal wheel is normally driven by an electric motor, which introduces a small energy penalty, and this should be considered in the assessment of the overall thermal energy recovery efficiency. Rotary thermal wheels require little maintenance and tend to be self-cleaning because the direction of air flow through the matrix is reversed as each section of the wheel rotates from the supply to the exhaust air stream.

9.2.3 *Run-around coil systems*

Run-around coils or loops are heat recovery devices which employ finned tube water heat exchangers (coils) in the supply and exhaust air streams of buildings or processes, as shown in Figure 9.3. The coils in the two air streams are connected in a way that forms a closed loop through which water or antifreeze solution is pumped in a counterflow arrangement with the air flow. Precautions must be taken at the design stage to ensure that water vapour does not condense and freeze on the coil in the supply air duct.

Run-around coils are very flexible devices and can be used where supply and exhaust air ducts are not located close to each other. They can also enable heat to be transferred from a variety of sources and uses. To accommodate changes in the fluid volume during operation at different temperatures, an expansion vessel should be provided in the closed loop. Typical effectiveness values for run-around coil heat recovery systems range between 45% and 65%.

9.2.4 *Heat pipe heat exchangers*

Heat pipe heat exchangers are two-phase, closed-loop thermal energy transfer devices that contain no moving parts. A volatile fluid in the pipe (Figure 9.4) vaporises at one end by absorbing heat from the air flowing over the evaporator section. The pressure that is created from the evaporation process transfers the vapour to the condenser end of the tube, where

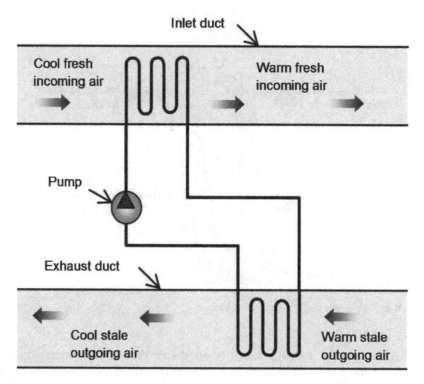

Figure 9.3 Run-around coil thermal energy exchanger

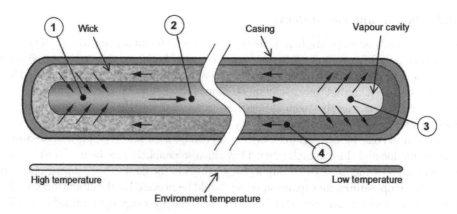

Heat pipe thermal cycle:
1) Working fluid evaporates to vapour absorbing thermal energy
2) Vapour migrates along cavity to lower temperature
3) Vapour condenses back to fluid and is absorbed by wick, releasing thermal energy
4) Working fluid flows back to higher temperature end

Figure 9.4 Principle of operation of heat pipe

it is cooled by the cooling air flowing over the tube and condenses to liquid. The liquid then flows back to the evaporator by gravity, the pressure difference between the two sections and, in some cases, capillary action created by a wick-type material for the repetition of the cycle.

Because of the two-phase heat transfer, heat exchange with heat pipes can be many times higher than other heat recovery devices, reducing the size of the equipment required. The performance of heat pipes depends on many parameters, including the working fluid, orientation to facilitate vapour and liquid flow, as well as other design parameters such as pipe material, diameter, extended heat transfer surfaces employed (fins), wick structure, material and arrangement of heat pipes in a heat exchanger block, etc.

9.3   Heat pump systems

As described in Section 8.2.1, heat pumps are devices that extract heat from a low-temperature source and reject this heat at a higher temperature level. Heat pump systems can normally be classified on the heat source medium, such as air, water or ground, and the heat rejection medium, normally air or water. The vast majority of heat pump systems in the market today operate on the vapour compression cycle, but sorption systems are also commercially available. Extensive research and development, such as more efficient heat exchangers and controls, including inverter control of the compressor to better match the heating capacity of the system to the load, and intelligent controls, have resulted in substantial improvements in the coefficient of performance (COP) of these systems. The COP is a function of the difference between the heat source and sink temperatures, and for moderate temperature differences, high values of COP can be obtained, making heat pumps attractive in terms of primary energy savings compared to other heating equipment.

Because of the flexibility to reverse the refrigeration cycle, many heat pump systems can provide cooling in the summer and heating in the winter. When air is used as a heat source, the COP of the system will be a function of the ambient temperature. At low-ambient temperatures the evaporator surface temperature will be below 0°C, and this will cause moisture condensation and freezing on the evaporator surface. Frost accumulation on the coil will reduce heat transfer performance, and to maintain satisfactory performance at low-ambient temperatures, the coil needs to be defrosted periodically.

The use of the ground as a heat source can overcome some of the disadvantages of air as the heat source, as the ground temperature beyond a certain depth remains fairly constant throughout the year. At 10 m below ground depth in the UK, the ground temperature remains approximately constant at around 10°C. The ground can also be used as a heat sink when the heat pump operates in the cooling mode in summer. Typical arrangements of a ground heat exchanger for ground source heat pumps are shown in Figure 9.5.

Ground heat exchange systems consist of two main sub-systems: *the ground loop* and the *thermal energy distribution system*. There are two basic types of loops: closed and open. In open-loop systems ground water is abstracted from an aquifer through one borehole, passed through the heat exchanger and then discharged to the same aquifer through a second borehole some distance away. In the UK, abstraction and discharge of groundwater requires a licence from the Environment Agency. In closed-loop vertical systems, water or antifreeze (brine) is circulated through a U-tube in a borehole or a series of boreholes spaced uniformly in a grid. In closed-loop horizontal systems, the fluid is circulated in a pipework laid horizontally in a trench and exchanges heat indirectly with the ground.

Ground heat exchange systems have been extensively and successfully used worldwide for many years. Most applications have been in North America, Northern Europe and Japan

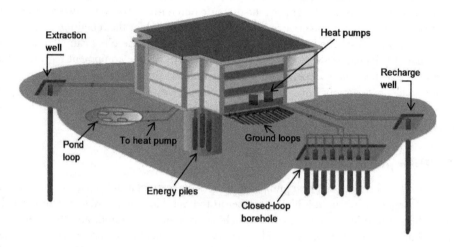

Figure 9.5 Possible arrangements of ground heat exchangers for ground source heat pumps

Source: www.geothermalint.co.uk

for space heating in the domestic sector in conjunction with heat pumps. There have also been successful applications in the commercial sector in schools, hotels and office buildings. Ground heat exchange systems have been less successful where cooling was the predominant load.

Water can also be used as a heat source and/or heat sink with heat pump systems. Water in rivers or lakes will have a more constant and, in most cases, higher temperature than air during most of the heating season and can be a good heat source if rivers and reservoirs are in close proximity to the use of the heat. Water can also provide cooling directly or with minimum cooling by the heat pump when the two are used in combination in cooling applications.

The installation and use of water-source heat pumps requires compliance to specific environmental regulations and considerations. Depending on the type of system and quantity of heat to be abstracted or added to the water, licences from the local environment and water authorities may be required for their application to ensure minimal impact on the environment.

9.4   Solar thermal technologies

Solar thermal technologies transform solar radiation into useful thermal energy. The solar yield replaces conventional sources of heat, mainly fossil fuels or electricity. The simplest way of utilising solar energy for the heating of buildings is through passive means. The primary elements in passive solar heating systems are windows. Glass has the beneficial property of transmitting solar radiation, allowing energy from the sun to enter the building and warm the interior spaces. Clearly, the larger the windows, the more sunlight will enter the building. Unfortunately, though, windows are not as thermally insulating as the building walls, and a passive solar design will optimise window surface area, orientation and thermal properties to increase the energy input from the sun and minimise heat losses to the outside while ensuring occupant comfort.

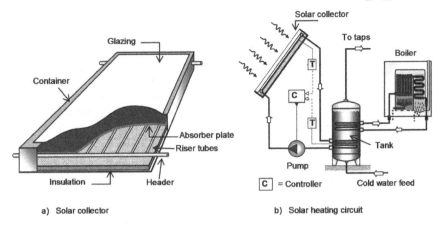

Figure 9.6 Solar thermal heating system

Passive solar heating is best applied to buildings where the heating demand is high relative to the cooling demand. Low-rise residential buildings in moderate to cold climates offer the greatest potential. Passive solar heating is more difficult to apply to office and other commercial or industrial buildings where there are high internal heat gains, especially during the day.

An indirect way of collecting and utilising solar energy is through solar collectors which collect solar energy on absorber plates, as shown in Figure 9.6. Selective coatings are often applied to the absorber plates to improve the overall collection efficiency. A thermal fluid absorbs and transfers the energy from the collector plates to a storage tank from where it can be used for domestic hot water heating, space or process heating.

Several types of solar collectors are used to heat liquids. The selection will depend on the temperature of the application being considered and the climate in which the system will operate. Non-concentrating solar collectors are the most common type of collectors and are normally used for domestic hot water applications or space and process heating. Concentrating solar collectors use reflectors to concentrate sunlight onto the absorber area. These collectors are used for high-temperature applications, and in particular power generation.

Flat-plate collectors can have many different designs but generally all consist of a) a flat-plate absorber which intercepts and absorbs the solar energy, b) a transparent cover that allows solar energy to pass through but reduces heat loss from the absorber, c) a heat-transport fluid flowing through tubes to remove heat from the absorber and d) a heat-insulating backing. An evacuated tube collector differs, as it uses a vacuum between the absorber and the glass surface to minimise heat loss.

9.4.1 Costs

Investment costs of solar thermal systems consist of the cost of hardware (collector, tank, piping and, where appropriate, the control unit and pump) and the cost of installation. Solar thermal systems are sold in a wide range of sizes and applications, and the cost of the hardware therefore varies substantially. This also depends on quality criteria. The cost of installation also varies depending very much on the timing; it is much cheaper to install a solar thermal system during the construction or refurbishment of a building than at a later time. For small systems, installation typically accounts for 20–30% of the total investment costs.

The system price for large collector fields (thousands of square meters), as used for industrial-process heat or district heating, is approximately £175/m². The initial investment constitutes by far the largest part of heat production costs. Modern, good-quality solar thermal systems have a lifetime of 20–25 years with very low maintenance requirements. As with investment costs, the final heat production costs vary greatly depending on the type and size of the system, the location, the timing of the installation and several other factors.

Annual operation and maintenance costs are usually below 1% of the investment. The simplest systems hardly require any maintenance; more complex systems need regular monitoring and some maintenance to keep up high productivity throughout their lifetime.

9.5   Evaporative cooling

Evaporative cooling utilises the evaporative capacity of water to cool air supplied to the conditioned spaces. Evaporating cooling can be achieved through direct air cooling, indirect air cooling, a multi-stage combination of both or through a combination with existing mechanical refrigeration systems and desiccant technologies.

In direct evaporative cooling systems, water evaporates directly into a supply air stream, producing both cooling and humidification. The principle of operation is shown in Figure 9.7.

Indirect evaporative air cooling evaporates water into a secondary air stream flowing in one direction through channels of a heat exchanger. This sensibly cools a primary (supply) air stream flowing through other channels in the heat exchanger in a crossflow direction to the secondary air, as shown in Figure 9.8. This results in 'sensible' cooling only. The secondary air can be all outside air or all exhaust air, or a mixture of the two. In the case of exhaust air, the same heat exchanger can become a pre-heater for the outside air in the heating season. Except in extreme dry climates, most indirect systems require several stages to further cool the primary air entering the conditioned spaces.

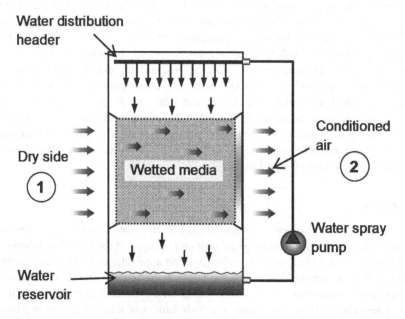

Figure 9.7 Schematic diagram of direct evaporative cooling

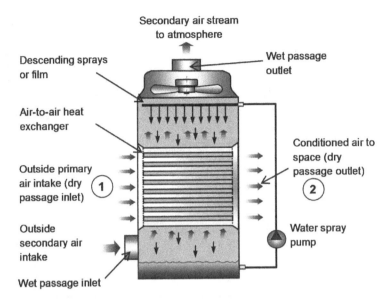

Figure 9.8 Indirect evaporative cooling system

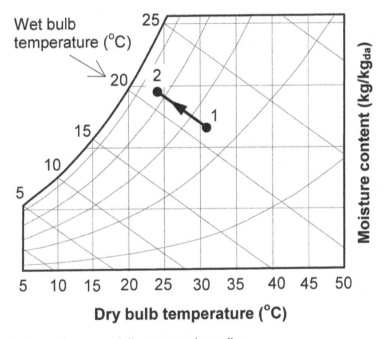

Figure 9.9 Psychrometric process of direct evaporative cooling

The psychrometric processes for direct and indirect evaporative cooling are shown in Figures 9.9 and 9.10, respectively.

Most common system layouts combine indirect with direct evaporative cooling. When the design temperatures required are lower than those available from indirect/direct cooling, a

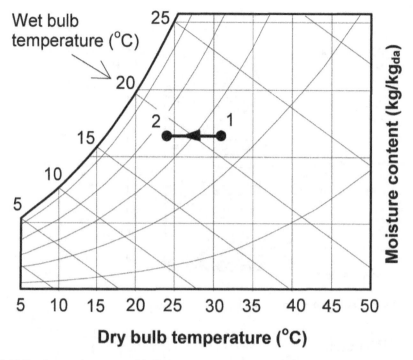

Figure 9.10 Psychrometric process of indirect evaporative cooling

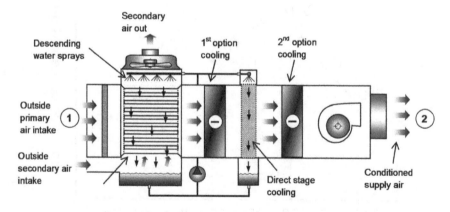

Figure 9.11 Indirect/direct evaporative cooling diagram

third cooling stage may be used, provided by a small-sized refrigerative direct expansion or chilled water coil, as shown in Figure 9.11.

In direct, indirect or indirect/direct stage systems, the capability of cooling relies on the outside air climate. In general, some benefits of evaporative cooling technology are:

• providing comfort cooling in arid and semi-arid regions or relatively dry environments

- improved indoor air quality by introducing high ventilation rates, diluting certain indoor airborne contaminants
- easily combined with existing air conditioning systems, integrated with other cooling technologies or operated with heat recovery.

9.5.1 Applications and performance

Evaporative cooling is effectively employed in arid climates. It can provide relief as well as comfort cooling, depending on regional weather conditions and type of building. In very humid areas or when a lower supply air temperature is specified, there are several approaches, such as integration into other cooling systems or multi-stage evaporative cooling. Applications can be found in industrial, commercial and residential buildings. Some examples are factories, power plants and warehouses.

Direct evaporative cooling systems are used in residential cooling applications. The once-through air flow principle is normally employed without air recirculation.

Indirect or indirect/direct evaporative cooling systems can be coupled to central fan systems for humidity control in commercial buildings. Also, because of the constant moisture content of indirect evaporative cooling, the principle can be used to pre-cool without the addition of water moisture. The effectiveness of indirect systems can be in the range between 50% and 70%.

Table 9.1 shows the performance of evaporative cooling systems for selected locations around the world. The assumptions are 65% of cooling effectiveness for the indirect systems and 85% for the direct systems (Foster 1996).

Since evaporative cooling systems use evaporation of water to cool large quantities of ambient air, they work best in arid regions where water can be very expensive. Water usage

Table 9.1 Evaporative cooling performance for selected regions

Locations	1% Design DB/WB Temperature (°C)	Direct Supply Air DB Temperature (°C)	Indirect/Direct Supply Air DB Temperature (°C)
Asia/Pacific		22.9	17.6
Alice Springs, Australia	39.4/20.0	19.3	16.4
Christchurch, New Zealand	27.8/17.8		
Middle East		23.6	17.1
Riyadh, Saudi Arabia	43.9/20.0	19.6	15.0
Jerusalem, Israel	33.3/17.2		
Africa		25.6	22.1
Cairo, Egypt	38.9/23.3	23.1	19.0
Casablanca, Morocco	34.4/21.1		
Europe		22.3	18.2
Madrid, Spain	35.6/20.0		
South/Central America		21.4	17.9
Santiago, Chile	32.2/19.4	21.8	19.7
Caracas, Venezuela	28.9/20.6		
North America		22.4	16.1
Las Vegas, Nevada, USA	42.2/18.9	17.6	13.9
Mexico City, Mexico	28.9/15.6		

Source: Foster 1996

is highly dependent on the air flow, the effectiveness of the wetted media and the wet-bulb depression (ambient dry-bulb temperature minus wet-bulb temperature) of the intake air.

The energy savings of evaporative cooling over conventional cooling systems will depend on the climatic conditions and the type of evaporative cooling system employed. Experimental observations for Dallas, Texas, showed that the seasonal energy efficiency ratio (SEER) of an indirect evaporative cooler could be 70% higher than that of a conventional air conditioning system and that it could displace nearly 12% of the conventional air conditioning capacity (Hunn and Peterson 1991). The two-stage systems can provide energy savings in the order of 80% in very dry regions to 20% in humid regions. Where the mean coincident wet-bulb design temperature is 19°C or lower, the average annual cooling power consumption of indirect/direct systems may be as low as 0.04 kW/kW(cooling); by comparison, the power consumption of a conventional air conditioning system utilising air-cooled condensers can be greater than 0.3 kW/kW(cooling) (ASHRAE 2019b). Further energy cost advantages can be obtained when the indirect/direct system is employed in the heat recovery mode.

9.6 Desiccant cooling

Desiccants can be solid or liquid materials that can dry either liquids or gases, including ambient air, and can be employed in many air conditioning systems. To demonstrate the principle of operation, a typical solid desiccant cooling system is illustrated in Figure 9.12 and the psychrometric cycle in Figure 9.13.

The ambient air (point 1) is first dried as it passes through the dehumidifier, where it comes into contact with the matrix surface, which is coated with a solid adsorbent. The air is heated up during the adsorption process. Then the air passes through a heat exchanger, where it transfers heat to the regeneration air and its temperature is reduced. The pre-cooled dry air then passes through an evaporative cooler, where it is cooled and humidified before it is supplied to the conditioned spaces (point 4). Because of the direct removal of moisture from the air without cooling it, this technology can perform independent control of humidity (latent cooling) and temperature (sensible cooling), if desired, and reduce a large portion of the cooling load and the size of the air conditioning system.

Solid desiccants include silica gel, zeolites, molecular sieves, calcium bromide, lithium chloride, lithium bromide, carbons, activated aluminas, titanium silicate and polymers. Polymer materials have some commercial potential and are regarded as advanced desiccants for the future. Solid desiccants can be non-regenerated, disposable packages; periodically

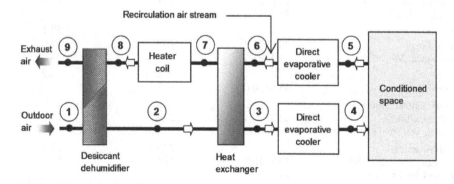

Figure 9.12 Schematic of a solid desiccant cooling system

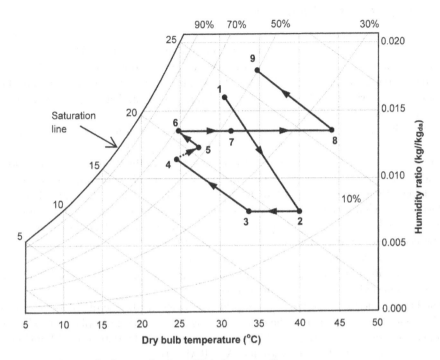

Figure 9.13 Psychrometric diagram for the solid desiccant cooling system

regenerated cartridges; and continuously regenerated. The selection is dependent on the size of the moisture load and the application. Rotary solid desiccant systems constitute the most common arrangement in the continuous removal of moisture from the air. The desiccant wheel rotates through two separate air streams: the process air is dehumidified by adsorption, which does not change the physical characteristics of the desiccant, while the reactivation or regeneration air, which is first heated, dries the desiccant.

For liquid desiccant systems, the moisture is removed by absorption through spraying a concentrated solution of desiccant into the incoming air. The diluted solution is then passed over the recovery units, where it is heated by a return air stream, reconcentrating the liquid desiccant to absorb moisture again. Liquid desiccants in common use are triethylene glycol and solutions of lithium chloride and lithium bromide (Pearson 1997).

In practice, the adsorption performance of solid-based desiccants can be controlled by their total contact area, the total volume of their capillaries and the range of their capillary diameters. The absorption of liquid desiccants can be regulated by their concentration, their temperatures or both.

The benefits of desiccant cooling technology are:

* ability to maintain lower humidity or satisfy high latent loads
* indoor air quality can be improved by handling large quantities of fresh air, drying the process air, removing certain airborne contaminants and minimising health risks
* ability to use various energy sources such as waste heat, solar power and natural gas
* reduction or elimination of the use of hydrofluorocarbon (HFC) refrigerants
* independent temperature and humidity control.

9.6.1 Applications and performance

Desiccant-based systems have been employed in industry for over 50 years but have only recently been developed for heating, ventilation and air conditioning (HVAC) systems for commercial and residential uses, as well as for new and retrofit applications. Other uses showing promise include supermarkets, retail stores, restaurants, hotels, hospitals, etc. Thermal sources most widely used for desiccant reactivation in the market are gas, condenser heat and solar energy.

Starting from the Pennington cycle in 1955 (Figure 9.12), several configurations have been developed, together with various definitions of the thermal COP, which is used to evaluate the performance of alternative designs. In general, though, the COP of this technology is around 1.0. Heat sources for regeneration can be solar energy, heat pumps and gas-fired systems. Electricity is still necessary to rotate the desiccant wheel, fans and pumps. The energy performance is highly dependent on the system configuration, geometries of dehumidifiers, types and properties of the desiccant material, degree of desiccant degradation, etc.

The economics of desiccant applications depend largely on the energy cost savings in dehumidification and the relative utility charges for gas and electricity. Depending on the installation location, the application and relative energy costs, paybacks of three to five years have been reported for dehumidification applications.

The life of a desiccant cooling system primarily relies on the life of the desiccant itself and is a function of the material, the manufacturing process and the application. The life of commercial desiccant materials can be between 10,000 and 100,000 hours. Manufacturers normally anticipate 10–20 years' plant life (ASHRAE 2017a).

9.7 Slab cooling

Slab cooling can be classified as a thermal storage system. The technology provides short-term storage and sensible cooling only, but it has greater potential for either the cooling or heating of buildings in moderate climates. Using the building's thermal mass as the storage medium, cooling or heating is implemented by bringing air (the approach in Figure 9.14) or water (the approach in Figure 9.15) into contact with the slabs in the building envelope.

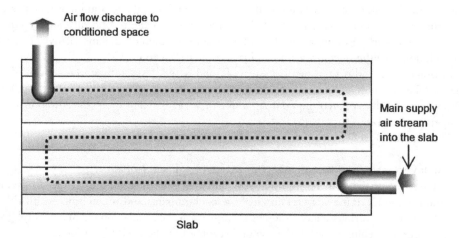

Figure 9.14 Schematic diagram of air slab cooling

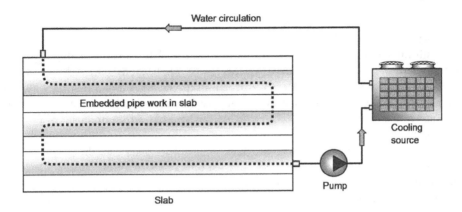

Figure 9.15 Schematic diagram of water slab cooling

The slabs can be the floors, walls or ceilings. At present, most systems utilise floor and ceiling slabs.

9.7.1 *Air slab cooling*

During the summer, cool night outdoor air is brought into contact with the slabs (Figure 9.14) at low velocity before entering the occupied space, cooling down the slabs. During the day, the cooling energy stored in the slabs is used to cool the air supplied for day ventilation.

During winter, the outdoor air can be heated up and then brought into contact with the slabs at night. The heat energy stored in the slabs is then used to warm the supply air to the indoor spaces in the daytime.

The fabric energy storage (FES) slab (trade name: TermoDeck, *www.termodeck.com*) is the most widely used system. It is a pre-fabricated rectangular concrete block 1200 mm wide with a variable length/depth determined at the design stage. Each block contains a minimum of three and a maximum of five cores of sufficient size and shape to allow the required air volume through one inlet hole. The air path is established by interconnecting the hollow-core slabs, and the cooled or heated air is discharged via ceiling diffusers to the indoor spaces.

In general, the merits of the air slab cooling technology are:

* reducing or eliminating mechanical cooling
* use of off-peak electricity
* avoidance of suspended ceilings
* low capital and operating cost
* high levels of human comfort.

The drawbacks are:

* slow thermal response to indoor variations in load
* possibility of air leakage at connections between cores
* extra access holes may be required for maintenance of dead regions at the core junctions.

Applications and performance

The technology is not suitable for regions where high levels of humidity control are required, as it provides sensible cooling only.

As with the night cooling technology, the potential of slab cooling is highly dependent on the level of overnight ambient air temperatures. For applications in the UK climate, this should not be a severe problem. Recent research in the field has found that the bulk temperature of the concrete slabs appears to remain within the accepted range under a daily maximum air temperature of 28°C. Overcooling can be avoided by introducing appropriate night control strategies.

Since the channels of the FES slabs are used as air distribution ducts, the system pressure drop should be minimised. The system would still require space for air terminal devices in conditioned rooms and air handling units in the plant room.

Apart from the integration with mechanical supply and exhaust ventilation night cooling, air slab cooling can work with evaporative cooling and displacement ventilation.

From past experience with TermoDeck, the cooling load applicable to air slab cooling technology is around 40 W/m^2 without exposing the lower slab surface to the indoor air and 60 W/m^2 with the lower surface exposed. The COP (night cooling/fan energy) of slab technology is strongly dependent on system pressure loss. Typical COP values of 2.5 and 3.0 have been estimated for pressure losses of 1,000 Pa and 100 Pa, respectively. High-pressure drops can also induce fan heat pick-up that will subsequently lower the cooling capability.

The air slab technology also provides the flexibility to work with other technologies, like natural, mechanical or mixed-mode ventilation cooling and evaporative cooling.

Costs

This technology is an integral part of the building structure in which the slab channels serve as air distribution ducts. In addition to short-term cooling storage, the advanced FES slab cores have been used for heating energy storage and accommodating the utility services such as water pipework and electric cables. As the advanced FES slabs are mainly pre-fabricated concrete blocks, a large portion of the installation costs related to the ductwork system is transferred to the builder's work accounts, which may need skilled labour, special handling equipment and may have to deal with safety issues during system construction. Although the capital cost of the FES slab technology is difficult to establish in the traditional sense, when directly compared to the capital cost of conventional air conditioning systems, it should be less.

Because the characteristics of FES slabs can allow the use of off-peak electricity at night, energy savings are greatly dependent on the quality of slabs, overnight climate, air flow rates and control strategies. The life span of FES slab cooling systems is normally expected to be equivalent to that of the building.

9.7.2   *Water slab cooling*

Similar to the principle of the air slab cooling technology, where the thermal storage features of the building mass are utilised to provide cooling in the summer and heating in the winter, a water slab heat transfer system relies on the mechanism of temperature difference between circulating water and the slab. The technology employs embedded water tubing or attached piping in the slab where surface temperatures are maintained by circulating water. The water pipework can be in a layer of screed of about 75 mm thick or a layer of solid concrete slab exposed to the conditioned space, as shown in Figure 9.15 (ASHRAE 2016b). The cooling source for the embedded water circuit can be:

- a heat pump with a reversing cycle
- a heat exchanger
- a chiller.

The merits of water slab cooling can be:

- reduction or elimination of mechanical cooling
- the use of off-peak electricity
- low capital and operating cost, if radiant heating is employed.

The drawbacks are:

- possibility of water leakage at embedded piping connections under cyclic cooling and heating during inter-seasonal changes
- difficulty in replacing damaged embedded pipework due to corrosion and erosion problems.

The cooling capability of this technology is approximately between 30 W/m^2 and 50 W/m^2 with cooling water temperatures of about 19°C and occupied space temperatures of less than 28°C.

9.8   Thermal energy storage with phase-change materials

Slab cooling systems utilise the sensible heat principle to store thermal energy in the building fabric. With sensible thermal energy storage, however, the quantity of energy that can be stored is limited by the specific heat capacity of the material used. An order of magnitude increase in thermal energy storage capacity can be achieved if the energy is stored and released by materials during phase change, i.e. freezing or melting. Apart from their high thermal energy storage capacity, phase-change materials (PCMs) also offer the advantage of near-isothermal energy discharge.

Ice has been employed as a PCM for air conditioning applications for many years. A variety of storage methods are available, including encapsulated ice storage systems where water is stored in small polymeric vessels, which are placed in tanks through which a secondary fluid circulates, alternately freezing and melting the water in the vessels; ice on coil systems, where coils are immersed in tanks of water and a secondary fluid flows through the pipe, forming ice on the coil; ice harvesting systems, where the ice forms on plate or circular tube evaporator coils to a thickness of 6–10 mm, melted through a coil defrost system and then gathered in a storage tank below the coil; and ice slurry systems (ASHRAE 2016c).

An ice slurry is a mixture of ice 'micro-crystals' (typically 0.1–1 mm in diameter) formed and suspended within a solution of water and a freezing point depressant. The ice slurry can be formed by a variety of methods, including scraped surface generators, direct contact generators and supercooling generators. The slurry can be stored in tanks and used as a static storage medium. It can also be pumped and used as a secondary two-phase heat transfer fluid in heat exchangers (IIR 2005). The high-energy content of the slurry allows significant reductions to be made in the size of pipes and storage tanks compared to single-phase energy storage and heat transfer with water. A number of energy storage applications of ice slurries are described by Bellas and Tassou (2005).

One disadvantage of ice thermal energy storage is the narrow temperature range of operation, around 0°C, and the significant change in volume with the change of phase from liquid to solid. A number of other PCM materials have been investigated in recent years for thermal energy storage at temperatures both below and above 0°C. The most common of these materials can be classified into two groups: organic and inorganic compounds. Inorganic materials such as hydrates and eutectics have low cost but exhibit problems of phase separation, incongruent melting and subcooling or superheating. To overcome these problems these materials are normally packaged in small capsules (0.02–0.1 m diameter) and are referred to as encapsulated PCMs. Organic materials such as alkanes, waxes and paraffin do not have the disadvantages of organic materials but have lower thermal conductivity, higher cost and can be flammable.

Significant research has been conducted on PCMs for the space heating and cooling of buildings in recent years, but at present there are only a small number of demonstration systems in use and limited experimental data from real installations. By embedding PCMs in gypsum board, plaster or structural elements such as the floor or roof, the building can store large amounts of energy while maintaining the indoor temperature within a relatively narrow range (Iten *et al.* 2016).

9.9   Summary

In this chapter we have considered a number of approaches that can be employed in isolation or in combination with traditional technologies to satisfy the heating and cooling requirements of commercial buildings. One such low-energy technology that can be employed to provide both heating and cooling is the heat pump. Heat pumps remove thermal energy from a low-temperature source and supply it to a higher temperature sink. In recent years the ground has become a popular energy source for space heating, as well as a sink for heat rejection, due to its relative constant temperature throughout the year. Heat pumps can also be used to recover and upgrade heat from the exhaust air from buildings. Other thermal energy recovery systems include fixed plate and rotary heat exchangers, run-around coils and heat pipes.

A number of low-energy cooling technologies can also be considered alongside or in place of conventional refrigeration technologies in appropriate climatic conditions. These include evaporative cooling and desiccant cooling. Thermal energy storage can also provide a means of reducing peak heating and cooling loads and shifting these loads to off-peak periods where the cost of electricity and environmental impacts may be much lower than in peak demand periods. Energy can be stored in the building fabric or in specially designed thermal stores that employ single-phase or two-phase fluids as storage media. Examples of these are water, ice and ice slurries. In recent years, a number of organic and inorganic compounds have also been developed and applied for thermal energy storage in the building fabric.

10 Energy-efficient electrical systems, controls and metering

10.1 Introduction

This chapter describes some of the key design features that can improve energy efficiency and reduce carbon emissions arising from the main energy-consuming electrical systems, together with the controls and metering systems that can enable effective energy management across all active building services systems. As with thermal systems, the desired carbon reduction features are a mix of those that control demand to achieve comfort or function and those directly related to minimising losses in distribution between plant and terminal devices, accepting that there is some overlap between these. The key design decisions are always related to selection of appropriate systems, equipment and distribution arrangements, as well as inclusion of features to limit demand to that necessary to achieve the required criteria at any point in time.

It must be emphasised that the primary consideration when designing electric power systems is always to address the safety aspects and design in accordance with the relevant codes and regulations, such as BS7671 (IET 2018). This will include design considerations such as current-carrying capacity; voltage drop; overload, overcurrent and Earth-fault protection; discrimination, or selectivity, of circuit-protective devices; disconnection time; and earthing and bonding. The energy efficiency aspects covered here should be considered alongside design to achieve code compliance.

10.2 Energy-efficient power distribution arrangements

10.2.1 Optimising sub-station locations

A brief outline of the typical arrangement and principal components in a building's power distribution system is provided in Chapter 12. This is highly relevant to this chapter also, so that the potential distribution losses can be understood and systems can be designed to minimise such losses. It is recommended that Section 12.2 be read prior to reading this chapter.

As outlined in Chapter 3, key design measures for reducing carbon emissions in any electric (or thermal) system are the location of main plant items close to the principal load centres and energy-efficient distribution. Any power distribution arrangement that can minimise energy losses will save energy and costs and will have a beneficial impact on the overall carbon performance. For large sites and systems with widely dispersed loads, the most important design consideration is to plan the location of the substations on the site so that they are close to the main load centres. This will mean that the lengths of low-voltage (LV) distribution cables are kept short, thereby reducing I^2R power losses on cables. Where a single sub-station is to be provided for a group of buildings on a site, the location should be the optimum for minimising losses, taking into account the load profile for each building.

The selection of a location for a sub-station must take account of many factors, including the generic issues for space planning outlined in Chapter 13. The sizing of the distribution cabling will take into account the relevant electrical design factors, including protective device rating, design current, prospective fault level and voltage drop, together with any group rating applicable to the relevant installation arrangement. However, for optimum energy performance, the location should be selected so that energy losses on the LV distribution systems are minimised. This is a different (although similar) approach to optimising sub-station location by reducing losses solely in relation to the maximum demand at each load centre. Because loads are dynamic, it is the summated load profiles over a whole year – taking account of the diurnal and seasonal load patterns – that are important. The situation is, of course, more complicated because the losses are proportional to the square of the load. The only way to obtain a complete figure would be from computer modelling with a full annual load pattern.

To minimise energy consumption, transformers should, ideally, be selected to achieve optimum energy efficiency for the anticipated duty cycle and foreseeable load growth; however, transformer selection is a complicated issue. The step-down high-voltage/low-voltage (HV/LV) transformers will have losses, which can be separated into iron (no-load, magnetising) losses and copper (running, I^2R) losses. The EU Ecodesign Directive for transformers has set new design specifications to address losses in transformers to improve energy efficiency and reduce avoidable CO_2 emissions, so transformers should be selected in line with this directive (noting that there might be similar standards applicable elsewhere). The value of total losses, and the proportional split between iron and copper losses, will vary between transformer types. The operating efficiency will be a function of the losses and the load profile and could be determined by calculations from published losses for different levels of load. In its simplest sense, where the transformer will be running on low load for much of the time, the iron loss value becomes more important, and where the transformer will be running on a low load for a minimal amount of time, the copper loss value becomes more important. It is not necessarily (or often)the case that transformers are most efficient at full load, as the optimum efficiency point can sometimes be at about two-thirds of the rated load, so transformers might have to be somewhat oversized (Court 2011). This aspect of the transformer rating selection cannot be considered in isolation, but must be considered alongside other aspects, including prospective fault level, physical size, cost and the criteria for resilience and redundancy. It should also consider the level of harmonics present in the load and a sensible longer-term margin for load growth.

10.2.2   Optimising LV distribution arrangements

As well as optimising sub-station locations, the lengths of LV cables from the transformer to the main LV switchboard, and from the main LV switchboard to distribution equipment, motor control centres and other items, should be kept to a minimum to limit the losses. This will require attention to the distribution routes as part of the space planning exercises described in Chapter 13. It could also be useful to consider the arrangement of LV distribution components in relation to both operational and embodied energy as for other components (as part of a whole-life impact assessment).

Distribution components used for LV mains and sub-mains power distribution – cables and busbars – can be represented by resistance of the conductors, plus, to a lesser extent, reactance. The resistance will reduce if the temperature is lower and will increase if the temperature is higher, so the I^2R power loss will increase as temperature increases. Therefore, a cable can carry more current in a lower-ambient temperature and less current in a higher ambient temperature. To make the most efficient use of the cost and embodied energy in the distribution system, any factors that will limit the cable or busbar current-carrying capacity should be avoided. So, first, as a simple and fairly obvious measure, power distribution should be routed away from areas of high-ambient temperature.

It is common practice to group power cables on cable ladders, cable trays, in ducts and so on to make economic use of the cable management system and to reduce the space required for distribution. There are various standard installation arrangements for cables that are touching or spaced apart. To take account of the influence of heat emitted by adjacent cables on the rating of an individual cable, a group rating factor (always less than unity) is used as a multiplier to determine its current-carrying capacity. Thus, two detrimental energy issues arise from a group of cables touching, or any other arrangement that increases the ambient temperature for cabling. First, the cable in question is likely to have a higher power loss per unit length; and second, the current-carrying capacity of all of the cables in the group will be reduced. Although these are interrelated, they are better seen as separate issues. The first is an operational energy issue and will increase carbon emissions per unit of useful energy consumed by the loads. The second is an embodied energy issue, as it will result in an unnecessary additional embodied energy (and capital cost) in the cabling system. In general, better energy utilisation can usually be achieved by spacing cables farther apart.

A balance has to be made, however, between the additional one-off capital cost and embodied energy associated with an increase in cable management materials (tray/ladder) and the ongoing operational energy impact. This is mainly a consideration for larger groups of larger-sized cables that will be carrying significant loads for long periods, where it is worth making an engineering judgement on the cable management arrangement. More generally, it is good design practice to keep loads balanced across phases and provide a symmetrical formation where it is necessary to use single-core cables for distribution.

There has been much recent interest in the energy benefits that might be available from adjusting the voltage level at the origin, or at other locations, in an LV distribution system through incorporation of 'voltage optimisation' equipment. While lowering the voltage could potentially save energy in certain types of loads in some legacy installations, it could give rise to risks and may not be the most appropriate energy reduction approach (Court 2011). Such measures always require careful technical assessment to assess their viability.

There has been relatively little research or monitoring on the proportion and pattern of energy losses arising from power distribution within buildings, but this is clearly a worthwhile area of research to inform design decisions. This might become more important as designers seek to further minimise energy losses and carbon impacts.

10.2.3 *Power factor correction and harmonic filtering*

It is normally the case that the electrical loads within buildings will have a variety of levels of reactance, and hence a variety of power factors. Some of the larger-magnitude loads – such as those for the induction motors associated with chillers, pumps and fans – will be inductive and have lagging power factors. As a result, the overall power factor will tend to be lagging.

For power distribution losses to be minimised, power factors should be close to unity, so that the magnitude of current (and hence I^2R losses) is close to the theoretical minimum for the value of 'real' power flow. Power factor correction (PFC), using capacitor banks with automatic switching in stages, is a well-established technology that offers a simple way to reduce energy losses in a distribution system. Correction can be applied at the main switchgear and/or at appropriate load centres to reduce current and thus I^2R power losses in cables, busbars and transformers. In reality, there are diminishing economic returns as power factors are corrected close to unity. While each case has to be assessed separately, there is often a strong environmental case for power factors to be kept at about 0.95 lagging. In Building Regulations Part L, PFC is recognised as an enhanced management and control feature.

In the UK, electricity suppliers normally encourage customers to maintain high power factors, typically no lower than 0.95, and customers can be charged for reactive power when the average power factor is less than 0.95 lagging. Reactive power changes are for excessive kVArh (typically based on a one-month period). There is usually also an 'availability' charge based on the contractual reserved (or authorised) supply capacity in kVA. The cost payback will vary depending on the load pattern, but it can be low, typically in the order of 1.5–2.0 years.

The benefits of reducing the kVA load are:

- lower energy costs and reduced carbon emissions due to lower distribution energy losses
- an increase in the efficiency of the supplier's distribution system due to better usage of the cost and embodied energy and less need for infrastructure upgrade, which means that the consumer can prevent the potential business impact(s) that could arise from insufficient service capacity.

Figure 10.1 shows typical locations for PFC equipment in an LV distribution system. Where multiple motors are fed from a motor control centre but only a proportion of the motors will run at any one time, the most beneficial arrangement is likely to be a common multi-stage

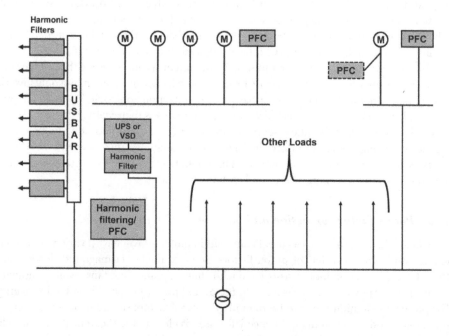

Figure 10.1 Typical locations for PFC and harmonic filtering equipment

PFC located at the motor control centre. Ideally, it should be sized to provide correction for the maximum cumulative reactive power of the motors, with switching in equal stages to match the load variation. In these cases there would be no benefit in providing a separate PFC for each motor. For large individual motors, such as chiller compressors, a separate local PFC cubicle can be provided for each chiller machine (or this may be integral), sized specifically to provide the required correction and only operating when the chiller operates.

The selection of PFC equipment should not be considered in isolation, as it needs to take account of the presence of harmonic currents. The types of non-linear power loads in buildings that can give rise to harmonic currents are outlined in Chapter 12. Harmonics can have a detrimental carbon impact in two ways:

- equipment will need to be sized for increased capacity, resulting in an increase in embodied energy (and cost)
- an increase in distribution energy losses in cables, busbars, switchgear and transformers.

The extent to which harmonics will have an impact will depend upon the magnitude and the harmonic pattern or signature. The most appropriate approach is to reduce the level of harmonics at the source. There are methods for reducing the level of harmonics, including filtration equipment, that can (and in many cases should) be incorporated at suitable parts of an installation. It is always preferable to reduce harmonics rather than having to design oversized system components (with the resulting embodied energy and cost) to address their impact.

In the UK, regulations are in place to limit the impact of an electricity consumer on the power quality of the supply to other adjacent consumers. The limitations relate to the point of common coupling, which is the point in the public supply system nearest to the consumer (in the electrical sense) at which other customer loads could be connected. If a building has multiple LV supplies (as is sometimes the case for large buildings in city centres), each supply will have its own point of common coupling. Harmonics can cause problems with capacitors resulting in overheating, over-voltage and failure. If resonance occurs at a harmonic frequency (say fifth or seventh), the magnitude of the harmonic current (and hence voltage) will be magnified. Problems often arise from the magnitude of the fifth harmonic (ABB 2010). Therefore, where non-linear loads are significant, PFC capacitors should be combined with harmonic filters.

Passive harmonic filters are normally selected so that they are de-tuned (i.e. the tuned frequency is below the frequency of the lowest-order harmonic generated) so that the magnitude of fifth, seventh and higher harmonics is reduced (ABB 2010). This can improve the power factor. Active filters work on the principle of an active device generating harmonics that are equal but opposite in phase to those generated by the non-linear load and injecting these between the supply and the load. This results in cancellation, so that the current waveform from the supply to the point of filter connection contains the fundamental component only. This approach can also contribute to balancing the loads. Active and (de-tuned) passive filters are often used in combination (ABB 2010). Other solutions are available for different harmonic patterns, such as isolating transformers. These and other technical aspects of PFC and harmonic filtration selection require detailed assessment and reference to specialist equipment suppliers.

Figure 10.1 shows typical locations in a system where PFC and/or harmonic filtering are regularly used. It should be understood that while the use of filters at the main LV switchboard might be necessary in order to meet regulatory criteria, it will only reduce energy losses on the supply side. Filters should be located as close to the load as possible to reduce losses in the internal LV distribution. They can be provided for individual distribution boards feeding loads with significant harmonic content. In addition, suitable space should be allocated in planning for the required PFC and/or harmonic filtering equipment. In cases where the nature of the load pattern is less certain, it can be prudent to allocate provisional space and undertake post occupancy measurements to determine the specific requirement. This can be particularly useful in situations such as multi-tenanted high-rise buildings.

The widespread usage of variable-speed drives (VSDs) (described in Section 10.3) is likely to reduce the need for PFC (as the drives will have power factors close to unity) compared with conventional drives. But it is likely to increase the need for active filters to counteract the harmonics arising from the drives. Due to the growth in utilisation of VSDs and other switched-mode-type power supplies for most loads within modern buildings, it is likely that the need for PFC will increasingly be primarily for legacy installations.

10.3   Motor power for HVAC equipment

In buildings with air conditioning or mechanical ventilation, the motors driving chiller compressors, fans and pumps are likely to be responsible for a significant proportion of the energy consumption and carbon emissions; they are therefore a major focus for energy efficiency. In buildings which have heating generated from heat pumps, the motors driving the compressors will also be significant for energy and carbon. This section covers motors for fans and pumps. The general points on motor efficiency are also relevant to chillers and heat pumps.

Because fans and pumps are components of heating, ventilation and air conditioning (HVAC) systems, it is necessary to understand how both the mechanical and electrical aspects contribute to energy consumption and identify the key measures that can reduce energy usage. It is useful to consider the whole-system approach, as outlined by Stasinopoulos *et al.* (2009). Figure 10.2 shows a typical power-balance Sankey diagram covering components

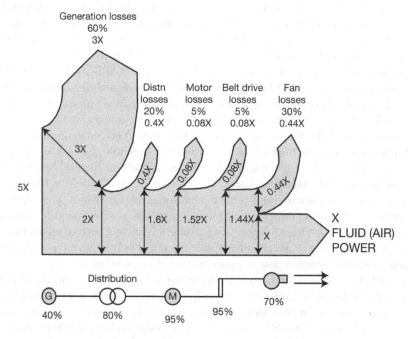

Figure 10.2 Sankey diagram for power to a fan or pump (based on conventional electricity generation – not to scale)

from electricity generation to the fan or pump (and relates to a grid scenario with a high level of conventional fossil fuel generation). Serial efficiencies for each component must be considered in determining the primary power requirement or energy to deliver power to the air or fluid from the fan or pump. This simplistic representation, with the whole of the electricity transmission and distribution losses up to the motor conveniently lumped together (shown here for convenience as a transformer symbol), indicates that the primary power (which relates to the fuel input and hence CO_2 emissions) could be about five times the actual power imparted to the fluid (air) stream. As the decarbonisation of the grid continues, the generation losses shown in the diagram will reduce, and hence the primary power input. It should be noted that the efficiencies in this diagram are nominal only, and in many older installations the overall efficiency could be much lower. Moreover, there will be additional losses in the fluid system that will further reduce the useful power delivered (von Weizsäcker *et al.* 1998). From this awareness, any improvements in the efficiency of motors can provide significant benefit. There are international classifications for LV motors, together with standard methods for calculating efficiency. A motor's efficiency can be defined as:

$$Motor\ efficiency = \frac{mechanical\ output\ power\ at\ the\ shaft}{electrical\ power\ input\ to\ motor} \tag{10.1}$$

The three international classes for the efficiency of motors are defined by the International Electrotechnical Commission (IEC) and have International Efficiency (IE) references (IEC 2014):

IE1: standard efficiency
IE2: high efficiency
IE3: premium efficiency
IE4: super premium efficiency.

The EU has minimum energy performance standards for motors. This requires that fixed-speed LV motors (of two, four- and six-pole construction) must conform to IE3, or IE2 (for motors from 0.75 kW to 375 kW) if equipped with a VSD. IE4 provides further improvement and is most suited to large motors running for a high proportion of the time.

Motor efficiency does not vary significantly between 50% and 100% of the load, but can fall considerably below 25% of the load. The power factor also reduces at low loads. Motors should therefore not be significantly oversized for their required duty (CIBSE 2004a).

The fluid power requirement necessitates that, at a minimum, fans and pumps should have high efficiency plus appropriate and rigorous time control arrangements to suit occupancy in order to manage the demand. The power requirement is the product of flow rate and pressure. Under the fan/pump laws, there is a cubic relationship between flow rate and motor power. Therefore, any reduction in the flow rate will significantly reduce the power consumption. Figure 10.3 shows a range of flow rate and energy consumption percentages for a pump for different hours of operation. This illustrates that considerable energy savings are possible by reducing the flow rate. For example, it can be seen that by reducing the flow rate to 30%, the energy consumption can be reduced by 90%. VSDs using inverters have become a standard feature for variable flow systems. They eliminate the inefficiency of fixed-speed belt drives and improve overall efficiency by continually adjusting to balance flow rate against demand. VSDs sometimes have settings to optimise energy efficiency and normally provide a high power factor at all loads (CIBSE 2004a), providing further improvements.

Hour	1	2	3	4	5	6	7	8	9	10	Average
Flow rate (%)	100	90	70	55	40	30	30	32	38	45	53·0
Energy consumption (%)	110·0	82·9	42·3	23·6	13·4	9·7	9·7	10·3	12·5	16·1	33·1
Energy saving (%)		−10·0	17·1	57·7	76·4	86·6	90·3	90·3	89·7	87·5	83·9	66·9

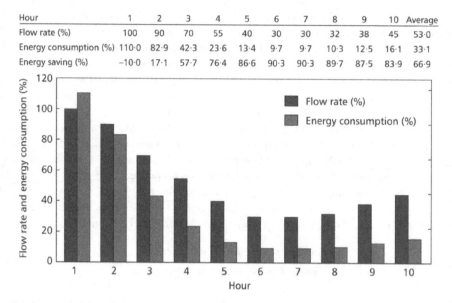

Figure 10.3 Typical savings from a VSD pump

Source: Reproduced from CIBSE KS14 (2009a) with the permission of the Chartered Institution of Building Services Engineers

For certain applications, the power requirement could be reduced by using appropriate hydronic circuits and pumping arrangements, or a wider temperature difference, that allow the flow rate to be reduced. It can also be reduced by minimising system resistance. A simple solution is to reduce the length and complexity of ductwork or pipework systems, which can significantly reduce losses and greatly reduce power requirements (CIBSE 2004a), and hence energy and carbon. This can be achieved byreducing the pressure drop arising from changes of direction, filters, fittings and attenuators; maintaining straight runs and suitable sizes; removing air and dirt from hydronic systems; minimising ductwork leakage; and other similar measures (von Weizsäcker *et al.* 1998).

Chapters 6–9 describe methods of reducing the energy for fans and pumps in specific HVAC systems. Motor power for lift drives and for fan coil units are covered in Sections 10.5 and 10.6, respectively.

10.4   Lighting

10.4.1   Design objectives

The overall objective for a lighting design is to achieve the desired lighting performance in terms of creating a satisfying visual appearance and facilitating the functional performance within the space, with the minimum use of energy. In the UK in 2012, lighting represented about 33% of final energy consumption in retail buildings, about 17% of final energy consumption in commercial office buildings and about 20% of final energy consumption for all buildings in the service sector (DECC 2012). Lighting would, of course, have represented a much higher proportion of carbon emissions in each case, due to the primary energy carbon factor for electricity. In 2004, lighting could account for over 40% of electricity costs in naturally ventilated offices (CIBSE 2004a). The proportions of energy consumption, electricity costs and carbon emissions are likely to have changed, and will continue to change, with grid decarbonisation and the development of new light sources, specifically light-emitting diodes (LEDs) (and the proportions will also depend on developments for other active energy systems). Nevertheless, lighting is likely to remain one of the significant energy consumers amongst electrical services, so it merits a strategic approach to minimise impact.

It is essential to appreciate at the outset that, unlike the other main energy-consuming systems in buildings, lighting performance has an aesthetic dimension alongside those aspects with physical parameters. For many types of spaces there is a paramount objective to create a visual appearance that is both satisfying to the occupants and supports their functional activities. This aspect is an essential feature of user-centred design, and of health and wellbeing in buildings, as outlined in Chapter 5. Therefore, the strategic energy efficiency approach to lighting design described here is only one aspect and should be considered along with the user-centred aspects to create a balanced holistic solution.

As with all active engineering systems, the first priority for the energy strategy is to reduce the demand. The primary requirement is therefore to seek a suitable level and quality of daylight so that this can satisfy the lighting requirements for the optimum portion of time. The artificial lighting design should complement the daylighting provision. The energy efficiency of the artificial lighting will primarily be a function of the efficiency of the equipment but also controlling the usage of the equipment so that it matches the need at the times required for operation. As with the other engineering systems, the operation throughout its life will be the determining factor. As such, the engagement of the operational staff and occupants will be necessary. A key design challenge, therefore, is to understand the occupancy patterns and the need for flexibility and then select the most suitable mix of automatic and manual controls to provide a practical solution.

Daylight potential can be optimised through liaising with the architect to influence decisions on the shape and orientation of the building and the glazing arrangements and characteristics at the initial design stage. This is a key part of passive design activity – see Section 3.4.5.

10.4.2 Design criteria

A key consideration for the artificial lighting system is to critically examine the design criteria. This is an area of building services design that can benefit most from exploring the potential relaxation of design parameters with a client, as outlined in Chapter 3. In particular, it is necessary to agree on realistic and appropriate overall illuminance levels to meet the needs of the routine activities undertaken in the space and avoid over-illuminating through an inappropriate and unnecessarily high level of illuminance.

While the primary focus will be on the general illumination for a space, the design approach should also consider the extent to which functional needs can be realised through lighting planned to satisfy illuminance of the task. This could be satisfied through the arrangement of the general lighting provision or through a separate system of task lighting that supplements the lighting in the area where the task is undertaken. It is obviously beneficial in energy terms to use task lighting for localised demanding tasks, rather than increasing the level of general illuminance over a wider area. The proposed usage pattern for the space should be determined at the briefing stage so that a suitable balance can be selected between task and general illuminance.

Design illuminance levels are defined as a maintained illuminance value, E_m, in lux. This is the average value that the illuminance has fallen to at the time that maintenance has to be undertaken. The maintenance factor (MF) is defined as the ratio of maintained illuminance to initial illuminance (CIBSE 2002). The MF takes account of all the losses that contribute to a reduction in illuminance level over time:

$$MF = LLMF \times LSF \times LMF \times RSMF \tag{10.2}$$

Where:
LLMF = lamp lumen maintenance factor
LSF = lamp survival factor
LMF = luminaire maintenance factor
RSMF = room surface maintenance factor

The higher the MF, the better the energy performance. Hence, each term in the equation should be as high as possible. Two of these terms are functions of the lamp performance. The lamp lumen MF is an indication of the extent to which light output deteriorates with time, and lamp survival factor is an indication of lamp life expectancy. The selection of light sources should consider these lamp parameters to achieve good whole-life performance. The luminaire MF and the room surface MF are both related to the quality of the maintenance regimen and should form part of early discussions with the client.

Lamp manufacturers usually provide rated LLMF figures for different periods of usage, such as 2,000 hours, 5,000 hours and up to 20,000 hours. For a T5 fluorescent lamp, typical LLMF ratings would be 0.96 at 2,000 hours, falling to 0.9 or below after 20,000 hours of operation. Rated lamp survival factors vary considerably between lamp families.

An unfortunate reality of lighting design from an energy perspective is that to achieve the required maintained illuminance level, spaces end up being over-lit for the majority of the life expectancy period of the lamps. Dimming controls can help to improve the overall energy efficiency for certain lamp types, as outlined next.

10.4.3   Energy-efficient light sources and luminaires

The energy efficiency of light sources will depend upon the luminous efficacy of the light sources and the effectiveness of the luminaire in directing the lumen output towards the surfaces to be lit. In order to achieve energy-efficient light sources, lamps should be selected with the highest luminous efficacy (lumens/watt) that is appropriate for the application, taking into consideration the colour rendering and colour appearance of the source. For any particular family of lamps, the luminous efficacy would usually increase with the power rating of the lamp. In addition, luminaires should have a good light output ratio (LOR), typically greater than 0.6. Figure 10.4 shows the range of luminous efficacies for the main families of lamp types used in buildings.

In the UK and in many other parts of the world, the usage of LED light sources, which have high luminous efficacies, has become so widespread that they have come to dominate the market. Their dominance is such that other light sources, such as linear and compact fluorescent, have a much lower presence in the market, although they are still likely to be used for legacy and specialist installations. As shown in Figure 10.4, LEDs typically have luminous efficacies in the range of 80–120 lumens per watt and can have even higher efficacies. LEDs will inevitably be the light source for most buildings. It is anticipated that their present high

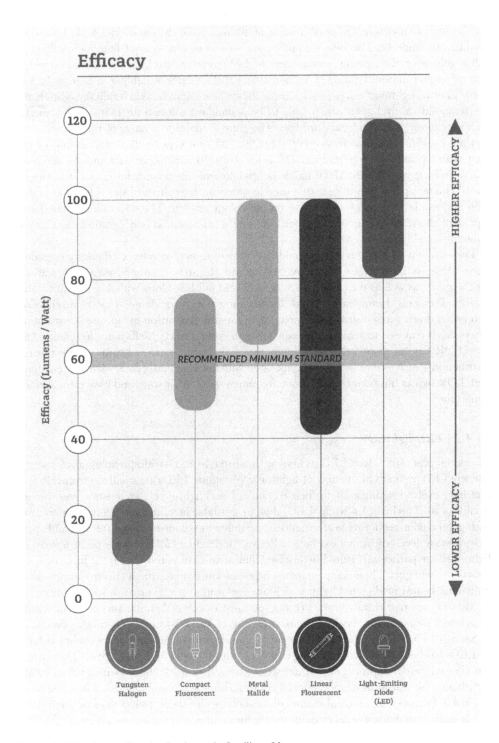

Figure 10.4 Luminous efficacies for the main families of lamp types

Source: Courtesy of Hoare Lea

luminous efficacies will rise with further improvements in the technology, and there may also be improvements in the performance of fittings. As such, the overall LED luminaire product may improve. The focus for optimising the energy efficiency of lighting installations will largely be on the lighting management of LED installations. LED lighting is covered in general and for interior lighting in Section 10.4.4 and for exterior lighting in Section 10.4.6.

For internal lighting, fluorescent T5 linear fluorescent lamps used in luminaires with high-frequency (hf) control gear, which used to be a standard solution for many non-domestic buildings, is now largely a legacy product. These are available in a range of tube lengths and typically provide luminous efficacies of about 90–100 lumens per watt. Compact fluorescent lamps can typically provide luminous efficacies of 50–70 lumens per watt and are also used in a wide variety of spaces. Metal halide lamps can typically provide luminous efficacies of 60–100 lumens per watt and are often used in spaces such as airports, sports halls, assembly halls and exhibition spaces, depending on the colour criteria. The selection of these lamp types will also need to consider whether dimming is proposed, as only certain lamps have a dimming capability.

The selection of the luminaire should not be determined by energy efficiency consider-ations alone, as it will involve design integration with the architectural layout and aspirations for the space, as well as the ceiling arrangement and finishes, along with any relevant statu-tory requirements. Luminaires should, for the most part, provide good internal reflection and broad downwards distribution, including a modest proportion of upwards distribution where this is required to achieve a visually satisfying solution. High-efficiency luminaires can have LOR figures up to 0.8–0.85. Luminaires that provide narrow beams of downwards distribution – such as recessed downlights, often with low luminous efficacy lamps – can have high LOR figures but poor overall luminaire lumens per circuit-watt, and have more limited application.

10.4.4 LED lighting

The development of 'white' LED technology has brought about widespread usage of energy-efficient LED products in a variety of lighting applications. LEDs are small electronic devices that have evolved significantly in their technology and lighting characteristics over the last decade or so. This includes 'white' LEDs that are available in a range of colour temperatures with good colour-rendering performance. The different colour temperatures available pro-vide different levels of luminous efficacy. Because individual LEDs provide point sources of light, they are particularly suited to localised illumination requirements in the form of spot-lights or downlights. Their compact form and point source make them effective in providing individual beams of directed light, which are useful in applications with specific direction needs. They are particularly suited to more specialist aspects of lighting, such as architectural or accent lighting. A specific application is the use of LED lighting in lift cars, as described in Section 10.5. To meet the needs of general illumination of building spaces, a linear form of LEDs has been produced providing a linear source of light, or 'light engine'. This format has allowed the development of bespoke styles and shapes of luminaires with a broader light distribution (similar to those for a fluorescent tube/luminaire combination). These are more amenable for providing general illumination, such as for offices, rather than through using much higher numbers of smaller point-source luminaires.

The emergence of white LEDs has stimulated the development of products providing either direct replacement of conventional lamps or new types of luminaires designed to uti-lise the particular features of LED light sources. Because of the high lighting load densities of display lighting, a particularly useful development is the packaged LEDs that contain all the components that would normally be within the lamp/luminaire combination, including heat management for the LEDs. In display lighting installations where these integrated sources can be used as direct replacements for conventional display lamps and where the quality of their lighting output is considered satisfactory, they can provide both significant reductions in energy consumption and much longer lamp life expectancy (Carbon Trust undated a).

LED light sources have many features that can contribute to sustainable whole-life performance. They perform well at low temperatures, so they have potential applications in a variety of spaces and locations where other light sources are not appropriate. Where spaces have mechanical cooling, LEDs can help to achieve a good overall energy performance, as they reduce the unwanted heat gain from the lighting system, and hence reduce the cooling load. A notable whole-life feature of many LEDs is their longevity. Typical life expectancies for LEDs can be in the order of 50,000 hours, but this is dependent on the application and the thermal management being effective. This means they can be a good solution in parts of buildings where light sources would be in awkward locations and would pose difficulties for replacement. This could help to meet health and safety (H&S) requirements under the construction and design (CDM) regulations. A further sustainable feature is their ability to undergo regular switching with no significant reduction in performance or life expectancy (Carbon Trust undated a).

Alongside the significant benefits of LED lighting described earlier, some particular factors need to be recognised and addressed to obtain effective energy performance and satisfying lighting outcomes. These are outlined briefly next.

As with all light sources, individual response to the quality of light can be highly subjective, but some people tend to find LED light less satisfying than other light sources. The quality of light within an installation can vary, as can the consistency of colour temperature and colour rendering. This can make it difficult to achieve appealing visual environments in certain applications.

It is essential to maintain good heat management for LEDs. This is a key issue that is related to the junction temperature on the LED chip and can affect colour temperature. If the heat management deteriorates with age, this can cause a shift in colour temperature, which can be larger than that for fluorescent tubes. This can be a problem in fittings with many LEDs, as the temperature is cumulative.

The MF needs to be good for the life expectancy of the product, and to obtain effective whole-life performance, it should not be below 0.7. For LED products, the MF is described as a combination of life (i.e. lumen depreciation) (L) and failure rate within a fitting (B). For good performance, it is useful to aim at a minimum standard, typically L70/B10, based on 50,000 hours life expectancy.

Numerous issues with glare may need to be addressed to a greater or lesser extent, depending on the location, application and likely user response.

To provide effective control, all LED drivers should be dimmable. Drivers can be pre-programmed off-site to provide constant illuminance control. This will dim the LEDs to maintain constant illuminance. However, for the luminaire to achieve the same illuminance at the end of its life as at the beginning, the power consumption is likely to increase.

Most LED drivers operate at frequencies that are significantly lower than for hf fluorescent lamp drivers. This can give rise to 'flicker', to which some people are susceptible, and a much smaller proportion can be highly susceptible.

While LEDs have very good performance at low temperatures, the performance can deteriorate at high temperatures, making them unsuitable for certain locations and applications.

10.4.5 *Lighting controls and management*

The basic requirement is for the control system operation to provide acceptable lighting levels for occupants throughout the period of usage and hence avoid wasting energy. Properly designed lighting control systems can provide a much wider benefit by allowing the potential of the lighting system to be fully realised. This can address such aspects as the visual comfort of the occupants and the satisfying appearances of spaces. Control systems can also be used to assist in the maintenance scheduling and life cycle management of the installation. The flexibility of the controls should match the anticipated usage pattern of the building. In many commercial premises, a considerable amount of energy is wasted due to the continued usage of the lighting system outside of the main operational periods. There is also, of course, a practical need for partial usage for a proportion of the non-operational time, for example, for cleaning purposes and to a lesser extent for security. However, due to poor controls and management, energy usage is often well above the level required for these purposes (CIBSE 2002). From an energy perspective, the need can be satisfied with suitable controls that provide the required flexibility so that individual luminaires, or groups of luminaires, can be dimmed or switched off when their light contribution is not required. Such controls should always have override facilities to allow usage for security or emergency purposes (CIBSE 2002). It is likely that the controls will be one or more of the types listed next, and could be a combination of these, although some of these control types and features may not be applicable to certain light sources.

1   Local switching, through distributed groups of multi-gang switches located to control distinct zones or discrete spaces. This might be appropriate in some locations, provided it has a logical division and labelling. However, it has limited benefit, as lighting does not switch off when people are absent. Under Part L2A, local manual switching has a standard definition and limitation.

2   Time switching, either based on fixed time schedules by room or zone, or pre-set values related to daily, weekly or annual usage patterns.

3   Automatic switching planned to respond to the occupancy patterns in spaces, used in conjunction with passive infrared (PIR) movement detectors, which sense the presence or absence of people. Absence detection is preferred, in which the luminaires are switched off when the space becomes unoccupied and switching on is manual. Presence detection switches the luminaires on when someone enters the space and switches them off when the space becomes unoccupied. Such control systems should include time delays to limit the switching operations. For certain lamp types, this will avoid the reduction in lamp life expectancy that can be a result of excessive switching. It will also avoid the nuisance to occupants of premature switching off. For certain types of lamps, the time delay features should relate to the switching characteristics of the lamp for the run-up to full output and the re-strike time. This will vary, depending on the characteristics of the lamp type.

4   Automatic switching with daylight linking used in conjunction with light sensors, also known as photoelectric cells (PECs) or photocells. This can be used for areas close to window walls, with good daylight penetration, preferably to provide dimming, but it could just be for switching. Dimming not only reduces energy consumption but also allows flexibility for change of usage (although in some cases this may detract from the scope for energy efficiency). Suitably designed dimming control can achieve a number of objectives. It can continuously match the illuminance provided to the needs

at any point in time, which not only reduces unnecessary energy usage but also helps to maintain visual comfort. Moreover, it does this through continual and user-friendly seamless change rather than less satisfying switching on and off.

To provide the best scope for energy reduction, it is preferable to locate light sensors in zones corresponding to the areas of best daylight distribution.

It should be noted that the luminous efficacy of the lamp circuit would normally reduce as the lamp is dimmed, so the relation between reduction in illuminance and reduction in energy is not necessarily linear.

For many buildings, a mix of manual and automatic controls should be utilised to suit particular spaces. The proposals should be discussed in detail with users' groups so that the arrangement selected matches their needs. In some cases, a balanced choice will be required between traditional hard-wired switching, centralised control or localised stand-alone controls. As with most controls, it is often better to select for simplicity rather than technical sophistication. The unfortunate experience from many completed projects is that controls that are too complicated for the users to understand are simply not used, and this rejection of controls can happen quite early in the occupancy period. This is obviously a wasted opportunity to both minimise energy usage and provide the intended lighting performance and can have serious consequences for meeting regulatory criteria in practice and achieving whole-life targets.

Automatic controls, as described in items 3 and 4 earlier, are most effective if they allow for a degree of individual occupant control, but this should not conflict with any central control features. Automatic systems of this type have the flexibility to allow for changes in the layout or usage of spaces. They can also have benefits in maintenance planning through monitoring of the system (CIBSE 2004a).

There is an anomaly in lighting performance versus design requirement that can be resolved through dimming controls (CIBSE 2002). This relates to the need to design to achieve maintained illuminance, which results in the initial illuminance level being much higher than the design illuminance, as outlined earlier. Modern fluorescent luminaires with dimmable hf control gear can reduce the lighting level to correspond to the design-maintained illuminance level. As the lamps age, their lumen output reduces. As a consequence, the level of dimming reduces with ageing and the power consumption increases (CIBSE 2002). At the time related to the requirement for maintenance attention, the power consumption will be at the maximum level, as designed to overcome the MF reduction (CIBSE 2002). One effect of this would be that the range of dimming achievable to respond to daylight levels would be reduced by its usage to offset the initial over-illuminance. Conversely, as the lamp ages, the effective range of dimming achievable will be higher. The accumulated energy consumption can be reduced by increasing the frequency of maintenance attention, which should be addressed in the operational regimen (CIBSE 2002). Control arrangements that under-run the lighting level initially and then, as the lamps age, increase the level until maintenance is required come under the 'constant illuminance factor' control method in Building Regulations Part L2A.

Specific technical aspects of lighting control are described in detail in guidance publications such as the Society of Light and Lighting (SLL) Lighting Guide LG14: Control of Electric Lighting (CIBSE 2016).

Examples of two very different solutions to achieving successfully lit and energy-efficient interior spaces are in Figures 10.5 and 10.6. Figure 10.5 shows a solution to office lighting where a variety of light fittings/systems have been used to provide a balance between

Figure 10.5 Example of a successfully lit and energy-efficient office space
Source: Courtesy of Hoare Lea (credit: Don Paton)

Figure 10.6 Example of a successfully lit and energy-efficient foyer area
Source: Courtesy of Hoare Lea (credit: Don Paton)

energy efficacy, natural light and user comfort. The installation uses a range of product types, including direct/indirect pendants over the office desks and downlights over the thoroughfare corridor. These are all on separate lighting controls so that, as the daylight entering the space varies over the course of the day, artificial light fittings nearer the windows are dimmed down, while those farther from the windows remain at a higher output. This is good practice in user-centred lighting design, as occupants need to feel the spaces they work in are 'well-lit', and sometimes, if this 'variance' is not considered, high contrast ratios across a floor plate can make some areas appear 'well-lit' while others can appear gloomy. Figure 10.6 shows an office reception area where the availability of versatile and compact energy-efficient light sources (most notably LEDs) has made it possible to create a more 'integrated' lit impression. Light fittings with long lamp life and low power consumption enhance the architectural volume of the space. This is achieved by adding light as part of the architectural fabric, rather than having 'visible' fittings which might have been a distraction.

10.4.6   *Legislation for lighting energy efficiency*

Because lighting systems can be one of the biggest energy-using (and carbon-emitting) systems in buildings, many countries and jurisdictions are likely to have legislation in place to minimise impacts as part of their regulations for buildings. In England and Wales, there is guidance for interior lighting for new and existing non-domestic buildings to help comply with the energy efficiency requirements of the Building Regulations Part L2A and L2B (HMG 2013). This sets out recommended minimum energy efficiency standards, including the use of controls. For new buildings, the standards are design limits, whereas for existing buildings, the standards represent a reasonable provision for compliance. However, in most cases it is necessary to aim for higher standards, as these might be necessary to achieve compliance, depending on the carbon emissions of the overall design. Selected aspects are described next, but reference should be made to the guidance document (HMG 2013) for the full criteria, definitions of terms, reference tables and ways to seek compliance with the regulations.

For lighting in new and existing buildings, there are two methods for meeting the standards, based on either a) the efficacy and controls or b) the Lighting Energy Numeric Indicator (LENI). In addition, the lighting should be metered and the controls should follow the guidance in BRE Digest 498 *Selecting lighting controls*. There are further specific requirements for display lighting. The LENI is a measure of the performance of lighting in terms of energy per square metre per year (kWh per sq.m per year).

To meet the standards based on efficacy and controls, the general lighting in offices, industrial and storage spaces should have a minimum lighting efficacy for the relevant controls arrangement. This is averaged over the whole area of the applicable type of space in the building. The efficacies are expressed as luminaire lumens per circuit-watt. The initial luminaire lumens per circuit-watt criteria is 60, but this is reduced for the particular controls arrangement. So, for example, a daylit space with photoswitching with or without override has a control factor of 0.9 applied, so the minimum lighting efficacy to be achieved is reduced to 54 luminaire lumens per circuit-watt. There are also combinations of control arrangements. For example, a space that is not daylit, has dimming for constant illuminance, but is unoccupied with auto on and off control has a control factor of 0.8 applied, so the reduced efficacy to be achieved is 48 luminaire lumens per circuit-watt. For display lighting, the efficacy is expressed as lamp lumens per circuit-watt. The efficacy standard relates to the total figure as averaged over the whole area of the relevant types of spaces in the building. This allows design flexibility to vary the LOR of the luminaires and the luminous efficacy of the lamps, and use different lamps and luminaires to meet the needs in different spaces.

The LENI is the total of the sum of the daytime, night-time and parasitic energy use divided by the area and is expressed in kWh per square metre per year. To meet the standards based on LENI, this should not exceed a specified maximum lighting energy limit for a given illuminance and hours run. The maximum LENI is shown in the guidance document for the illuminance required and the relevant day and night hours. So, as an example, from the guidance document reference table, a room with an illuminance of 500 lux, used for 4,400 hours (3,621 hours during the day and 779 hours during the night) a year would have a lighting energy limit of 37 kWh per square metre per year.

The actual LENI for an area A can be calculated (HMG 2013):

$$\text{LENI} = (Ep + Ed + En)/A \tag{10.3}$$

Where:
Ep is the parasitic energy use
Ed is the daytime energy use
En is the night-time energy use

The data for Ep, equations for Ed and En and definitions of the terms are given in the guidance document.

10.4.7   External lighting

External lighting is required on and around buildings and their associated grounds for a mixture of purposes, including access, safety, security and amenity. The extent of the external lighting provision will depend upon the types and numbers of buildings and their layouts on a site, as well as the facilities to be provided for access, roads, car parking, pedestrian and cycle routes and amenity areas. As with internal lighting, the starting point should be to properly explore the brief and determine the specific functional requirements during times when daylight is insufficient. External lighting has the potential to provide wide-ranging benefits beyond the obvious functional needs of safety and security. It is therefore worthwhile undertaking a full engagement and consultation, involving all parties, to help establish appropriate design objectives at the concept stage. This will be essential, as a merely functional brief could yield a significantly different solution from a brief seeking social enlivening or a particular aesthetic treatment of an external space. External lighting can enhance the experience of the architecture and places more generally and create safe, accessible and satisfying community areas that allow civic and social interaction to continue for extended hours.

As part of the brief development the design proposals must be suitable for the specific location, including any distinguishing features of the urban or rural environment in the vicinity. Particular influencing factors could be the architectural character, history and style and whether the site is in a conservation area. If the site has listed buildings or conservation status, there might be specific criteria imposed by the planning authorities to achieve the relevant permissions. There could be specific considerations, including the extent to which the proposed lighting might infringe upon authority spaces, rural environments, waterways and the environments of protected species (Carbon Trust undated b).

The brief development should lead to a scoping of the extent, function and aesthetic outcome of external lighting on, around and between buildings. The scoping could include these types of spaces and features:

- Internal roads within a site
- Car parking areas and other hard surfaces
- Footpaths, passages, alleyways and other pedestrian routes
- Squares, piazzas, gardens and other amenity spaces
- Cycle routes
- Covered walkways, pedestrian bridges and subways
- Architectural features, artworks, murals and water features.

Because of the potential energy impact of long hours of operation, it is important to critically question whether external lighting is required throughout the night. If extensive operation is considered necessary, it is worthwhile discussing the needs with the client to explore the possibility of providing two or more levels or stages at different times, so that energy consumption levels can be reduced for a part of the time. It is also worth exploring whether lighting could be localised in certain areas, rather than providing a continuous widespread coverage. This is a more strategic aspect of the brief.

As with internal lighting, it is important to select efficacious lamp and luminaire combinations with appropriate colour rendering for all applications. There is a trend towards using 'white' light sources rather than potentially higher luminous efficacy discharge lamps which have poorer colour performance. This could allow a reduction in the design illuminance due to a perceived achievement of safety and security at a lower illuminance with good colour rendering, compared with poorer colour rendering, as well as providing a more satisfying appearance. In situations where closed-circuit television (CCTV) is in use, external lighting using 'white' sources is likely to be more effective, so this is a worthwhile discussion point with those planning the security strategy.

As outlined in Section 10.4.4, LED light sources have evolved to become a practical solution for many energy-efficient internal lighting applications. These sources also have several technical features that make them highly suitable for a wide range of external lighting applications. Their small size means they can be used in bespoke luminaire shapes and forms that would not be practical for other lamp types, allowing the development of new types of luminaires. They can continue working effectively in low temperatures. In addition, LEDs provide the full level of light output when switched on, whereas certain other lamp times take a defined period to reach their declared light output. This allows LEDs to be used with time switches to give the intended illumination immediately and to provide an effective solution with movement detection where a delay in illumination of the desired level would not be acceptable. The compact size of LED sources allows usage for architectural feature lighting in locations where other technologies, with larger luminaires, would be considered to be intrusive in appearance. The colour rendering of some LEDs used in exterior lighting products may not be as good as that for interior lighting products.

In the UK, many authorities have undertaken widescale replacement programmes for their streetlighting installations, replacing lighting schemes using discharge lamps with new installations using LED light sources. For various technical reasons, it is usually necessary to replace the whole column and lantern with new column and lantern types specific to LEDs, rather than retrofit LED light sources to existing installations. A case study of Central Bedfordshire Council's LED luminaire replacement of traditional street lighting showed a reduction in annual electricity consumption and annual CO_2 emission reductions of more than 50%, along with reduced light trespass (Carbon Trust undated b). However, it should be noted that the light distribution pattern and perception may be diminished or considered less satisfactory to some familiar with the spread of luminaire coverage provided by previous installations.

Particular consideration should be given to luminaire mounting heights in relation to the architectural and planning objectives, as well as the maintenance aspects. While mounting luminaires at increased heights can provide broader coverage, it also reduces the illuminance level and so may not result in an overall reduction of installed power. Figure 10.7 shows a successfully lit and energy-efficient exterior space utilising a mix of column-mounted luminaires and recessed luminaires.

A key factor in the selection of luminaires and their locations is to avoid 'light pollution'. One aspect is 'encroachment' or 'trespass', which is light entering other properties with an

Figure 10.7 Example of a successfully lit and energy-efficient external space: Highbury Square, London
Source: Courtesy of Hoare Lea

unwanted impact. This is both intrusive and a potential subject of conflict with neighbours, as well as a waste of energy. A different aspect of waste is 'sky glow', which is light directed upwards into the sky rather than on the surfaces that require to be lit, and is also wasteful. Another aspect to be avoided is glare. The locations of luminaires and the directions of their light distribution should be carefully selected to avoid or minimise any form of light pollution or glare, and hence avoid the associated energy wastage. There is increasing societal pressure to address light pollution and encourage opportunities to appreciate the dark night sky.

The light output characteristics of lanterns should direct the light downwards, both to minimise upward light spill and to achieve good downward illuminance. Luminaire shapes will depend on the lamp sources being used, but can be most efficient with a flat or slightly curved lower surface.

Time controls should be an essential feature of any external lighting system and should be set to restrict usage to only those times when security, access or amenity usage is required or anticipated. The hours of usage of floodlighting should be sensible, recognising the diminishing benefits of usage for extended hours when the number of people who can appreciate it is reduced. In simple terms, the hours of usage should be limited to meeting the civic objectives. However, usage needs change with time and client policies. All controls should therefore be flexible and easily adaptable so that they can be adjusted if the planned usage changes in the future. It could be worth exploring ways in which a mixture of time settings and staged reductions in coverage would be appropriate. For certain types of areas, carefully selected combinations of time controls and movement detection could provide adequate coverage to satisfy safety and access needs.

In large open spaces where appearance is less important and there is no public amenity purpose, high-pressure sodium (SON) lamps could provide adequate colour performance and be more energy and cost-effective than metal halide or LED sources (Carbon Trust undated c). The lighting of industrial compounds, marshalling yards and outdoor sports facilities often have highly specific design criteria and standards, so reference should be made to the relevant guides and publications, but many of the principles outlined earlier are likely to be relevant.

10.5 Lifts

The planning of a lift installation forms an integral part of the building design process and has a fundamental impact on the design development of the space planning, and hence the architectural and structural engineering solutions. The planning of the numbers, capacities and characteristics of passenger lifts is an outcome of a detailed analysis of passenger traffic performance to satisfy the design criteria. With proper selection of the lift passenger installation parameters, the lifts should provide an efficient solution to the building's passenger traffic needs in terms of capital cost, space requirements (for lift shafts, lobbies and machine room space, if required), energy and running costs (CIBSE 2004a). The planning of lift shafts and lobbies (and lift motor rooms, where required) is a fundamental consideration when designing the building's cores, as outlined in Chapter 13. For many offices and other buildings up to medium height, 'machine room–less' lifts can provide a space-efficient solution, as the machine is located in the shaft, so they do not require a separate machine room. In some cases this could result in lower embodied energy when compared with the lift shafts and machine room for a traditional lift arrangement.

The energy consumption of lifts will vary considerably between different building types and patterns of vertical passenger movement and will be most significant in tall buildings (Russett 2010). The energy consumption of lifts and escalators in some buildings can be

5–15% of total energy costs (CIBSE 2004a). However, the energy efficiency of lift installations has received little attention and has not featured directly in the UK regulatory framework. Lift installations can incorporate simple features that reduce energy consumption while having negligible effect on traffic performance.

Lifts use energy for these items:

- lighting in the lift cars (the single largest energy consumer)
- drive machinery
- control systems
- ancillaries.

The most significant energy efficiency improvement is likely to be the use of LED lighting in the lift car, lift shafts and other areas associated with the lift installation and for lighting to be controlled in addressing the operational state, as outlined later.

The most efficient drives are electric traction. For new lift installations, the motor drives should be of the variable voltage, variable frequency (VVVF) type and can be regenerative. Regenerative drives have high levels of energy efficiency and recover energy from the out-of-balance condition by regenerating power back into the mains electricity supply. They should be used when they are available for the lift drives being used (BCO 2019). It should be recognised that existing older lift installations could have other types of motor drives that may be much less energy-efficient.

The lifts in most buildings are mainly used for three relatively short periods during the day. However, in most multi-lift installations, all of the lifts normally remain in a continual operating mode throughout each 24-hour period, resulting in a considerable wastage of energy. The key to reducing energy consumption in a lift installation is therefore to reduce the operational state of lift cars to a level consistent with the variable traffic level, as outlined by Barney (2006). This is a form of demand matching and covers two separate aspects:

1   *Reducing the number of lifts in service outside of the peak operational period.* Outside of the peaks, lift traffic is primarily inter-floor. Barney (2006) has suggested that the number of lifts in service could possibly be reduced and still provide a satisfactory level of traffic performance in these periods of low traffic demand. The UK British Council for Offices (BCO) specification (BCO 2019) proposes that during off-peak periods, lifts should operate in 'standby' mode. The specification requires that when lifts are in this mode, they should be idle, with the lighting, ventilation and display equipment switched off and the following items powered down: drive controls, traffic control system and door operator equipment. This would reduce energy consumption from lift controllers, lighting and ventilation and from the motor drives due to overall reductions in motor starts and motor running time for the multi-lift group. Because individual car loading is likely to increase, lifts will operate closer to their optimum balanced operating condition, and therefore in a more energy-efficient mode. At a minimum, one lift would be left in operational mode. Multiple lift installations therefore require suitable shutdown control protocols (Russett 2010) for programming lift management. The BCO specification (BCO 2019) states that the operational arrangement should be configured to provide a phased sequence for lifts to enter and exit standby mode. This can provide the appropriate number of lifts in operational mode to satisfy the demand pattern. The specification also recommends that after an idle period of 30 minutes, the energy consumption for a lift on standby should be less than 50 W.

2   *De-activating energy-consuming equipment in lifts that are idle but still in service.* Barney (2006) has suggested that individual lifts in a multi-lift installation can be idle for considerable periods, possibly up to 60% of the time over a whole year. This provides an opportunity to switch off the car lighting and auxiliaries in these lifts after a predetermined period of idleness (say five minutes). Lift controllers consume energy in standby mode when a lift is idle. There is an opportunity to switch off the power side of controllers (isolation transformers and pre-energising motor winding power) while the electronic systems for control operation would remain in standby mode. For safety purposes, switching and timing arrangements would obviously have to be such that lighting would be operational prior to a passenger entering the car, as would any other safety-related features (Barney 2006).

The lift installation should achieve the relevant energy efficiency target of the environmental assessment methodology being used (e.g. Building Research Establishment Environmental Assessment Method [BREEAM] in the UK), and the energy consumption should be included at the design stage when estimating the overall operational energy use (BCO 2019).

It may be worthwhile considering transferring one or more escalators in a group to an 'out of service' mode during times when traffic demand is low, but this is always subject to the overall passenger handling criteria, scenarios for maintaining traffic flow and the prevailing safety considerations.

10.6   EC/DC fan coil units

The factors that should be considered in the selection of an indoor climate control system are described in Chapter 7. While it may not be the preferred system in terms of energy performance, there is often a strong economic case to use a traditional fan coil unit (FCU) solution, particularly in commercial offices. Although FCUs may not be the favoured choice for energy efficiency in most circumstances, they can be made much more efficient by using fan motors that are electronically commutated (EC) instead of conventional single-phase motors. EC motors are also known as direct current (DC). In an EC/DC motor, power electronic devices create a rotating magnetic field in the stator winding. The motor contains a permanent magnet. For FCUs using this drive technology, carbon emissions can be reduced in the following ways (Blackwell 2010):

- they are more efficient than an equivalent single-phase alternating current (AC) motor; typically up to 90% efficiency, compared to in the order of 50%
- variable-speed operation can be achieved without any reduction in efficiency, allowing matching of the motor speed to the required air flow rate at the commissioning stage
- the fan speed can be varied in response to different load conditions, using suitable controls based on demand to provide variable-volume operation
- they do not incur the problem associated with heat transfer to the air from a less efficient motor as with a conventional FCU, which avoids the increase in load on the cooling coil, and hence reduces the cooling energy requirement.

Therefore, EC/DC FCUs should be the preferred choice when a decision has been made to use them in areas with high cooling loads (Blackwell 2010), subject to the electrical design considerations described next.

10.6.1   Electrical design considerations

There are two electrical aspects of EC/DC FCUs that require careful consideration in the planning of the electrical distribution system. The first is their poor power factor, which is typically about 0.6 (or above) for larger units, but decreases considerably under part-load conditions to 0.3–0.4 (Blackwell 2010). The second is the generation of harmonics from the power electronic devices. Such systems may require suitable PFC and/or harmonic filtering at relevant locations in the electrical distribution system. Where widespread use of an EC fan coil system is proposed for a refurbishment project, the impact on the existing electrical distribution system could be considerable. This might necessitate some upgrading of the distribution system, which could be a determining factor in the viability of this technology in some refurbishment projects.

10.7   Key operational decisions for electricity usage

A key operational consideration is that the demand for cooling should be minimised by ensuring that there are no unnecessary heat gains to cooled spaces from electrical equipment. It is useful to consider the compounded energy impact of unnecessary electric heat gains to a space that requires cooling to maintain its temperature criteria. Carbon emissions will arise from:

- electrical energy for the unnecessary electrical load
- electrical energy for the chiller to remove this heat gain
- electrical energy for the associated fans and/or pumps for heat rejection.

Consequently, it is essential to exercise rigorous demand management to avoid any unnecessary usage of electricity in spaces requiring cooling. This is particularly important where the electricity has a high carbon factor.

10.8   Unregulated loads: small power equipment

The types of unregulated loads and their relevance to overall energy consumption have been described in Chapter 3. The use of personal computers (PCs), monitors, laptops and tablets in particular has a significant energy impact. There are various power-limiting features and settings, power management equipment and controls for PCs and other equipment to reduce energy consumption. Such features should be discussed with clients at the initial stage, alongside behavioural aspects, to address energy usage from process loads as part of the overall energy management strategy. Similar considerations apply to all other electrical and electronic process equipment. Such equipment often has a load that is non-linear, so it gives rise to harmonics, which can incur additional energy consumption, so policies to limit usage can play a useful part in carbon mitigation.

There has been much focus on the wasted energy associated with electrical and electronic equipment that is not in active operation but is in 'standby' mode. This relates to both business equipment and to consumer electronics in residential buildings. In some cases (particularly for older equipment), this can consume power at a rate that is a considerable proportion of the operating value. This represents avoidable energy consumption, which will increase if, in turn, this leads to an additional cooling energy requirement.

There are EU Ecodesign requirements for standby, off-mode power consumption of household and office equipment under Regulation 801/2013.

10.9   Process loads supported by UPS systems

Data processing and other loads supported by uninterruptible power supply (UPS) systems can represent a significant proportion of the energy consumption in certain types of buildings. Chapter 12 provides a brief outline of these systems in the context of load assessment. In order to provide the required level of resilience, these systems often comprise multiple parallel UPS modules of equal size sharing the load. Individual UPS module efficiencies could be high (perhaps 96%) at full load, but they will normally reduce as the load reduces. For UPS systems with high levels of resilience, individual modules will operate at a lower load level and the overall group of modules will be less efficient, with heat loss to the space. There will be further energy usage for the cooling system to control the space temperature. This will result in additional energy consumption and carbon emissions. While a client's primary consideration for their UPS-supported load is likely to be the resilience, there is an increasing focus on the energy efficiency of such systems. The selection of the sizes and numbers of parallel modules sharing load to meet resilience criteria should therefore also consider the resultant overall energy efficiency and carbon impact of the UPS system and associated cooling.

10.10   Enabling energy management through controls, monitoring and data collection

The need for rigorous demand management has been emphasised in Chapter 3. For all the active energy-consuming systems described in this chapter and in Chapters 6–9, a key design feature should be the inclusion of facilities to enable effective operation and energy management throughout the life of the building. This will provide the most appropriate automatic and/or manual control. It will also provide the operational engineering team with meaningful data on the performance so that they can understand where energy is being consumed and thereby make suitable interventions to reduce consumption, as outlined in Chapter 3. It is likely that, in the future, artificial intelligence (AI) algorithms will be used to respond to data collected and maintain efficient operation.

These facilities are not active energy-using systems delivering a functional requirement, but the means through which the active systems can perform in an optimal way. So the starting point for developing the physical system requirements for the data collection, controls and monitoring should be the energy management strategy and the proposed operational regime. This should be an early design consideration, to follow the energy strategy report, and should include a strategy for data collection. While the controls and metering facilities are covered separately in Sections 10.11 and 10.12, it is best to think of them together as an integrated system.

10.11   Controls and building management systems (BMS)

When considering the requirements for the controls, it is perhaps tempting to commence with an engineering systems approach. This would typically cover aspects such as controllers (sometimes known as outstations), communications networks and interfaces, user interface (sometimes known as supervisor) and so on. Similarly, considerations for the controls systems objectives will often focus on the control of the comfort parameters and functional operation within the occupied spaces. These engineering aspects are, of course, essential and will need full consideration within the eventual controls provision.

From a perspective of focusing on energy performance optimisation, however, other factors need to be addressed within the wider initial objectives. It is necessary to stand back from the engineering perspective and consider what contribution the overall controls provision could provide, both in the shorter and longer terms. In this sense the controls requirement is integrally related to the energy management strategy and the wider operational strategy for the building, as shown in Figure 3.8. It is best to think about what the facilities could provide above and beyond mere control of the plant and equipment. The automatic and manual controls will be both the principal human–machine interfaces (HMIs) of the active (and, to a limited sense, passive) systems and, in a Building Energy Management System(BEMS), an information and data resource that is an essential requirement for allowing operators to monitor and improve performance.

In essence, the control facilities provide a mixture of automatic/digital and manual control actions and data allowing operator intervention that can assist in optimising the energy performance. It will include a need to allow the moderate level of local manual intervention from occupants that is necessary to help make adjustments to suit localised conditions and meet their psychological need to have some personal measure of control. This would be for selective aspects of systems, but can provide a beneficial sense of wellbeing (Figure 10.8). However, the most effective energy management is likely to be through the primary energy management controls being automated. The automatic controls for HVAC systems can have a major beneficial impact in reducing energy consumption, provided they relate to practical spatial zone allocations and have appropriate levels of dynamic responsiveness. The emphasis is always on honing controls so that they can respond at a local level to match the varying needs at any point in time. This will then allow the central plant to adjust accordingly. Controls should be amenable to adjustment to suit the likely flexibility of space usage during operation.

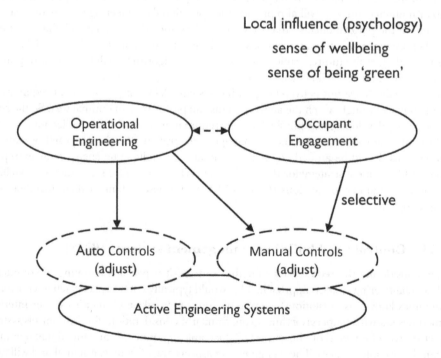

Figure 10.8 Concept for operator intervention

There are obviously benefits from controls facilities that provide enhanced capability and flexibility, particularly in relation to lifetime adaptability and 'future-proofing'. However, it should always be recognised that greater so-called 'intelligence' and technological sophistication in the systems does not, by itself, contribute to energy-efficient performance. It could, in some cases, be detrimental if the perceived complexity and/or the lack of user-friendliness prevent operators from intervening to the extent required. The key is always to select an appropriate and suitable controls capability through considered planning. There is no reason for controls to be any more complicated than required. Instead, the controls system should be of the simplest form and range of flexibility that can deliver the required level of operation, in an intuitive way, through the anticipated usage pattern of the building (CIBSE 2004a; Horsley 2010). There are ongoing developments in user interfaces, with touch-screen graphics becoming easier to use and helping to make systems easier to understand and operate.

The key functional requirements for the control systems are:

1   The primary requirement is to regulate the various conditions in accordance with the design intention and maintain the condition of the controlled parameters. This is achieved by measuring and adjusting system variables, which will then alter the system output through control loops incorporating the appropriate feedback arrangement. The mode of control will depend on the system dynamics, as briefly outlined later.
2   Data collection from metering and other means to provide information on patterns of energy usage with the facility to create a variety of outputs to represent meaningful aspects of energy performance. Data collection should include metering of electricity, water and gas, together with heating and cooling (and air flow, where worthwhile).
3   A range of monitoring, alarm and control facilities specifically to restrict operation to within suitable parameter boundaries to maintain good energy performance, such as automated data collection and alarms.
4   A range of facilities to assist in the maintenance of active engineering systems. This is to ensure that suitable maintenance regimens can be instigated, such as planned preventative maintenance (PPM), requiring appropriate predictions and planning schedules.

These items can overlap and are integrally related to the data collection and metering provision.

For item 1, it should be understood that the HVAC systems involved have a dynamic response, with inherent time lags. Achieving and maintaining the desired condition will require feedback and a control loop with parameters to suit the system dynamics and user comfort needs. The nature of control provided for such closed-loop systems is that a controller seeks to control a variable, the value of which is continually measured by a sensor. Control loops with feedback can utilise different control modes, depending on their dynamic response and the level of accuracy required for the controlled parameter. The most widely used control mode for HVAC systems is proportional plus integral (PI) control. This mode can provide a stable level of control of desired parameters, without any error or offset, for many of the system types used. All such systems have time delays or time lags, so that responses to a drift of the measured parameter will take a defined period of time. There are two types of delays in a feedback control loop. A distance-velocity (or transport) lag relates to the distance between the controlled entity or medium and the location of the controlled space. A transfer lag relates to the time for a change in the controlled space due to the inertia of the various components (such as thermal capacities). The effect of lags in the loop is that there is 'dead time' that leads to poor control. PI control with suitable settings can reduce the error. (CIBSE 2009c)

To achieve the desired functionality will require suitable settings and adjustments, including timed switching profiles and schedules. Operational staff should be able to (and be encouraged to) fine-tune settings (particularly time settings) to match active systems to the actual dynamic performance in practice. The same adjustments should apply to occupancy patterns in practice, so that there is continual matching. Together, these can reduce energy and carbon while maintaining comfort criteria as part of an active participatory plan, which can help during settling in and optimisation.

It should be understood that the building services designer is not responsible for the detailed software and engineering design for the BEMS/BMS and automatic control systems, control panels and controllers. Instead, the engineer creates the controls concept (i.e. description of operation) and provides an outline of the performance required. The appointed specialist contractor undertakes the detailed design, including the integration, interfaces and other compatibility aspects with the plant and equipment (Horsley 2010), and has a system integrator role.

The scope and extent of controls provision are important design decisions. It is useful to discuss this with the client as part of the design development process. It might be appropriate in some smaller buildings just to use direct digital control (DDC) controllers on a stand-alone basis. This would be sufficient to achieve the required controls function, but would not provide any information for management. The next level for extending the automatic controls capability would be to use a network of DDC controllers. This would allow sharing of data and, through a suitable user interface, would provide a facility for management intervention through monitoring and adjustments. The first level of integration would be to have a number of distributed DDC controllers linked together on a local area network (LAN) for sharing information (CIBSE 2005b). However, for most medium and large projects, it will be appropriate to utilise a BEMS. While a BEMS monitors and controls the active energy-using building services systems, it is likely that its functions will form an integral part of a full building management system (BMS). A BMS has a wider monitoring and control coverage that usually extends from the energy-related aspects to include non-energy engineering systems, such as fire alarm and detection systems and security systems. The terms 'BEMS' and 'BMS' are often used interchangeably. A BMS provides the functions of both control and monitoring for the engineering systems via a networked system, with one or more user interfaces provided by operator terminals. A BMS has a multitude of beneficial functions, but its core usage can be described simply as a system with inherent intelligence that can control and monitor the engineering systems. By doing this, it facilitates active energy management (CIBSE 2005b). It is also possible to have an energy monitoring system (EMS) alone, without any wider building management capability.

The main components of a basic BEMS network are shown in Figure 10.9. A BMS would usually have features such as PC-based graphical user interfaces connected to a communications network of DDC controllers. In a BMS, the linked-up DDC controllers will be connected to a central PC, usually known as a 'supervisor' or 'head-end', and will have suitable graphical capability for information displays and facilities for logging and recording. This will allow the operator to integrate the information; see records, fault reports and trends; and adjust settings to improve energy performance. The system can be considered to have four levels: management (supervisor), network, control and field.

The communications system will usually comprise one or more LANs and/or wide area networks (WANs). This will allow adjustments to the control strategy to be made from the user interface and data to be logged and stored, and therefore extend the flexibility and the potential range of interventions to improve the energy performance. A variety of network options exist, including structured cabling and wireless systems. Systems can be integrated

System Components

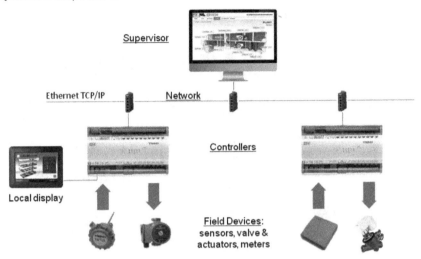

Figure 10.9 Components of a basic BEMS network

Source: By permission of Trend Controls Ltd

(or converged) with other data systems, including communications, fire alarm and detection and security (CIBSE 2005b) and can form part of wider 'smart building' features. An early aspect of design integration is often with the communications systems designer. It is also possible to provide integration with other buildings and sites as part of the energy monitoring for an estate portfolio. Figure 10.10 shows a typical full network for a BEMS.

The core components are:

Controllers: These are intelligent devices using DDCs. Controllers contain the embedded software programs and algorithms and the required numbers of analogue and digital inputs and outputs. DDC controllers allow a variety of control strategies and control modes to be pre-programmed into the software, which are suitable for the majority of situations. They can be programmed to undertake a variety of simple control functions, such as optimum start; temperature compensation; and one-, two- or three-term proportional, integral and derivative (PID) feedback control, together with a range of monitoring functions. The controller size will vary depending on the number and range of plant items to be controlled, and hence the number of digital and analogue inputs and outputs. Most motor control centres will contain one or more outstations providing control of fan or pump motor drives. See Figure 10.11. VSDs for motors usually have field bus connections that provide useful data about the motor's performance and can include alarm conditions. Control actions could include varying the flow rates via control of the motor VSDs in conjunction with relevant analogue inputs representing the controlled parameter.

Operational devices: Control is effected through operational devices such as actuators for motorised valves for heating and cooling systems and other field devices. There might also be a need for other auxiliary devices, such as relays, to achieve the necessary interfacing.

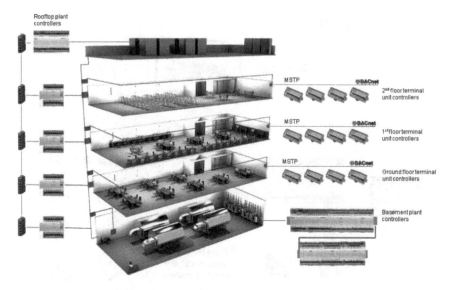

Figure 10.10 Typical full network for a BEMS

Source: By permission of Trend Controls Ltd

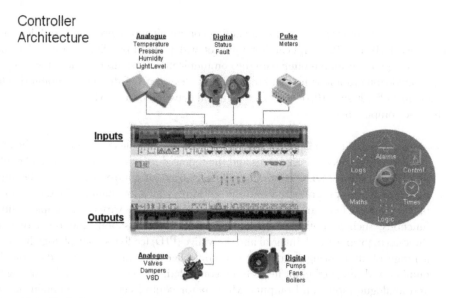

Figure 10.11 Typical controller architecture

Source: By permission of Trend Controls Ltd

It is normal for individual stand-alone items of equipment, such as boilers, chillers, generators, cooling towers, air handling units and combined heat and power (CHP) plants, to be packaged with their own integral control panels and cabling. This allows pre-site testing of the controls operation at the manufacturers' works. The BMS would therefore only have to interface with the local control panel. Network interfaces allow the integration of other plants or systems, which can be 'seen' as another controller (Horsley 2010).

Sensors: These are used to provide signals for parameters such as temperature and humidity. Sensors should be of high quality and allow regular calibration to reduce the drifting of set points. Control set points should be monitored on a regular basis. Sensor locations must be selected so that they give a truly representative indication of the controlled parameter.

The control strategy should define the specific performance of each controller, sensor, actuator and all other control system devices in relation to all conceivable operating conditions and scenarios. Control loops should be selected to suit the required dynamic operation and tolerance required for the set point of the controlled parameter. Settings should take beneficial account of thermal lags where this can reduce energy consumption (CIBSE 2004b, 2005b).

While a BEMS can provide facilities for a multitude of useful operational and management functions, the focus here is on the benefit that is provided for energy management to reduce fuel usage and carbon emissions. Consequently, as well as the BEMS capability to achieve and maintain the designed internal comfort conditions in all the spaces – itself a major contributor to reducing energy consumption – it can also provide direct energy management (Horsley 2010) through aspects such as:

*   identifying and reporting faults, which allows corrective action to be taken to return the plant to optimum operation
*   identifying unexpected conditions, including rises in energy usage in particular systems or at particular locations
*   verifying the performance of systems and allowing re-proving on a periodic basis
*   setting values for selected parameters, so that any parameter that goes out of range triggers an alarm.

Using information derived from the metering devices allows data logging by system, and overall combinations of systems, which could include:

*   summated consumption for each type of fuel
*   status, operational mode or cycle for plant and systems
*   time periods and total running hours for all main items of the plant
*   energy (kWh) delivered by all low and zero carbon (LZC) technologies
*   key parameters for LZC technologies, such as CHP running time, accumulated number of starts, power/heat ratios and key operating temperatures; heat pump operational parameters; peak power; and daily energy generated by wind power and photovoltaics
*   data required for compiling statutory certification
*   logs of all alarms and faults
*   trends, usage and optimisation.

The typical range of data outlined earlier can be analysed to provide an informed picture of the overall energy performance. The 'data analytics' should be specific for reporting and decision-making. For example, various types of data can be collected to support planned maintenance, and hence promote good whole-life performance for the active systems. These include accumulated plant running hours (which can be compared to typical periods for maintenance interventions) and monitoring of specific plant conditions (which can be related to trends for deterioration in performance). These data can assist in the production of planned maintenance schedules for individual items of the plant as part of the management regimen, and hence maintain energy-efficient system performance (CIBSE 2009c).

10.12   Metering and monitoring

10.12.1   *Usage for energy management*

The basic concept for using metering and monitoring for energy management is shown in Figure 10.12. The operational engineering team needs to engage with the occupants to jointly reduce energy consumption. However, at the outset of the project, they only have the theoretical knowledge of the building contained in the design information. Actual information on the true building performance comes from the data collected directly from the metering and the monitoring derived from metering, showing trends and patterns. This information provides operational engineering personnel with dynamic knowledge. If properly contextualised, this information can improve awareness and thus allow operational and occupancy sides to collaborate as a management regime, as outlined in Section 3.6. For nearly all buildings there will, of course, be a fiscal (i.e. used for billing) metering provision. To help effect the intended energy management strategy, the objective should be to consider suitable sub-metering or other portable data collection that can supplement the limited data from a building's fiscal meters with more meaningful consumption data (IET 2016).

A metering strategy should be formulated to provide the proposed range of data collection and should be included as part of the overall energy management strategy. It should be a consideration at an early stage in the design process and thereafter developed as an integral part of the mechanical, electrical and public health (MEP) systems design. An essential feature is that a metering strategy should be integrated with control systems planning. Metering provides essential information about the pattern of energy usage, without which management activities would be hampered by being based only on assumptions. It is

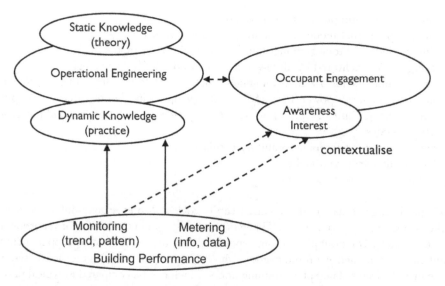

Figure 10.12 Metering and monitoring concept for awareness and engagement

therefore useful to allocate specific areas and responsibilities for different parts of the management structure (Carbon Trust 2012). The ways in which metering data can be utilised for energy management should be established through agreement as part of the liaison with operational engineering and should be set out in the building logbook. This information in itself will not save energy; however, suitable actions based on this information can form an important part of energy management strategies and can lead directly to considerable reductions in energy consumption. Metering is usually required to meet regulatory criteria and is specifically required to comply with Part L and the EU Energy Performance of Buildings Directive 2018 and Energy Efficiency Directive. At its simplest level, it can allow individual areas or end-use functions to be benchmarked against standard target consumption limits for comparison. By extending the information to convey an indication of the building's carbon impact and footprint, the metering strategy can provide a much wider benefit by stimulating occupant interest and involvement – and hence participation – in sustainability issues. In certain building types, it could also be used as an educational tool; for example, in schools, colleges and universities. A suitable metering and monitoring arrangement can help to create a 'smart' building and will be an essential feature to move towards net zero carbon buildings.

Case studies have shown that operational energy consumption can be reduced through behavioural change (CIBSE 2009b). The key to maintaining good energy performance is likely to be collection and logging of meter readings on an automatic basis. This needs to be in a suitable form for direct interpretation and can usually be provided via the BMS software (CIBSE 2004a). This would ideally be based on demand profiling derived from half-hourly data (CIBSE 2009b). The operational engineering regimen should provide regular reviews of the collected data and ongoing intervention strategies to reduce carbon emissions.

This section is about the types of metered quantities required to provide the monitored data to achieve effective energy management. It does not cover the specific technical details of secondary or sub-metering for electricity, gas, water and heat or the associated metering technologies, equipment and protocols. For these technical details, reference should be made to other documents, such as the IET Guide to Metering Systems (IET 2016) and relevant guidance on measuring and reporting metering (BBP undated).

10.12.2 *Metering coverage and selection*

For metered information to be meaningful, a primary requirement is that it should accurately represent the consumed energy (in kWh) of each identifiable service or function at a particular location within the building. As well as electricity metering, this could include metering for gas and water and heating and cooling systems. It is not necessarily the case that all services of interest need to be metered directly through fixed metering, as some energy consumption data could be obtained by calculations from other metered readings (CIBSE 2009b). Figure 10.13 shows a 'metering pyramid' with the variations in performance for different metering coverage. This indicates that while the preference is for direct metering to provide improved energy performance, a more economic approach might be necessary in certain cases, with a resultant reduction in reliability. A further requirement is that the measuring arrangements for multiple meters should be synchronised, with all readings being taken at the same points in time (CIBSE 2009b). This is so that meaningful comparisons and relationships can be deduced from the output information. In some cases, the fixed metering could be supplemented by portable meters where continuous readings are not so important.

The metering provision must, of course, comply with relevant regulatory criteria, standards and guidance. The selection of meters should address all necessary aspects of accuracy

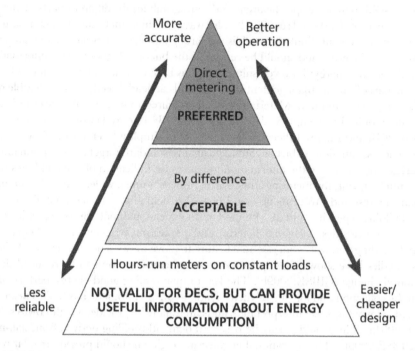

Figure 10.13 Metering pyramid

Source: Reproduced from CIBSE TM39 (2009b) with the permission of the Chartered Institution of Building Services Engineers

and linearity over the required range of measured quantity, so that metered data are reliable and truly representative. Metering equipment should be selected for compatibility with communications and other interfaces. As with any engineering system, metering systems should undergo planned pre-commissioning checks, tests and activities so that the system is operating correctly from the outset (IET 2016). This should include checks on the identification of channels from each meter, which should be uniquely described in the software, and checks that scaling factors for each meter have been set up correctly (CIBSE 2009c).

For most buildings, metering should be considered for the following functions, which include aspects described in Part L Guidance (HMG 2016):

1   In tenanted areas, it is likely that regulatory criteria will determine the requirements for sub-metering for billing purposes. In addition, for energy purposes metering should be provided for electricity for areas above 500 m², together with heating and cooling provisions where appropriate.
2   At least 90% of each service by end use, or so that at least 90% of the estimated annual energy consumption of each fuel can be assigned to the relevant end-use function.
3   Each major item of equipment, such as chillers or motor control centres.
4   The output from LZC technologies, including CHP. This is so that their individual contribution and performance can be monitored, which will aid the process of optimising performance. There will also be export metering to the grid.
5   District heating and cooling distribution.

6   Electric humidifiers, above 10 kW, because they can be high energy users.
7   Dedicated lighting distribution boards and lighting controllers.
8   Particular usages that might need to be subtracted so that consumption in other areas can be evaluated in relation to relevant benchmarks.
9   For areas above 1,000 m^2, automatic reading and data collection. In some cases this could be justified for smaller areas with high-energy usage density.

It should be noted that hours-run meters are only suitable for items where it is anticipated that the load will largely remain constant and the magnitude of the load is known.

It is also worth considering extending the coverage for particular buildings and systems, depending on the nature and magnitude of loads, and the anticipated extent of usage, including:

- data processing and IT systems, such as data centres and main and secondary equipment rooms, including loads supported by UPS systems
- small power systems, specifically in areas where particular attention is merited to monitor the pattern of usage to help control energy consumption
- specialist lighting where the function is beyond general illumination and is of relatively higher power density, such as in retail and display areas
- for lighting management systems, logging of the dimming levels and the running hours to allow calculation of consumed energy related to the installed load
- external lighting, particularly for car parks and floodlighting
- all catering installations, so it is worth considering sub-metering of separate sections of a catering facility or large items of equipment with high-energy usage
- other process loads related to the business function
- all significant load centres, switchgear and distribution boards
- other identifiable services or functions with the potential for high-energy usage without management intervention.

A judgement always needs to be made between the benefit and the economics of extending metering coverage. It is highly unlikely that metering will be of benefit in final circuits from distribution boards, where occasional use of portable meters is usually more appropriate (CIBSE 2009b).

10.12.3 Metering types

The type of metering should be selected to suit the most appropriate usage of the data, and this will often require a mixture of manual reading meters and automatic meters. Automatic meters can have three main types of output data: pulsed output, typically 0–10 V; analogue output, typically 4–20 mA; or a communications output that can be used with open protocol systems (CIBSE 2009b; Horsley 2010). The facilities should be planned so that output data are in a form that can be connected to a BMS or to another collection system, such as serial communications, Internet monitoring or wireless access (IET 2016). Where used in an energy management system, the hardware for metering will usually include a meter module and display module, together with a display logger. The logger communicates with the computer, which then analyses the data. It is often the case that the energy management function is an integral part of the BMS, where the meters provide pulsed outputs to the intelligent controllers (CIBSE 2009c). This information can then be used for the automatic monitoring and targeting of energy data. A useful method for encouraging positive occupant behaviours to reduce energy consumption is to link metered data to an 'energy dashboard', which is often a large TV screen in a prominent location (IET 2016). This can show a wide range of energy usage and related data that provide context, broken down for different floors, zones, sectors, tenancies, departments, buildings or areas as required to suit management responsibilities and accountabilities (IET 2016; Carbon Trust 2012).

10.12.4   Smart metering

A development in metering technology that has potential for energy and carbon reduction in the UK is the programme for the widescale introduction of 'smart metering' to consumers. Smart meters communicate data on energy consumption to the energy supplier for the purposes of monitoring and billing. They also provide an in-home display (IHD) for the consumer and can facilitate two-way communication between the meter and the energy supplier. However, there have been concerns about the extent to which it actually aids reducing energy consumption and also concerns about the security of information on consumers and their lifestyles that could be derived from load profiles and patterns.

It was thought that the deployment of smart meters would be an important step toward a more flexible and more efficient supply infrastructure for energy and assist the overall objective to introduce 'smart energy grids'. Numerous ways have been identified in which smart meters could help to reduce carbon emissions (IET 2009). These include the management of peak energy demand and management of imbalances in supply and demand in the short term. Smart meters should also allow energy systems to be operated more efficiently and more securely. For electricity, this should allow the security of the supply to be improved by facilitating adjustments to reserve and reactive power and frequency response. It should also assist the management of energy networks. By educating customers about energy usage, it should allow them to become engaged through seeing their direct interrelationship with the supply and demand of energy.

In the UK, the government proposed that every home be offered smart meters by 2020 on a voluntary basis, and a roll-out programme was launched. However, the take-up rate has been slow and there have been numerous issues raised about the effectiveness and security. There are concerns at the limitations of the advice given via IHDs, the extent to which customers check their meters and whether customer behaviours are actually changed to reduce energy. There are also concerns about data privacy – related to patterns of consumption and hence potentially inferred behaviours – and security of the system, including protection from malicious cyber interference. The energy and carbon effectiveness will, to some extent, depend on ways in which the two-way communication can help to interactively manage peak loads to improve the efficient running of the overall energy networks and reduce carbon impacts.

10.13   Renewable electricity generation: wind power and photovoltaics

10.13.1   Renewables in perspective

The need to develop a logical energy strategy was outlined in Chapter 3. The measures outlined in the preceding sections, covering a mixture of demand management and energy efficiency, will directly reduce energy consumption and carbon emissions and should therefore be seen as a higher priority than the incorporation of renewable electricity generation.

For renewable technologies in general, it is important to recognise that there are often other more appropriate, and less costly, measures for reducing carbon. Incorporating building-integrated renewables usually involves most of the issues listed in Section 3.5.13, which is not the case with routine energy efficiency measures. The viability of renewables should always be assessed in the wider context of the range of potential carbon reduction measures and the overall implications for the building design.

Two renewable technologies for on-site electricity generation for buildings – wind power and photovoltaics – have traditionally in the UK often been, respectively, the most favoured and least favoured in the renewable assessment hierarchy when based on the cost per unit of carbon reduction due to the previous high carbon content of the electricity grid. As the electricity grid continues to decarbonise, the cost per unit of saved carbon will rise, affecting viability. Both of these technologies have potential hazards that must be addressed within the design and as part of the designer's obligations under CDM as outlined in Chapter 2. These two technologies are discussed next.

10.13.2   *Wind power*

A: Renewable energy availability and generation technology

It is estimated that the proportion of the planet's incident solar radiation that is used to maintain winds is only about 1% (Coley 2008). Wind power is a form of kinetic energy and can reach much higher power densities than solar irradiance, so it has high potential as a renewable energy source (Quaschning 2005). Winds arise from variations in the solar heating in different locations, resulting in differences in pressure in the atmosphere and hence movement of air masses (Boyle 2004). Wind patterns are influenced by geographical location, elevation and terrain. In coastal locations, the potential wind resources are usually much higher than for locations that are deep inland. Sea breezes are generated due to the variation in heat capacities of the land and sea. Because of its lower heat capacity, the land heats up quickly in the daytime, but cools more quickly at night. Cooler air flows from the sea to the land in daytime, replacing warm air rising from the land. At night, cooler air flows from the land to the sea in a reverse relationship (Boyle 2004). Thus, a regular feature in coastal location is 'compensating' winds from the sea to the land during the day and from the land to the sea during the night. Winds can move easily across the relatively smooth surface of the sea, which offers little surface resistance compared with the land. In a coastal area, mean wind speeds are typically about 6 m/s. In inland areas, mean wind speeds are often less than 3 m/s. Mountainous regions also provide good wind conditions. In mountain valleys winds travel upwards during the day and downwards at night, driven by similar differential heating and movement of air masses (Boyle 2004).

The British Isles include locations with some of Europe's higher mean wind velocities, and for the UK, studies have indicated that wind power – both offshore and onshore – is likely to have both higher potential and lower costs than other renewables (Boyle 2004). Britain has locations with highly favourable wind resources, particularly in areas close to the western coastline, and inland on many hills and mountains. The wind resource generally increases with altitude. In upland and coastal areas, the average wind speeds at 25 metres above the ground are often in the range of 7–10 m/s. In areas away from the western coast and uplands, average wind speeds are often in the range of 5–7 m/s. In all locations wind speeds tend to be higher in the winter and during the night (McCrae 2013).

Wind power has become more economically viable in recent years. Historically there was little wind generation in the UK; however, within the past 10–15 years the number and scale of wind farms have increased significantly. Large-scale wind turbines are a tried and tested technology, but planning issues have, to some extent, limited the growth of inland wind generation. There are now many rural wind farms in the UK, but the larger wind generation projects are mainly in offshore locations, where there are fewer restrictions. Some other European countries, including Germany, Spain and Denmark (Boyle 2004), have extensive wind generation capacity. The world's total installed capacity of wind power in 2018 was 597 GW, with 50.1 GW added in 2018 (WWEA 2019). In some countries wind power has provided a considerable proportion of their delivered electricity for many years, for example, in Denmark, where it has provided about 42% of their demand (Andrews and Jelley 2017).

It should be recognised, however, that as wind speeds are variable and can be still for prolonged periods, wind power is an intermittent resource. This has a major bearing on its viability for both on-site generation and larger-scale contribution to the grid, as it means that wind energy has minimal impact in addressing maximum demand aspects, its primary benefit being reduction in energy consumption (Boyle 2004) and helping to decarbonise the grid. The load factor for small wind turbines is about 15–20%, compared with about 25–45% for wind farms (McCrae 2013).

The estimated 'energy payback ratio' for wind power has been estimated as 80:1, which is much higher than for fossil fuel power stations; however, this figure is somewhat artificial, as it does not make allowance for the necessary back-up generation to address intermittency (Coley 2008).

The power generated by a turbine is proportional to the cube of the wind velocity, so the predicted annual energy generation can be severely reduced if the incidence of the mean wind speed is overestimated. The power generated is also directly proportional to the swept area of the blades, and so is proportional to the square of the radius (or diameter) of the blades. These cubic and square relationships mean that the best viability will be for large turbines in locations with high wind speeds for a significant proportion of time.

The mean wind speed alone does not fully describe the intermittent nature of wind on a site, but it is relatively easy to obtain and is often used as the measure of site quality or availability. Wind speed frequency distribution (ideally measured for a whole year) provides better information, but is time consuming and costly to record. Figure 10.14 shows a typical wind

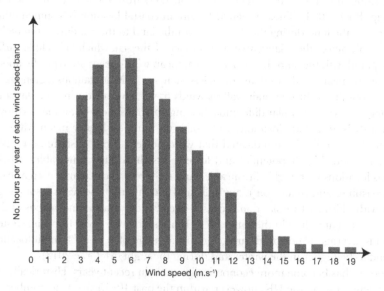

Figure 10.14 Typical wind speed frequency distribution

Source: Derived from Boyle (2004: Figure 7.30)

speed frequency distribution format. If wind speed is measured at wide intervals, this can result in incorrect estimates of the frequency of wind speeds (Quaschning 2005). Measurements are usually taken at a height of 10 m, corresponding to the hub height of a typical small-scale horizontal axis turbine.

The maximum power of a wind, P, in watts is given by (Quaschning 2005):

$$P = 0.5 \, \rho \, A \, v^3 \tag{10.4}$$

Where:

ρ = density of air in kg per cubic metre
A = swept area of the blades in metres
v = velocity of air in metres/second

As air density varies with altitude and temperature, the power generated will also vary with altitude and temperature.

A large proportion of modern wind turbines are of the horizontal axis wind turbine (HAWT) type, which provides good overall performance. Most HAWTs are three-bladed. This arrangement is considered to be slightly more efficient than two-bladed versions (Quaschning 2005). As the number of blades on a turbine increases, they tend to produce higher levels of noise (Boyle 2004).

The velocity reduces as wind passes through a turbine. The turbine utilises the power difference between the wind speeds on either side, while the mass flow rate of air passing through a wind turbine remains constant (Quaschning 2005). However, a wind turbine can only usefully extract a proportion of the total power content of the wind (Coley 2008). This utilisation capability is called the power coefficient of a turbine, Cp, and is defined as:

$$Cp = \frac{P_T}{P_O} \tag{10.5}$$

Where:
P_T is the power used by the turbine
P_O is the power content of the wind

So the power generated by a turbine, Pgen, in Watts is:

$$Pgen = 0.5 \, Cp \, \rho \, Av^3 \tag{10.6}$$

The theoretical maximum value for Cp is about 0.6. In practice, good-quality wind turbines usually have power coefficients of about 0.4–0.5 (Coley 2008).

So the power generated by a good quality turbine, Pgen, in Watts will typically be:

$$Pgen = (0.5)(0.45) \, \rho \, Av^3 = 0.225 \, \rho \, Av^3 \tag{10.7}$$

To obtain a good energy yield, it is necessary to have high mean wind speeds at the turbine height. Wind speeds can be influenced significantly by terrain, including changes in elevation. Wind speed will therefore increase with height. Wind is slowed down by the surface features that determine the roughness of the ground. The wind can also be slowed down considerably by obstacles in its path, such as buildings, trees or hills (Quaschning 2005). In this 'boundary layer', the energy of the wind is dissipated, creating turbulence and making the available wind power highly sensitive to height and location (King 2010b). If the total rotor area is more than three times the area of an obstacle, this usually causes no problem, although this depends on the distance between the turbine and the obstacle (Quaschning 2005). As a result, it is far less appropriate to locate wind turbines in built-up areas or in areas with significant tree cover or other obstructions. Due to the low mounting height of most small-scale and micro turbines, the wind power can reduce to about 12.5% of that available in open countryside, and there would be a further reduction in city centres (King 2010b). Micro wind turbines typically require wind speeds in the range of 6–8 m/s to generate a useful power output, but in many urban areas, wind speeds tend to be lower at rotor heights of about 10–12 metres (McCrae 2013). Small-scale and micro wind turbines mounted on buildings in urban and suburban locations therefore provide minimal energy contribution, and their usage is questionable. Further issues that make smaller turbines inappropriate and non-viable for non-rural locations are that planning authorities are often unaccepting of wind turbines that protrude significantly above building roof lines, and the nature of wind patterns, which includes regular gusts and eddies, can have a detrimental physical impact on small wind turbines, which can reduce their life expectancy (McCrae 2013).

The best viability is large-scale wind power in unobstructed rural locations with high levels of annual wind velocity distribution but located close to the load.

B: Energy yield in practice

When considering wind generation for a building project or site development, the key issues are location and likely energy yield. The site location will determine the wind pattern and hence the potential for wind generation. The proposed built form and planning considerations will influence the potential locations within the site for wind generation, using either free-standing or building-mounted turbines, and the acceptable rotor height and blade diameter. As well as the more general aspects of acceptability for planning and neighbourhood impact, it will be necessary to consider the potential impact on the site and occupants. Apart from visual impact, turbines create noise and can cause electromagnetic interference and other environmental impacts (Boyle 2004; Coley 2008). Because of the numerous issues involved, depending on the location, planning approval can be a major obstacle. All wind turbine proposals require careful assessment to address these issues. The likelihood of new structures being erected in the vicinity, and their potential effect on performance, should also be considered (Roe 2011). For example, a proposal to locate wind turbines on the roof of a medium-height building in a city centre might be a viable proposition at the time the building is being designed; however, the potential energy yield in the future could be considerably reduced if one or more tall buildings are erected in the vicinity.

The performance of an individual wind turbine will be represented by a wind speed power curve, usually in the typical form shown in Figure 10.15. A wind turbine only provides useful output power within an operational velocity band. Below a predetermined cut-in velocity, the power generated is negligible and operation is prevented. Similarly, operation is prevented above a predetermined high wind speed, or shutdown velocity, in order to protect the turbine from damage. The power generated is often poor at low operational velocities (the lower part of the sloping line in Figure 10.15). The wind turbine provides maximum efficiency at the design wind speed and generates rated power at the rated wind speed (Quaschning 2005).

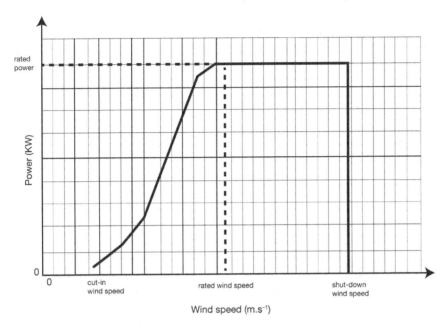

Figure 10.15 Typical format of a wind turbine wind speed power curve

Source: Derived from Boyle (2004: figure 7.29)

The typical operational velocity ranges are (Quaschning 2005):

cut-in wind speed: 2.5–4.5 m/s
design wind speed: 6–10 m/s
nominal wind speed: 10–16 m/s
shutdown wind speed: 20–30 m/s

The wind speed frequency distribution, in the form shown in Figure 10.14, can be used in conjunction with a wind speed power curve to provide the wind energy distribution. This can then show the summated energy production in relation to the band of operational velocities. There could be inconsistencies between the power curves of different manufacturers' products, as the test procedures may not be standardised (Roe 2011). At the lower end of the power output scale, turbines rated up to about 50 kW output have often failed to deliver the predicted energy yield, particularly in urban environments (King 2010b).

The power output generated by the turbine must be converted to a fully synchronised electricity supply via an inverter and synchronised with the mains frequency. An inverter will be required to convert the output from the wind turbine (covering voltage, phase and frequency) to allow connection to the grid. The selection of the inverter should seek the optimum efficiency, taking account of the range of operating power levels. The inverter's power loss must be included when assessing the delivered energy from the turbine. There will also be a power loss for distribution from the turbine to the inverter and from the inverter to the point of connection on the LV distribution system; therefore, the locations should be selected to minimise cable lengths.

Where the power generated is above the level of usage within the building or site, the excess can be exported to the grid, in which case the appropriate protection, metering and approvals would be required. For connection to the grid, the electricity provider will usually have relevant regulations, recommendations or guidelines that must be followed and will normally require a connection agreement to be signed.

C: Practical considerations for building integration

For roof-mounted wind turbines, the selection of the location on the roof area must consider H&S, installation and maintenance requirements. The need to provide the necessary clear space can cause limitations in the planning of other roof-mounted services, which could have a major impact on the overall planning of spaces for services. This could impact on the viability of wind turbines in certain cases. Suitable structural support will be required, so design liaison with the structural engineer should commence at an early stage, as transmission of vibrations to the structure can be an issue.

Connecting wind turbines to power systems can cause voltage fluctuations, which can be perceived as changes in the luminance of certain types of lamps (Quaschning 2005). As a general rule, the higher the power rating of the turbine, the more likely that such disturbances and fluctuations would become problematic to occupants. This is more likely to be the case where the grid-derived supply has a high source impedance (Quaschning 2005), although it will depend, to some extent, on the point in the system where the supply from the turbine is connected.

There are other potential issues related to the integration of wind turbines into building projects due to daylight interference caused by shadows and reflections. This phenomenon is sometimes called 'shadow flicker'. It occurs where sunlight hitting the turbine blades produces shadows, and they rotate or flicker, causing visual disturbance across windows or buildings. This can be a source of annoyance to occupants, so consideration is required as to the siting of the turbine, or some form of shading would be necessary, in relation to occupied spaces and the sun path, to limit potential annoyance (McCrae 2013)

So, overall, considering the many factors outlined earlier, by far the best utilisation of wind power is using large-blade-diameter turbines in locations with minimal obstructions and high levels of annual wind velocity distribution, and preferably locations that are some distance from buildings (to avoid noise intrusion) and more amenable to planning acceptance. Hence, offshore wind farms are an ideal way to utilise the wind resource (as have been used for decarbonising the grid). Certain rural locations can also be favourable. For on-site generation, a balance is required between having a sufficiently large free-standing turbine size to provide useful energy yields and satisfying the various issues related to acceptability for the site layout, occupants and planners. Wind turbines located on buildings in urban areas generally have limited and unpredictable energy yields and would usually need careful assessment and justification to form part of an energy strategy.

10.13.3 *Photovoltaics (PVs)*

A: Renewable energy availability and generation technology

The incoming solar radiation does not hit the Earth's upper atmosphere in an equal manner, and it is not spread equally at different places on the Earth's surface, as it is dependent on latitude and cloud cover; the average net incoming radiation for the planet is 240 W/m^2

(Coley 2008). However, this varies considerably with location, so information on the average accumulated radiation over a period of time is required to determine the potential viability of PV generation. Solar radiation will be much higher closer to the equator and will also increase at higher elevations, so the viability for any site will be considerably influenced by a site's location and orientation. A useful metric is 'capacity factor', as for a given location, this considers the amount of time the sun is shining. In 2015 in Europe, systems had an average capacity factor of 12%, compared to 20% in the United States and 26% in the Middle East. The global average is 15% (Andrews and Jelley 2017). Figure 10.16 shows a map of the UK with the variation in average annual total solar radiation in kWh per unit area. This varies from about 800 kWh/m² in the north of Scotland to about 1,050 kWh/m² on the south coast of England. This solar radiation is a mix of direct and diffuse.

A PV or solar cell is a large-area semiconductor that uses the 'photovoltaic effect' to absorb the energy of the sun and cause current to flow between two oppositely charged

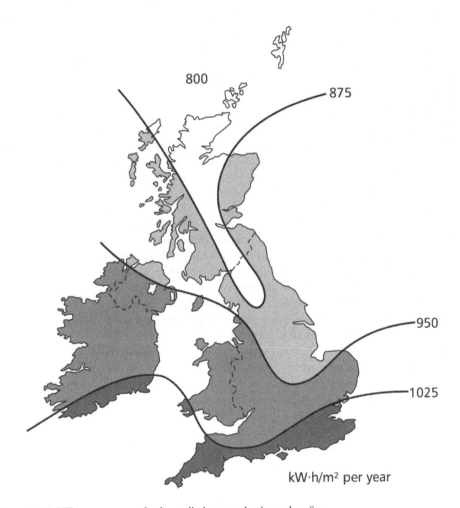

Figure 10.16 UK average annual solar radiation on a horizontal surface

Source: Reproduced from CIBSE TM25 (2000b) with the permission of the Chartered Institution of Building Services Engineers

semiconductor layers. A PV cell is, in effect, a semiconductor junction, usually formed from two differently 'doped' crystals of silicon. One-half will be an n-type layer, or negatively doped, with a surplus of electrons compared with holes; the other half is a p-type layer, or positively doped, with an excess of holes compared to electrons; and a voltage is created in a boundary layer at the junction (IET 2007). The current generated varies with solar intensity, while the DC voltage remains constant. The upper p-type layer is translucent to sunlight and has an anti-reflection surface on the covering of dielectric material (IET 2007; Quaschning 2005). The semiconductor physics and operation are discussed in detail in many other books (Andrews and Jelley 2017; McCrae 2013).

The most common types of cells are silicon (monocrystalline or polycrystalline) and thin film. For silicon cells, the majority are monocrystalline, which are formed from a single crystal and are considered to be efficient and reliable, but they are relatively expensive. A polycrystalline cell is formed from several crystals and is less efficient but more economic, while thin film cells are more costly with lower efficiency but use fewer materials (CIBSE 2000b; IET 2007). Cell efficiencies – and hence module efficiencies – have been improving over the last 10–15 years. The best conversion efficiencies for modules (using different silicon or thin film materials at different sizes) are typically between 16% and 22%, and for commercial silicon modules, typical efficiencies were about 17% in 2015 (Andrews and Jelley 2017). The semiconductor technologies are continually changing, and this is a major area of research, particularly related to thin film devices.

The duty of PV devices is expressed as peak power output, Wp, which is the power output under an illumination of 1,000 W/sq.m. PV cell efficiency is peak power output divided by incident solar radiation (1,000 W/sq.m). In northern Europe, solar incidence on sunny days is closer to 900 W/sq.m at ground level (McCrae 2013). Individual cells would typically generate a voltage of about 0.5–0.6 V and have a power rating of about 1.5 Wp. The most practical way to protect the cells and harness energy is to form them into modules, typically containing 30–70 cells connected in series and embedded into plastic (CIBSE 2000b; IET 2007). The front surface is glass, while the back is made of glass or plastic. The typical module characteristic is an open circuit voltage of about 20 V, with a short-circuit current of about 5 A, but this can vary (CIBSE 2000b). A module would typically have a rating of up to 100 Wp. The usual physical arrangement is for a group of modules to form a panel and an interconnected set of panels to make an array.

Modules are connected electrically in series as a string, with the number in the string being that required to provide the desired voltage level. The number of strings that are connected in parallel will determine the current. If just one cell in a module is shaded, the module performance can be significantly reduced, as the cell can act as a load and dissipate heat (CIBSE 2000b). Each module usually has a bypass diode to provide protection, which is more economic than a separate diode for each cell (CIBSE 2000b), and each string of modules usually has a blocking diode to prevent reverse current flow.

In its simplest sense a PV system can be considered in a block form, as shown in Figure 10.17. The PV array generates DC power, and this is converted to AC in a power conditioning unit, which also contains control and protection equipment (CIBSE 2000b). The system will require the necessary protection equipment for grid connection to ensure disconnection on loss of mains and DC and AC isolation devices and should not result in disturbance beyond established limits (CIBSE 2000b). Where appropriate, suitable meters register the amount of energy imported or exported. There is no storage of electricity in grid-connected systems.

PVs are an established technology that is reliable and requires minimal maintenance, as the modules are largely self-cleaning. They have widespread usage for local power applications

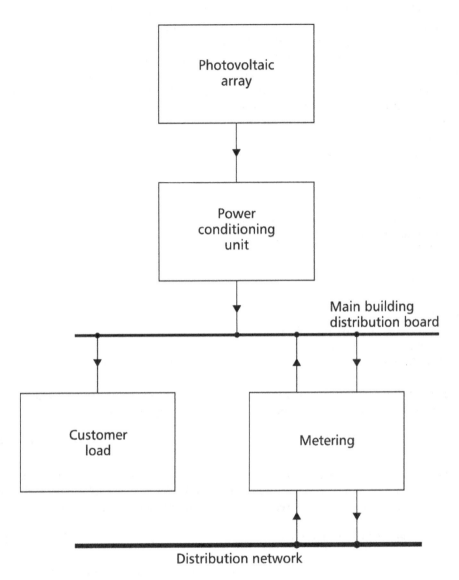

Figure 10.17 Block diagram of a grid-connected PV system

Source: Reproduced from CIBSE TM25 (2000b) with the permission of the Chartered Institution of Building Services Engineers

in remote locations, such as for communications, control, data logging, sensing and monitoring equipment. Small-scale panels have been a familiar sight alongside railway lines for many years. Their wider usage has been dependent on locality issues, primarily costs, and the energy yield from annual solar radiation. In locations where there is no local grid system, PVs are economically very competitive, on a par with grid systems in many sunny regions, and they could be much more cost-effective than a grid system in some developing countries (Andrews and Jelley 2017).

Solar PV use in buildings and in stand-alone PV generation facilities has been growing fast in recent years. More recently, there has been a major expansion of 'solar farms' in the UK, with solar PV contributing 12.9 TWh to UK electricity generation in 2018 (compared to 30.2 TWh for onshore wind and 26.7 TWh for offshore wind). However, the use of PVs in buildings is undergoing a revaluation for cost payback and carbon benefit due to the decarbonisation of the grid. The situation will be influenced by anticipated changes in Building Regulations Part L. It is likely that it will be considered more advantageous for further PV growth to be at grid level rather than at building level.

There is likely to be a growth in PV usage in some countries with high solar incidence and limited wind resources.

B: *Energy yield in practice*

To assess the annual energy yield of a PV system, it is necessary to consider the theoretical yield and then the likely yield in practice. The theoretical annual energy yield of an array, E, in kWh is given by:

$$E = A \, n \, H \tag{10.8}$$

Where:
A is the area of the active part of the array in m^2
n is the efficiency of the array
and H is the annual solar irradiation per unit area in kWh (Quaschning 2005)

So, for a PV array with an inclined area of 25 m^2, an efficiency of 14% and an annual solar irradiation of 950 kWh/m^2, the theoretical annual energy yield of the array would be:

$$E = (25).(0.14).(950) \text{ kWh} = 3325 \text{ kWh} \tag{10.9}$$

The actual energy yield will be lower in practice due to shading, dirt on the modules, installation factors and operating temperature (Quaschning 2005). A system might have losses of about 15–20% for these aspects. So, in the earlier example, a realistic annual energy yield might be:

$$E = (3325).(0.8) = 2660 \text{ kWh} \tag{10.10}$$

There will be additional losses external to the array due to the inverter, transformer (if required) and distribution cabling. The performance ratio (PR) of PV systems (ratio of actual to theoretical annual energy production) is about 80–90%. This is a measure of all the system losses (Andrews and Jelley 2017). Coley (2008) notes these figures for losses for an example application in the UK:

Multiplier to account for losses from cables: 0.98
Efficiency of inverter: 0.85

The resultant annual energy yield would then be:

$$E = 2660 \times 0.98 \times 0.85 = 2216 \text{ kWh} \tag{10.11}$$

There are useful guidance figures on energy yield related to either module area or peak output. Boyle (2004) states that UK tests have indicated annual yields of 700–1,000 kWh/kWp and 22–120 kWh/m^2 for unshaded optimal orientation and that a practical assumption in the UK would be yields of 750 kWh/kWp per year for crystalline cells. 'Rules of thumb' for the UK of 90–110 kWh/m^2 per year for crystalline cells with reasonable tilt, orientation and system efficiency and 700 kWh/kWp per year for a roof-mounted grid-connected system have been provided by CIBSE (2000b). In the British Isles in mid-July, the energy available on a horizontal surface is about 14.8 kWh/sq.m, compared to January, where it is about 0.5 kWh/sq.m (McCrae 2013). It should be noted that cloud cover affects the output, so it can change rapidly between clear skies and cloud cover.

In the UK, the best power densities are about 120–140 W/m^2. Quite obviously, for the best solar collection, the positioning of cells should be selected so that they are tilted towards the sun. To achieve optimum performance in the Northern Hemisphere, cells should be orientated from southeast to southwest and tilted at about 30–45 degrees above horizontal. The tilt angle is not too critical to the output. If an array is vertical, orientated between southeast and southwest and unobstructed, it will receive about 70% of the available maximum annual energy (CIBSE 2000b). If the angle of tilt is more inclined to the horizontal, summer solar radiation collection can be maximised, whereas if the angle of tilt is more inclined to the vertical, winter solar radiation collection can be maximised. An approach that is effective is to have a tilt angle that is the same as the location's latitude angle, or about 10 degrees less than the latitude angle. In northern latitudes, tilt angles can be about 30 degrees to the horizontal (McCrae 2013). Also of importance is the need to avoid shadows that would reduce the power output and hence energy yield. Photovoltaic power generation continues even when skies are overcast or cloudy, but the power generated reduces, so it is an intermittent resource. Power is generated in daylight hours, with peaks in the summer, but in the UK power demands are often at a peak in winter and at times with little daylight. The use of PVs can be particularly beneficial where there is an electrical load throughout the year during daylight hours.

Key issues with PV viability have related to the area required for a meaningful level of power generation in relation to the total building electrical load, and, most of all, the very high cost (per unit of carbon offset) of PVs compared with some other renewable technologies or other carbon reduction measures. The carbon reduction aspect will change with decarbonisation of the grid. In the past, cost payback was often estimated to be in the order of 70–80 years (and sometimes longer), although payback can reduce considerably where grants are available. The cost scenario is complicated and has been changing with the widespread use and growth in production of PVs. However, grants are based on political policies prevailing and can be withdrawn. The lifespan of PV cells is usually estimated at about 25–30 years. The average embodied energy for PV modules varies from 1,305 MJ/m^2 for thin film devices to 4,750 MJ/m^2 for monocrystalline devices (Hammond and Jones 2011). The emissions for manufactured PV systems are about 44 gCO2e/kWh, which compares to a global average of about 600 gCO2e/kWh (Andrews and Jelley 2017). The energy payback time in northern Europe has been estimated at about 2.5 years compared to the south of Europe, where it has been estimated to be about up to about 1.5 years (Andrews and Jelley 2017). It has been estimated that a 2kWp system on an average domestic home should save approximately 800 kg of CO_2 emissions per year, which would amount to about 30 tonnes over the anticipated life of a system, and can provide energy return ratios in excess of 30 (McCrae 2013), but the potential carbon savings will change with grid decarbonisation. In the UK, the economic case related to exporting surplus energy to the grid is complicated due to the changing policies for feed-in tariffs.

It should be noted that the situation described here is primarily related to the UK. The energy yield, and hence the viability, would be quite different in countries with higher average solar intensities and longer annual hours of sunshine and/or where different grant arrangements may apply. PVs usually provide the most benefit in remote locations where establishing a power connection would be prohibitively expensive. This has led to their widespread usage to provide independent stand-alone, small-scale power supply in conjunction with suitable capacities of battery storage.

C: *Practical consideration of building integration*

There has been considerable development of the forms of PV modules to improve the options for integration into the envelopes of buildings. This includes glass modules, modules used in curtain-wall cladding systems, roofing tiles, brise soleils and other types of canopies. It is possible to have the whole weatherproofing roof system formed from PV cells. Where the PV format integrated into the building replaces the need for a conventional element of construction material, this can help to make an economic case. Examples would be where the PV cells provide the whole or part of a roof or where a canopy of PV modules also provides solar control shading. For many buildings there might be issues related to the extent of available south-facing (or southeast- to southwest-facing) space, particularly at the roof level, which might also be in demand for other MEP equipment. Where the geometric features of the roof and/or wall surfaces do not lend themselves to good solar collection, mounting panels on frames can provide improved height or orientation to maximise solar exposure. The incorporation of PV systems at roof level requires careful space planning and coordination along with other services equipment and must include considerations of H&S and access. A key objective is to select a location where the PV cells will not be overshadowed at any time of the year to avoid reductions in energy yield. The location should be such that it avoids shading from surrounding buildings, structures, MEP equipment, trees and other physical elements falling across PV modules. The selection of locations therefore needs to take account of the likely pattern of shadows from the solar path during different times of the day and for different seasons. Where there are parallel arrays of PV modules, the spacing should be sufficient to avoid one array shading another array during the solar path. The installation of modules should be such that the rear of the cells is properly ventilated to remove any generated heat. Figure 10.18 shows a visualisation of large arrays of PVs on the roof of a building which have been coordinated with other equipment to provide an integrated solution.

Special equipment is required to integrate with AC systems (inverter, controls, protection, etc.) (CIBSE 2000b). For connection to a building's AC grid-connected system, an inverter is required to convert the PV array's DC output to AC power at the required mains voltage and frequency. The grid-commutated inverters provide synchronisation and must operate the PV generator at the optimal operating point so that maximum power is generated. To do this, the inverters are often combined with a DC/DC converter, known as a maximum power point (MPP) tracker (CIBSE 2000b; Quaschning 2005). MPP trackers automatically vary the load seen by the PV cells in such a way that it is always operating at its maximum power point and so delivering maximum power to the load (McCrae 2013). For connection to the grid, the electricity provider will usually have relevant regulations, recommendations or guidelines that must be followed and will normally require a connection agreement to

Figure 10.18 Visualisation of photovoltaic arrays coordinated with roof plant layout

Source: Courtesy of Hoare Lea

be signed. For detailed guidance on connection arrangements, reference should be made to other documents, such as the IET Code of Practice for Grid Connected Solar Photovoltaic Systems (IET 2015).

The selection of the inverter should take account of the dynamic nature of the operation during changing levels of solar irradiance. It should be recognised that the inverters will operate at their rated power for just a few hours each year. For most of the time, they will be operating at part-load, so it is important for the inverter to maintain good efficiency levels for low loads. Inverters should not be oversized, should have minimal losses and should be switched off during hours of darkness (Quaschning 2005). The majority of inverters used for PVs generally have an efficiency of about 90% for an inverter with about 1-kW rating, and this efficiency is maintained even down to loads of about 10% of the inverter rating. Higher-rated inverters can have an efficiency of up to 96–97% (Quaschning 2005). The lengths of DC and AC cabling should be kept short to limit distribution losses.

10.14 Summary

The carbon content of electricity has been reducing and is continuing to reduce in some countries, but remains higher in some other countries. The carbon reduction due to energy savings in electrical systems will depend on the carbon scenario, which is likely to be going through a transition in coming years.

Electrical systems for buildings feed both direct electrical system loads and motive power for HVAC systems. Therefore, the efficiency of power distribution can be an important measure for achieving an energy-efficient building. Efficiency can be improved through the appropriate selection of sub-station disposition and transformer characteristics in relation to the load profile, together with cabling and cable management arrangements to minimise both operational and embodied energy impacts. Power factor correction and harmonic filtering can reduce energy losses in distribution components in many, primarily legacy, systems. The efficiency of internal lighting can be addressed through optimising the usage of daylighting, together with attention to design criteria. The use of LEDs, which have high luminous efficacies, and the application of suitable lighting controls and management can further improve energy effectiveness. Design approaches for energy effectiveness in motor drives for HVAC systems, lift installations and other electrical systems have been outlined. Process loads can give rise to significant energy consumption and should also be designed for energy-efficient performance.

Energy management and maintenance of functional performance for all active energy-using systems can be enabled through suitable data collection, automatic controls, metering and monitoring. Such systems should be considered alongside management involvement and occupancy engagement to create a successful energy management regimen.

Where the building location circumstances are appropriate, wind power and PVs may provide a useful on-site renewable energy contribution, but their viability will be dependent upon the costs, potential energy yield and the practical aspects of integration and should be a lower priority than reducing demand and improving efficiency.

11 Building thermal load calculations

11.1 Introduction

Building heating and cooling load calculations can only be undertaken if the heat transfer mechanisms between the building walls and the internal and surrounding environment are identified. Heat transfer through building walls is caused by a difference in temperature between the exterior and interior surfaces of the wall. During the day, solar radiation strikes the exterior wall surface. A small portion of this energy is reflected but the remainder is absorbed, thereby increasing the wall surface temperature. Because of the higher outer surface temperature of the wall, convection also takes place between the wall and the outdoor air. Heat received at the exterior wall surface is then transferred by conduction through the various material layers in the wall to the interior surface. It should be noted that the nature of this process is strongly affected by the number of wall materials involved, the thickness of each layer and the properties of each material, namely thermal conductivity, density, thickness and heat capacity.

The thermal interaction between the building and the environment is illustrated in Figure 11.1. At the interior surface of the wall, some of the heat is transferred to the room air by convection, while the remainder is radiated to the surfaces of other walls, floors, ceilings and furnishings within the room. At each surface that receives this radiant heat, a proportion is then transferred, through convection, to the room air, conducted into the material mass and stored, or re-radiated to other surfaces in the room. By the repetition of these processes over time, most of the original heat which entered through the wall is eventually transferred to room air.

Since the space heat gain by radiation is partially absorbed by the surfaces and contents of the space, it does not affect the temperature of the room air until sometime later. It follows therefore that the rate at which heat must be removed from the room air to maintain a constant temperature is not the same as the instantaneous rate of heat gain. In air conditioning systems design, it is therefore important to differentiate between related but distinct heat flow rates, each of which varies with time.

It is the radiant component of the total heat gain that falls on the internal surfaces of the building (and is partly absorbed into the structure as the surfaces are warmed) that appears as a cooling load through convection after a time delay.

This radiant component of the heat gain takes two forms, namely short- and long-wave radiation. Short-wave radiation is due to solar irradiation and part of the electric light spectrum, and long-wave radiation results from the remainder of the electric light spectrum, people and warm objects such as electrical equipment.

The heat gain due to convection within the indoor air volume takes place without a time delay and is known as the heat gain to the air node, and the heat gain due to the solar or

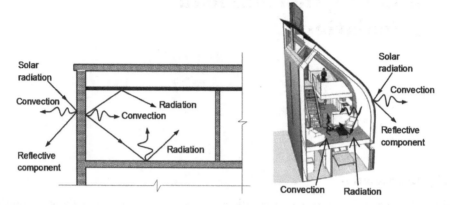

Figure 11.1 Thermal interaction between the building and the environment

thermal (long-wave) radiation is considered to take place to the environmental node – which is partly due to long-wave heat exchange between indoor surfaces. (The air node represents the air inside the room, and the environmental node represents a point just off the surface of the wall where convective and radiation heat transfers are taking place.)

Various heating and cooling load calculation techniques are in use; however, the emphasis here will be on the methods recommended by the UK Chartered Institution of Building Services Engineers (CIBSE).

11.2   The cyclic dynamic model and the admittance procedure

This method is based on the calculation of the thermal response of a building using the admittance procedure. It provides a manual method of calculating cooling loads for buildings by assuming a sequence of identical days when the external conditions repeat every 24 hours. This procedure estimates the proportion of the total heat gain that is absorbed by the internal surfaces of the building and therefore reduces the peak cooling load. Figure 11.2 presents the way in which heat energy interacts with the building structure and the way 'thermal latency' and 'thermal storage' affect the conversion of heat gains to cooling loads.

The conversion from heat gain to cooling load for a building is illustrated in Figure 11.3, where a building of a random structure is selected and the relation between the heat gain and the cooling load is plotted. It is clear that the component of heat gain that is stored in the building structure causes the peak cooling load to be lower than the peak heat gain.

In Figure 11.3, the upper curve shows the solar heat gain, and the lower curve represents the actual cooling load for a constant space temperature during the operating period of the equipment. The horizontally shaded areas represent the heat stored in the building structure and furnishings, while the vertically shaded areas represent the stored heat released to the building space. Since for a cyclical 24-hour operation, all the heat entering the structure must be removed, the two shaded areas must be equal. However, it can be seen that the peak load is lower than the peak heat gain and occurs after the peak heat gain has occurred by a noticeable time (time lag). And this time lag depends on the building construction, the external walls, floor, ceiling, internal walls or partitions and internal finishes.

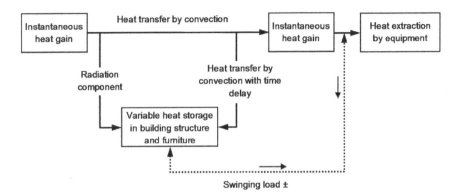

Figure 11.2 Difference between space heat gain and its cooling load

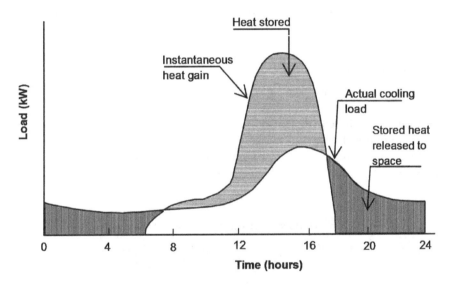

Figure 11.3 The relationship between the heat gain and the cooling load of a building

From fundamental heat transfer theory and for one-dimensional non-steady heat conduction through a homogeneous material of uniform thermal conductivity, the temperature distribution is given by:

$$\frac{\delta\theta}{\delta\tau} = \alpha\frac{\delta^2\theta}{\delta x^2} \tag{11.1}$$

Where:
θ is the temperature
x is the distance
τ is the time
α is the thermal diffusivity

given by: $\alpha = \left(\dfrac{k}{\rho c} \right)$

k is the thermal conductivity
ρ is the density
c is the specific heat

   This equation then has to be solved using the admittance method for sinusoidal variations. For practical building components such as multi-layered walls with materials containing different thermal properties, the analytical solution of this equation to sinusoidal variations is complex and cannot be easily solved by manual calculations. The solution requires computer calculation of the resulting matrix algebra or the use of numerical methods such as finite differences. For details of the mathematical solution of this equation for sinusoidal temperature variations, refer to BS EN ISO 13786 (1999) or Appendix 3.A6 of the CIBSE Guide A (2015), which also includes the mathematical derivation of thermal admittance.

   The steps to be followed in order to calculate building cooling loads using the admittance procedure will be detailed in a later section, but first the various types of building heat gains will be explained.

11.3   Building heat gains

The various types of building heat gains, illustrated in Figure 11.4, are:

*   heat gain through building fabric (sensible gains)
*   solar gains through windows and blinds (sensible gains)
*   heat gain due to air infiltration and ventilation (sensible and latent gains)
*   internal heat gains (sensible and latent gains).

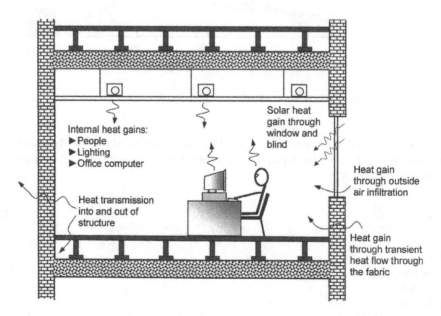

Figure 11.4 Heat gains to a conditioned space

11.3.1 Heat gain through the building fabric

The heat transmission through external walls, internal walls, roofs and a part of the heat transmission through the glazing takes place due to conduction of heat through solid structures and is calculated using the concept of the sol-air temperature.

The sol-air temperature is defined as the outside air temperature that, in the absence of solar radiation, would give the same temperature distribution and heat gain through the wall or roof as that which exists with the actual outdoor temperature and the incident solar radiation.

The sol-air temperature θ_{eo} is calculated from the following formula:

$$\theta_{eo} = \frac{\alpha I_t}{h_{so}} + \theta_{ao} \tag{11.2}$$

Where:
α is the absorption coefficient for solar radiation
I_t is the incident short-wave solar radiation
h_{so} is the outside surface heat transfer coefficient
θ_{ao} is the outside air temperature

Sol-air temperatures for 14 cities in the UK (Belfast, Birmingham, Cardiff, Edinburgh, Glasgow, Leeds, London, Manchester, Newcastle, Norwich, Nottingham, Plymouth, Southampton and Swindon) can be found in the supplementary files to CIBSE Guide A (2015) tables 2.14(a)–2.14(n).

The solar energy incident on an external wall or roof is periodic, so the variation in sol-air temperature is also periodic. The degree by which the amplitude of the external temperature variation is dampened at the internal surface is a function of the decrement factor, f.

The decrement factor is defined as 'the ratio of the rate of heat flow through the structure due to variations in the external heat transfer temperature from its mean value to the steady state conduction with the environmental temperature held constant' (CIBSE 2015) (see Section 11.5).

Figure 11.5 illustrates the temperature cycles induced by sinusoidal temperature variation at the outside surface. The external fluctuations give rise to temperature cycles of smaller

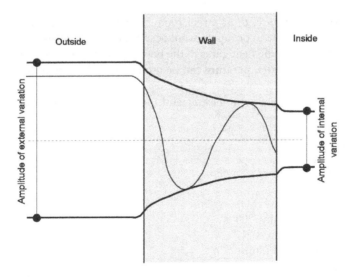

Figure 11.5 Temperature cycles induced by sinusoidal temperature variation at the outside surface

amplitude, which decay in an exponential way through the wall. There is also a time delay as the temperature wave passes through the wall.

In the cyclic dynamic model, the structural heat gain comprises both mean and alternating (or swing) components of heat gains through opaque and glazed areas and heat gains through internal partitioning walls where temperature differences exist across them.

Heat gain through opaque and glazed areas

A heat gain through opaque and glazed areas consists of mean and alternating heat gain components.

The mean heat gain through the fabric at the internal operative temperature is obtained by summing up the mean structural heat gains through the opaque and glazed surfaces:

$$\bar{Q}_f = \Sigma A_n U_n (\bar{\theta}_{eon} - \bar{\theta}_c) + \Sigma A_{gn} U_{gn} (\bar{\theta}_{ao} - \bar{\theta}_c) \tag{11.3}$$

Where:
$\bar{Q}_f$ is the mean fabric heat gain
$\Sigma A_n U_n$ is the sum of the products of the opaque surface areas and corresponding thermal transmittances
$\bar{\theta}_{aon}$ is the 24-hour mean sol-air temperature for the wall element n

$\Sigma A_{gn} U_{gn}$ is the sum of the products of the glazed surface areas and corresponding thermal transmittances
$\bar{\theta}_{ao}$ is the 24-hour mean air temperature
$\bar{\theta}_c$ is the 24-hour mean internal operative temperature
n is the number of wall elements on a particular facade
gn is the number of glazed elements on the same facade

The operative temperature, θ_c in a real room is defined as the air temperature in a hypothetical room such that an occupant would experience the same net energy exchange with the surroundings (i.e. the heat gains to the body equal the heat losses from it). At low air velocities, the operative temperature can be used as an index temperature for thermal comfort.

The operative temperature combines air and mean radiant temperatures into a single index temperature, as follows:

$$\theta_c = \frac{\theta_{ai} \sqrt{10v} + \theta_r}{1 + \sqrt{10v}} \tag{11.4}$$

Where:
θ_c is the operative temperature (°C)
θ_{ai} is the inside air temperature (°C)
θ_r is the mean radiant temperature (°C)
v is the air speed (m/s)

For indoor air speeds below 0.2 m/s, and when the mean radiant and air temperature difference is less than 0.4°C, the operative temperature becomes the average of the air temperature and the mean radiant temperatures:

$$\theta_c = 0.5\,\theta_{ai} + 0.5\,\theta_r \tag{11.5}$$

For differences smaller than 0.2°C, the operative temperature becomes so close to the air temperature that the two can be used interchangeably without any appreciable error.

The alternating heat gain through the fabric is obtained by summing up the alternating structural heat gains through the opaque and glazed surfaces:

$$\tilde{Q}_f = \sum f_n A_n U_n \tilde{\theta}_{eon} + \sum A_{gn} U_{gn} \tilde{\theta}_{ao} \tag{11.6}$$

Where:

$\tilde{Q}_f$ is the alternating fabric heat gain
f_n is the decrement factor for the opaque surface, n
$\tilde{\theta}_{eon}$ is the alternating component of sol-air temperature for the corresponding opaque surface
$\tilde{\theta}_{ao}$ is the alternating component of air temperature

The alternating component of sol-air temperature is given by:

$$\tilde{\theta}_{eon} = \theta_{eon} - \bar{\theta}_{eon} \tag{11.7}$$

where θ_{eon} is the sol-air temperature at the particular time the heat gain is determined. If the time is t, then θ_{eon} refers to $(\tau - \varphi)$, where φ is the time lag.

The alternating air temperature is obtained in the same way, setting the time lag to zero. Thus, the alternating component of air temperature is given by:

$$\tilde{\theta}_{ao} = \theta_{ao} - \bar{\theta}_{ao} \tag{11.8}$$

The values of decrement factors and time lags for typical walls and roofs can be found in CIBSE Guide A (2015), tables 3.48–3.54.

Heat gains through internal partitions

When an air-conditioned space is adjacent to a space which is maintained at a different temperature, heat transfer takes place with heat flowing from the hotter to the colder space through the partition materials. The heat gain can then be calculated, taking into account the steady-state conditions which exist in both spaces. The rate of heat transfer (heat gain or loss) to or from the conditioned space is given by:

$$Q_i = U\,A\left(\theta_{bi} - \theta_{ai}\right) \tag{11.9}$$

Where:
U is the overall heat transfer coefficient (W/m²K)
θ_{bi} is the temperature of air in the adjoining space (°C)
θ_{ai} is the room air temperature (°C)

Example 11.1 Heat transfer through wall

Location	London 51.7°N
Wall construction	105-mm dark-coloured brickwork
	50-mm air gap
	100-mm dense concrete
	13-mm dense plaster
Wall area	30 m × 3 m
Orientation	Facing southeast (SE)
Internal air temperature	21°C, held constant
Time	August at 9.30, 11.30, 13.30, 15.30, 17.30, 19.30

Assume: the indoor mean radiant temperature, $\theta_r = \theta_{ai}$, then $\theta_{ai} = \theta_c$

Solution:

Calculation of heat transfer through wall (data)

Item	Reference	Value	Unit
Date		August	
Orientation		SE	
U (thermal transmittance)	CIBSE table 3.48 (No 7a)*	1.77	W/m²
f (decrement factor)	CIBSE table 3.48 (No 7a)*	0.34	
φ (time lag)	CIBSE table 3.48 (No 7a)*	8.1	hours
$\theta_{sol\text{-}air}$ (24-hour mean)	CIBSE table 2.14(g)*	27.5	°C
Wall area	30 × 3	90	m²

Source: *CIBSE Guide A (2015)

Heat transfer through wall

Time of heat gain into room, τ (hour)	Time at which sol-air temperature should be considered, $\tau - \phi$ (hour)	θ sol-air at $\tau - \phi$ (°C)	24-hour mean heat gain (W)	Alternating heat gain component (W)	Total heat gain (W)
9.30	1.30	15.1	1036	−672	364
11.30	3.30	12.5	1036	−812	224
13.30	5.30	12.9	1036	−791	245
15.30	7.30	42.9	1036	834	1870
17.30	9.30	59.8	1036	1749	2785
19.30	11.30	54.1	1036	1294	2330

Note: It should be noted that to calculate the heat gain into the room at a given time, e.g. 5.30 p.m., the external sol-air temperature that occurred 8 hours earlier is used in the calculation, i.e. 9.30 a.m., as there is a time lag of 8 hours before the temperature wave passes through the wall.

11.3.2 Solar gains through windows and blinds

The response of a space to solar radiation transmitted through and absorbed by the glazing system is characterised by two parameters, the mean solar gain factor and the alternating solar gain factor. These are further divided into factors relating the heat gain to the room environmental node (causing an increase in the environmental temperature) and the room air node (causing an increase in the air temperature). The latter is only used where internal blinds are fitted – this is because increased convection from blinds significantly changes the proportions of long-wave and convective heat from the surface.

The environmental temperature is a hypothetical temperature determined from heat flow into a room surface by convection from the room and radiation from the surrounding surfaces. This temperature is traditionally given by:

$$\theta_{ei} = \frac{1}{3} \, \theta_{ai} + \frac{2}{3} \, \theta_r \tag{11.10}$$

a) The mean solar gain factors for the environmental and air nodes are given by (CIBSE 2015):

$$\bar{S}_a = \frac{\text{Mean solar gain at the air node per } m^2 \text{ of glazing}}{\text{Mean solar intensity incident on solar facade}} \tag{11.11}$$

$$\bar{S}_e = \frac{\text{Mean solar gain at the environmental node per } m^2 \text{ of glazing}}{\text{Mean solar intensity incident on solar facade}} \tag{11.12}$$

b) The alternating solar gain factors for the environmental and air nodes are given by (CIBSE 2015):

$$\tilde{S}_a = \frac{\text{Instantaneous cyclic solar gain at the air node per } m^2 \text{ of glazing}}{\text{Instantaneous cyclic solar intensity incident on solar facade}} \tag{11.13}$$

$$\tilde{S}_e = \frac{\text{Instantaneous cyclic solar gain at the environmental node per } m^2 \text{ of glazing}}{\text{Instantaneous cyclic solar intensity incident on solar facade}} \tag{11.14}$$

Where no shading devices are used, the alternating gain usually lags the solar intensity by between zero and two hours. The lag time depends on the surface factors (characteristics) for the internal surfaces. High surface factors (e.g. 0.8) give rise to delays of about one hour; low surface factors (e.g. 0.5) give rise to delays of about two hours (the surface factor is defined in Section 11.5).

Typical values of solar gain factors for various glazing configurations are given in Table 11.1. Glazing manufacturers do not usually provide values of these solar factors for their products, but the factors can be determined from fundamental principles and the properties of the glazing materials.

One or more of the following properties are normally provided by manufacturers:

*   properties at normal incidence (see Appendix 5.A11, CIBSE Guide A 2015)
*   shading coefficients
*   total solar energy transmittance ('G-value').

Table 11.1 Solar gain factors and shading coefficients for various glazing/shading configurations

Description (outside to inside)	Solar gain factor (environment node)[†]			Solar gain factor (air node)		Shading coefficient S^c	
	$\bar{S}e$	$\tilde{S}el$	$\tilde{S}eh$	$\tilde{S}a$	$\tilde{S}a$	Short-wave	Long-wave
Single glazing combinations:							
• clear glass	0.76	0.66	0.50	–	–	0.91	0.05
• absorbing glass	0.61	0.54	0.44	–	–	0.53	0.19
• clear absorbing slats	0.43	0.44	0.44	0.17	0.18	–	–
• clear reflecting slats	0.35	0.32	0.31	0.12	0.12	–	–
• clear glass with 'generic' blind	0.34	0.33	0.29	0.11	0.11	–	–
Double glazing combinations:							
• clear/clear	0.62	0.56	0.44	–	–	0.70	0.12
• reflecting/clear	0.36	0.32	0.26	–	–	0.37	0.08
• clear/low emissivity	0.62	0.57	0.46	–	–	0.62	0.18
• absorbing/low emissivity	0.43	0.38	0.32	–	–	0.36	0.15
• 'generic' blind/clear/low emissivity	0.15	0.14	0.11	–	–	–	–
• clear/clear/absorbing slats	0.34	0.36	0.37	0.18	0.21	–	–
• reflecting/clear/absorbing slats	0.19	0.19	0.19	0.12	0.13	–	–
• clear/low emissivity/absorbing slats	0.33	0.35	0.35	0.21	0.23	–	–
• absorbing/low emissivity/absorbing slats	0.22	0.22	0.22	0.16	0.17	–	–
• clear/clear/reflecting slats	0.28	0.29	0.26	0.15	0.16	–	–
• reflecting/clear/reflecting slats	0.17	0.16	0.16	0.10	0.10	–	–
• clear/low emissivity/reflecting slats	0.28	0.27	0.26	0.18	0.20	–	–
• absorbing/low emissivity/reflecting slats	0.18	0.17	0.17	0.14	0.15	–	–
• clear/low emissivity/'generic' blind	0.29	0.29	0.27	0.17	0.18	–	–
Triple glazing (combinations):							
• clear/clear/clear	0.52	0.49	0.40	–	–	0.55	0.17
• absorbing/clear/clear	0.37	0.35	0.29	–	–	0.33	0.15
• reflecting/clear/clear	0.30	0.28	0.23	–	–	0.30	0.09
• clear/low emissivity/clear	0.53	0.50	0.42	–	–	0.50	0.21

Source: Adapted from CIBSE Guide A (2015)

Note: [†]For $\tilde{S}e$, subscripts l and h denote thermally 'lightweight' and 'heavyweight' buildings, respectively.

The shading coefficient, S_c, can be defined as:

$$S_c = \frac{Solar\ gain\ through\ glass\ and\ blind\ at\ direct\ normal\ incidence}{Solar\ gain\ through\ reference\ glass\ at\ direct\ normal\ incidence} \tag{11.15}$$

The solar gain can be considered to mean the short-wave, long-wave or the total component of the two. The shading coefficient can therefore refer to any of the three gains, but in all cases the solar gain through the reference glass at normal incidence can be taken as 0.87 (CIBSE 2015).

After determining the solar gain factors, the solar heat gains can be calculated as described in the next sub-section.

Mean solar heat gains

Solar gains through glazing consist of solar radiation, which is absorbed in the glazing and transmitted to the environmental node, and also the transmitted solar radiation, which is absorbed at the internal surfaces of the room and appears at the environmental node.

The mean solar heat gain to the internal environmental node, $\bar{Q}_{se}$, is given by:

$$\bar{Q}_{se} = \bar{S}_e \bar{I}_t A_g \qquad (11.16)$$

Where
$\bar{I}_t$ is the mean total solar irradiance (W/m²)

The total solar irradiance I_t is the sum of the beam, I_{DV}, and diffuse, I_{dv}, solar radiation given by:

$$I_t = I_{DV} + I_{dv}$$

A_g is the area of glazing (m²)
$\bar{S}_e$ is the mean solar gain factor at the environmental node

For the case of internal shading (i.e. blinds), part of the solar gain will enter the air node and part will enter the environmental node.
The mean solar heat gain to the air node is:

$$\bar{Q}_{sa} = \bar{S}_a \bar{I}_t A_g \qquad (11.17)$$

Swing in solar heat input

The swing in solar gain to the environmental node is given by:

$$\tilde{Q}_{se} = \tilde{S}_e (\hat{I}_t - \bar{I}_t) A_g \qquad (11.18)$$

and that to the air node by:

$$\tilde{Q}_{sa} = \tilde{S}_a (\hat{I}_t - \bar{I}_t) A_g \qquad (11.19)$$

Where
$\tilde{Q}_{se}$ is the alternating solar gain to the environmental node
$\tilde{Q}_{sa}$ is the alternating solar gain to the air node
$\hat{I}_t$ is the peak total solar irradiance (W/m²)
$\bar{I}_t$ is the 24-hour mean total solar irradiance (W/m²)

There will be a time delay between the occurrence of the gain and the appearance of the solar cooling load due to the admittance of the room surfaces. This delay is one hour for spaces with a 'slow' response and zero for 'fast' response spaces.

Example 11.2 Calculate the peak solar heat gain and the solar heat gain at 17.30 for the following glazing configuration:

Location	Edinburgh
Glazing type	Reflecting slats (internal) with single clear glass
Window size	10 m × 2 m
Orientation	Vertical glazing facing southwest (SW)
Time	August
Space	Lightweight

Solution: The solution is detailed in the following tables.

Item	Reference	Value
Date		21 August
Orientation		SW
Glass area	$A_g = 10 \times 2$	20 m²
$\bar{S}_a$ Mean solar gain factor air node	CIBSE, table 5.A11.1*	0.12
$\tilde{S}_a$ Alternating solar gain factor air node	CIBSE, table 5.A11.1*	0.12
$\bar{S}_e$ Mean solar gain factor environmental node	CIBSE, table 5.A11.1*	0.35
$\tilde{S}_e$ Alternating solar gain factor environmental node, lightweight space	CIBSE, table 5.A11.1*	0.32

Source: *CIBSE (2015)

(a) Peak solar heat gain through the glazing

Item	Reference	Value	Unit
Time for peak heat gain	CIBSE, table 2.13(d)	13.30	
I_{Dv} (peak) – beam	CIBSE, table 2.13(d)	551	W/m²
I_{dv} (peak) – diffuse	CIBSE, table 2.13(d)	201	W/m²
$I_{t,\,max}$ (peak)	$I_{Dv} + I_{dv}$	752	W/m²
$I_{t,\,mean}$ Mean solar radiation (24 hours)	CIBSE, table 2.13(d)	156	W/m²
Mean solar gain at air node $\bar{Q}_{sa}$		374	W
Alternating solar gain at air node $\tilde{Q}_{sa}$		1430	W
Peak air node solar heat gain Q_{sa}	$\tilde{Q}_{sa} + \bar{Q}_{sa}$	1804	W
Mean solar gain at environmental Node $\bar{Q}_{se}$		1092	W
Alternating solar gain at environmental Node $\tilde{Q}_{se}$		3814	W
Peak environmental node solar heat gain Q_{se} @ 13.30	$\tilde{Q}_{se} + \bar{Q}_{se}$	4906	W
Total solar heat gain @ 13.30 Q_s		6710	W

(b) Solar heat gain through the glazing at 17.30

Item	Reference	Value	Unit
Time		17.30	
I_{Dv} (peak) – beam	CIBSE, table 2.13(d)	42	W/m²
I_{dv} (peak) – diffuse	CIBSE, table 2.13(d)	87	W/m²
$I_{t,\,max}$ (peak)	$= I_{Dv} + I_{dv}$	129	W/m²
$I_{t,\,mean}$ Mean solar radiation (24 hours)	CIBSE, table 2.13(d)	156	W/m²
Mean solar gain at air node $\bar{Q}_{sa}$		374	W
Alternating solar gain at air node $\tilde{Q}_{sa}$ @ 17.30		−65	W
Air node solar heat gain Q_{sa} @ 17.30	$\tilde{Q}_{sa} + \bar{Q}_{sa}$	309	W
Mean solar gain at environmental node $\bar{Q}_{se}$	$\bar{S}_e \bar{I}_t A_g$	1092	W
Alternating solar gain at environmental node $\tilde{Q}_{se}$ @ 17.30	$\tilde{S}_e (\tilde{I}_t - \bar{I}_t) A_g$	−173	W
Environmental node solar heat gain Q_{se} @ 17.30	$\bar{Q}_{se} + \tilde{Q}_{se}$	919	W
Total solar heat gain @ 17.30	$Q_{sa} + Q_{se}$	1228	W

Note: This calculation assumes that there is no shading of the glass area from external overhangs or deep window reveals, etc. In practice the shaded area of the glass would be calculated from the solar altitude and azimuth and deducted from the total area before the heat gain is calculated.

11.3.3 *Heat gain due to air infiltration and ventilation*

Infiltration is the uncontrolled flow of air through cracks and other openings around windows and doors and through floors and walls. Ventilation is the intentional displacement of indoor air by the air conditioning system or through specified openings such as windows and doors. The rate of infiltration or natural ventilation is a function of the pressure difference across the building envelope. This pressure difference is caused by:

- wind pressure
- difference in density between the indoor and outdoor air, often called the chimney or stack effect
- pressure generated by a mechanical ventilation system.

In recent years, several techniques for calculating infiltration rates have been developed. These are outlined in the ASHRAE *Handbook of Fundamentals* (2017b) and CIBSE Guide A (2015).

A simplified estimate of the instantaneous heat gain due to infiltration or ventilation to the air node of a building is given by:

$$Q = C_v(\theta_{ao,\tau} - \theta_c) \tag{11.20}$$

Where:
Q is the heat gain due to air infiltration/ventilation
$\theta_{ao,\tau}$ is the outside air temperature at time, $\tau(°C)$
θ_c is the indoor operative air temperature (°C)
C_v is the infiltration/ventilation conductance (W/K) and is given by:
$C_v = \rho c_p NV$

where N is the number of room air changes for air entering the space at the outside air temperature (per hour) and V is the room volume (m³). Values for N for various building types can be found in tables 4.16 to 4.24 in CIBSE Guide A (2015). Converting the number of air changes per hour to air changes per second and using $\rho c_p \approx 1200\,J/m^3K$, the following relation for the ventilation conductance in SI units becomes:

$$C_v = \frac{1200\,NV}{3600} = \frac{NV}{3} \tag{11.21}$$

Equation 11.20 can be used for simple calculation of infiltration gains. For more precise infiltration/ventilation heat gain calculations, the alternating component of the heat gain needs to be considered. Therefore, by the admittance procedure method, the mean and the alternating components of the infiltration/ventilation heat gains need to be calculated as follows:
The mean infiltration/ventilation heat gain component:

$$\bar{Q}_v = C_v \left(\bar{\theta}_{ao} - \bar{\theta}_c \right) \tag{11.22}$$

The alternating infiltration/ventilation heat gain component:

$$\tilde{Q}_v = C_v \tilde{\theta}_{ao} \tag{11.23}$$

Where: $\tilde{\theta}_{ao} = \hat{\theta}_{ao} - \bar{\theta}_{ao}$; $\hat{\theta}_{ao}$ being the air temperature at the peak hour

The total infiltration/ventilation heat gain can then be estimated from:

$$Q_v = \bar{Q}_v + \tilde{Q}_v$$

(11.24)

11.3.4   *Internal heat gains*

Internal heat gains arise from the heat released by the following elements into the environmental node within the considered space:

*   occupants
*   lighting
*   equipment.

Occupants give out heat and moisture in the conditioned space. The rates at which the heat and moisture are released depend on the different states of activity of the people within the conditioned space. Typical values of sensible and latent gains for various activities, adapted from the ASHRAE *Handbook of Fundamentals* (2017b), are given in Table 11.2.

Lighting

The heat gains from lighting are partly due to convection and partly due to radiation, which is first absorbed by the building structure and furnishings before it is released to the conditioned space. This creates a time lag, so that part of the absorbed energy is radiated back to the space after the lights have been switched off, as shown in Figure 11.6.

Table 11.2 Rates of sensible and latent heat gains from occupants for different levels of activity

Degree of activity	Typical building	Total rate of heat emission for adult male (W)	Rate of heat emission for mixture of males and females (W)		Percentage of sensible heat that is radiant heat for stated air movement (%)	
			Total	Sensible	Latent	High	Low
Seated at theatre	Theatre, cinema (matinee)	115	95	65	30	–	–
Seated at theatre, night	Theatre, cinema (night)	115	105	70	35	60	27
Seated, very light work	Offices, hotels, apartments	130	115	70	45	–	–
Moderate office work	Offices, hotels, apartments	140	130	75	55	–	–
Standing, light work: walking	Department store, retail store	160	130	75	55	58	38
Walking; standing	Bank	160	145	75	70	–	–
Sedentary work	Restaurant	145	160	80	80	–	–
Light bench work	Factory	235	220	80	140	–	–
Moderate dancing	Dance hall	265	250	90	160	49	35
Walking; light machine work	Factory	295	295	110	185	–	–
Bowling	Bowling alley	440	425	170	255	–	–
Heavy work	Factory	440	425	170	255	54	19
Heavy machine work: lifting	Factory	470	470	185	285	–	–
Athletics	Gymnasium	585	525	210	315	–	–

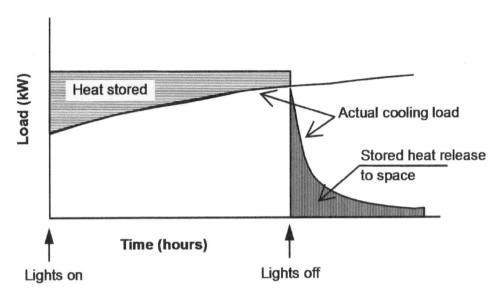

Figure 11.6 Influence of radiant energy from lighting on cooling load

Table 11.3 Nominal energy heat output from different types of lamps

Lamp type (common designations)	Heat output (%)		
	Radiant	*Conducted/ convected**	*Total*
Fluorescent (MCF)	30	70	100
Filament (tungsten, GLS, TH)	85	15	100
High-pressure mercury/sodium, metal halide (MBF, SON, MBI)	50	50	100

Source: Adapted from CIBSE *Code for Lighting* 2002

* The power loss from ballasts should be added to the conducted/convected heat.

Table 11.3 provides typical values of the convective and radiant heat outputs from various lamp types.

Equipment

The heat gains from the various types of equipment need to be considered when calculating internal heat gains. In buildings such as data centres and restaurants, equipment heat gains form the largest heat gain for the building. Typical values of heat gains from various items of office equipment are given in Table 11.4.

For building services engineers, it is normal practice to specify internal heat gains (including occupants, lighting and equipment) in the form of watts per square metre

Table 11.4 Typical heat gains from office equipment

PCs	Value for stated mode (W)	
Nature of value		
	Continuous	Energy saving
Average	55	20
Conservative	65	25
Highly conservative	75	30
PC monitors		
Monitor size	Value for stated mode (W)	
	Continuous	Energy saving
Small (13–15 inch)	55	0
Medium (16–18 inch)	70	0
Large (19–20 inch)	80	0

Laser printers			
Printer size	Value for stated mode (W)		
	Continuous	1 page/min	Idle
Small desktop	130	75	10
Desktop	215	100	35
Small office	320	160	70
Large office	550	275	125

Photocopiers			
Copier size	Value for stated mode (W)		
	Continuous	1 page/min	Idle
Desktop copier	400	85	20
Office copier	1100	400	300

Fax, scanner and dot matrix printer		
Device	Value for stated mode (W)	
	Continuous	Energy saving
Fax machine	30	15
Scanner	25	15
Dot matrix printer	50	25

Source: Adapted from Wilkins and Hosni (2000)

(W/m²). Typical values of these heat gains for different building types are given in Table 11.5.

In accordance with the admittance procedure, the total internal heat gain consists of the mean and alternating components as follows:

$$Q_{int} = \bar{Q}_{int} + \tilde{Q}_{int} \tag{11.25}$$

Where:

$$\bar{Q}_{int} = \frac{\Sigma_i^n \bar{Q}_{int,n} t_n}{24}$$

$$\text{and } \tilde{Q}_{int} = \hat{Q}_{int} + \bar{Q}_{int}$$

n refers to the heat gain source and t_n to the duration of the source (hours of operation). The internal gains can be a combination of convective (Q_{con}) and radiant (Q_{rad}) gains.

Table 11.5 Internal heat gains in typical buildings in W/m²

Building type	Use	Density of occupation (person/m²)	Sensible heat gain (W/m²)			Latent heat gain (W/m²)	
			People	Lighting	Equipment†	People	Other
Offices	General	12	6.7	8–12	15	5	–
		16	5	8–12	12	4	–
	City centre	6	13.5	8–12	25	10	–
		10	8	8–12	18	6	–
	Trading/dealing	5	16	12–15	40+	12	–
	Call centre (floor)	5	16	8–12	60	12	–
	Meeting/conference	3	27	10–20	5	20	–
	IT rack rooms	0	0	8–12	200	0	–
Airports/stations‡	Airport concourse	0.83	75	12	5	4	–
	Check-in	0.83	75	12	5	50	–
	Gate/lounge areas	0.83	75	15	5	50	–
	Customs and Immigration	0.83	75	12	5	50	–
	Circulation spaces	10	9	12	5	6	–
Retail	Shopping malls	2–5	16–40	6	0	12–30	–
	Retail stores	5	16	25	5	12	–
	Food court	3	27	10	†	20	§
	Supermarkets	5	16	12	†	12	§
	Department stores:						
	• jewellery	10	8	55	5	6	–
	• fashion	10	8	25	5	6	–
	• lighting	10	8	200	5	6	–
	• china/glass	10	8	32	5	6	–
	• perfumery	10	8	45	5	6	–
	• other	10	8	22	5	6	–
Education	Lecture theatres	1.2	67	12	2	50	–
	Teaching spaces	1.5	53	12	10	40	–
	Seminar rooms	3	27	12	5	20	–
Hospitals	Wards	14	57	9	3	4.3	–
	Treatment rooms	10	8	15	3	6	–
	Operating theatres	5	16	25	60	12	–
Leisure	Hotel reception	4	20	10–20	5	15	–
	Banquet/conference	1.2	67	10–20	3	50	–
	Restaurant/dining	3	27	10–20	5	20	–
	Bars/lounges	3	27	10–20	5	20	–

Source: Adapted from CIBSE Guide A 2015

Notes: The internal heat gain allowance should allow for diversity of use of electric lighting coincident with peak heat gain and maximum temperatures. Lighting should be switched off in perimeter/window areas (up to say 4.5 m) and or allowance accounting for any dimming or other controls.

† Equipment gains do not allow for heavy-duty local equipment such as heavy-duty photocopiers and vending machines.

‡ The exact density will depend upon airport and airplane capacity, the type of gate configuration (open or closed) and passenger throughput. Absolute passenger numbers, if available, would be a more appropriate design basis. Appropriate building-scale diversities need to be derived based on airport passenger throughput.

§ Latent gains are likely, but there are no benchmark allowances, and heat gains need to be calculated from the sources, e.g. for meals, 15 W per meal served, of which 75% is sensible and 25% latent heat (ASHRAE 2017b).

11.4 Total building heat gain

The total heat gain has two components, mean and alternating. These two components contribute to the air and environmental nodes as described in the following sub-sections.

11.4.1 Mean heat gain into the environmental and air nodes

This is the mean value over 24 hours and is given as follows for the air and environmental nodes:

$$\text{Environmental node: } \bar{Q}_{te} = \bar{Q}_{se} + \bar{Q}_{int} + \bar{Q}_{f} \tag{11.26}$$

$$\text{Air node: } \bar{Q}_{ta} = \bar{Q}_{sa} + \bar{Q}_{v} \tag{11.27}$$

11.4.2 Alternating heat gain into the environmental and air nodes

This is given as follows for the air and environmental nodes:

$$\text{Environmental node: } \tilde{Q}_{te} = \tilde{Q}_{se} + \tilde{Q}_{int} + \tilde{Q}_{f} \tag{11.28}$$

$$\text{Air node: } \tilde{Q}_{ta} = \tilde{Q}_{sa} + \tilde{Q}_{v} \tag{11.29}$$

11.5 Building classification and thermal response

Buildings are classified as having either a slow or a fast response to heat transfer. The response of a space to thermal input depends upon the:

- type of thermal input
- surface finishes
- thermal properties of the construction
- thickness of the construction
- furnishings within the space.

The heat input to the surfaces will be in the form of either:

- short-wave radiation (solar radiation and energy from electric lights)
- combination of long-wave radiation (from surfaces and other emitters) and convective exchange with the air.

The thermal response of a building is affected by:

- the thickness of the building's structural materials
- the properties of the structural materials (i.e. thermal conductivity, density and specific heat capacity)
- the relevant positions and orientations of the various construction elements of the building.

To determine the thermal response of a building, the following three parameters need to be known.

11.5.1 The admittance

Denoted by Y, this is the most significant parameter when using the admittance procedure. It is defined as 'the rate of flow of heat between the internal surfaces of the structure and the environmental temperature in the space, for each degree of deviation of the space temperature about its mean value' (CIBSE 2015) and it has the same unit as that of the heat transfer coefficient (U):

$$Y = \frac{\hat{q}_i}{\hat{\theta}_i} \qquad (11.30)$$

Where:
Y is the thet hermal admittance (W/m^{2}K)
$\hat{q}_i$ is the amplitude of heat entering the surface
$\hat{\theta}_i$ is the amplitude of temperature variation.

The associated time dependency of the thermal admittance takes the form of a time lead denoted by ω. The admittance equals the U-value of a building structure if this structure is thin (less than 100 mm in thickness) and consists of a single layer of material with a time lead equal to zero. In multi-layered constructions, it is the *inside* surface layer which primarily determines the admittance.

11.5.2 Decrement factor (f)

The decrement factor, as described in Section 11.3.1, is 'the ratio of the rate of heat flow through the structure due to variations in the external heat transfer temperature from its mean value (with the environmental temperature held constant), to the steady state conduction' (CIBSE 2015). The time dependency associated with f takes the form of a time lag denoted by φ. $f = 1$ and $\varphi = 0$ for thin structures of low thermal capacity value. With increasing thermal capacity of structure materials, f decreases as φ increases.

11.5.3 Surface factor (F)

The surface factor is defined as

> the ratio of the variation of radiant heat flow (from short-wave sources) about its mean value readmitted to the space from the surface, to the variation of heat flow about its mean value incident upon the surface. The associated time dependency takes the form of a time lag denoted by ψ.
>
> (CIBSE 2015)

The surface factor value decreases and its time lag increases with increasing thermal conductivity, but both remain virtually independent of the material thickness.

The physical process involved is that short-wave radiation is absorbed at the surface which, after a delay due to thermal storage, causes the temperature of that surface to rise. Heat is then transferred to the space in the form of long-wave radiation and convection. The effect is to raise the internal heat transfer temperature (i.e. environmental temperature). The response of a space to changes in environmental temperature is characterised by the admittance of the surfaces, which depends upon the long-wave emissivity, the surface heat transfer coefficient and the thermal properties of the structure.

Therefore, two time delays are associated with the thermal response of the space, one which applies to short-wave radiation and the other due to surface-to-surface and surface-to-air heat exchanges. Thus, it is possible for a space to be lightweight in terms of its response to solar radiation but heavyweight in terms of the change in temperature arising from other sources of heat input.

Buildings are categorised as fast or slow with regard to their thermal response to short-wave radiation as follows:

Fast: surface factor, $F = 0.8$ with a delay, ψ of 1 hour
Slow: surface factor, $F = 0.5$ with a delay, ψ of 2 hours

Response to the changes in the environmental temperature is characterised by the response factor, f_r, given by:

$$f_r = \frac{\Sigma AY + C_v}{\Sigma AU + C_v} \tag{11.31}$$

Where:
f_r is the response factor
$\Sigma(AY)$ is the sum of the products of surface areas and their corresponding thermal admittances (W/K)
$\Sigma(AU)$ is the sum of the products of surface area and corresponding thermal transmittance of surfaces through which heat flow occurs (W/K)
C_v is the infiltration/ventilation conductance (W/K) (see Eq.11.21)

Taking the response factor into account, buildings can then be classified as follows:

High thermal response ($f_r > 4$) → Slow response building (*heavyweight*)
Low thermal response ($f_r \leq 4$) → Fast response building (*lightweight*)

Nominal building classifications according to the response factor and values of the three parameters, Y, f_r and F, for buildings with slow or fast thermal response are provided in Table 11.6.

Table 11.6 Typical amplitude values of the admittance (Y), decrement factor (f) and surface factor (F) with time lag/lead values

Thermal Response Definition	Construction features (typical)	Response to short-wave radiation			
		Response factor, fr	Average surface factor, F	Time delay, φ (h)	Time lead for admittance, ω (h)
Slow	Masonry external walls and internal partitions, bare solid floors and ceilings	> 4	0.5	2	1
Fast	Lightweight external cladding, de-mountable partitions, suspended ceilings, solid floors with carpet or wood block finish or suspended floors	≤4	0.8	1	0

Source: Adapted from CIBSE Guide A 2015

11.6 Building cooling load calculations using the admittance procedure

The admittance procedure for cooling load calculation recognises that a person's feeling of thermal comfort depends on the heat exchanges between the body of a person to the indoor environment, by convective heat loss to the indoor air and by radiant heat loss to the indoor environment.

11.6.1 Operative temperature

To assess the cooling loads using the admittance procedure, the concept of *operative temperature*, as defined in Section 11.3, is used.

The mean operative temperature for a fixed ventilation rate is given by:

$$\bar{Q}_c = \frac{\bar{Q}_{ta} + F_{cu}\bar{Q}_{te}}{C_v + F_{cu}\sum AU} \tag{11.32}$$

Where:

$\bar{Q}_{ta}$ is the mean total heat gain at the air node, from Eq.11.27
$\bar{Q}_{te}$ is the mean total heat gain at the environmental node, from Eq.11.26
C_v is the ventilation conductance (W/K)
$\Sigma(AU)$ is the sum of the products of surface area and corresponding thermal transmittance over surfaces through which heat flow occurs (W/K)
F_{cu} is the mean room conductance factor with respect to operative temperature. Using standard heat transfer coefficient and emissivity values, F_{cu} is given by (CIBSE 2015):

$$F_{cu} = \frac{3(C_v + 6\Sigma A)}{\Sigma AU + 18\Sigma A} \tag{11.33}$$

The alternating value of the internal operative temperature caused by the building thermal storage is given by:

$$\tilde{\theta}_c = \frac{\tilde{Q}_{ta} + F_{cy}\tilde{Q}_{te}}{C_v + F_{cy}\sum AY} \tag{11.34}$$

$\tilde{Q}_{ta}$ and $\tilde{Q}_{te}$ can be obtained from Eqs.11.29 and 11.28, respectively. F_{cy} is the room admittance factor with respect to operative temperature and is given by:

$$F_{cy} = \frac{3(C_v + 6\Sigma A)}{\Sigma AY + 18\Sigma A} \tag{11.35}$$

$\Sigma(AY)$ is the sum of the products of surface area and corresponding thermal admittance (W/K).

The peak value of the internal operative temperature can then be calculated from:

$$\hat{\theta}_c = \bar{\theta}_c + \tilde{\theta}_c \tag{11.36}$$

Building overheating risk is detailed in CIBSE Guide A (2015).

11.6.2   *Building cooling load for a convective cooling system*

The cooling load is influenced by the characteristics of the heat emitter in the space and the temperature used for the control of the system. The following section outlines the method of calculating the cooling load for a convective cooling system. Equations for the calculation of cooling loads for a radiant or a combined convective and radiant system are given in Section 5 of CIBSE Guide A (2015).

For a convective cooling system (zero radiant component), the total sensible cooling required to control the operative temperature in a space is given by:

$$Q_k = \bar{Q}_a + \tilde{Q}_a + Q_{sg} + Q_v \tag{11.37}$$

Where:
Q_k is the total sensible cooling load (W)
$\bar{Q}_a$ is the mean convective cooling load (W)
$\tilde{Q}_a$ is the alternating component of the convective cooling load (W)
Q_{sg} is the cooling load due to windows and blinds (W)
Q_v is the cooling load due to air infiltration or ventilation (W)

11.6.3   *Mean convective cooling load*

The mean convective cooling load can be calculated as follows:

$$\bar{Q}_a = \bar{Q}_{fa} + F_{cu} \, 1.5 \, \bar{Q}_{rad} + \Sigma \bar{Q}_{con} - 0.5 \, \bar{Q}_{rad} \tag{11.38}$$

Where:
$\bar{Q}_{fa}$ is the mean fabric gain at air node (W) given by: $\bar{Q}_{fa} = F_{cu} \bar{Q}_f$
$\bar{Q}_f$ and F_{cu} can be determined from Eqs.11.3 and 11.33, respectively
$\bar{Q}_{rad}$ is the daily mean radiant gain (W)
$\bar{Q}_{con}$ is the daily mean convective gain (W)

11.6.4   *Alternating component of convective cooling load*

The alternating component of the convective cooling load is calculated from the following:

$$\tilde{Q}_a = \tilde{Q}_{fa} + F_{cy} \, 1.5 \, \tilde{Q}_{rad} + \Sigma \tilde{Q}_{con} - 0.5 \, \tilde{Q}_{rad} \tag{11.39}$$

Where:
$\tilde{Q}_{fa}$ is the alternating component of fabric gain at the air node, which is calculated from:
$\tilde{Q}_{fa} = F_{cu} \tilde{Q}_f$
$\tilde{Q}_f$ and F_{cy} can be determined from Eqs.11.6 and 11.35, respectively
$\tilde{Q}_{rad}$ is the alternating component of radiant gain (W)
$\tilde{Q}_{con}$ is the alternating component of convective gain (W)

11.6.5   *Cooling load due to windows and blinds*

The cooling loads due to windows and blinds can be determined in a simplified manner using tables 5.16(a – n) for unshaded and 5.17(a – n) for intermittently shaded glazing in

CIBSE Guide A (2015). These tables apply to certain locations in the United Kingdom as follows:

- tables 5.16(a) and 5.17(a) Belfast (54.7°N)
- tables 5.16(b) and 5.17(b) Birmingham (52.5°N)
- tables 5.16(c) and 5.17(c) Cardiff (51.4°N)
- tables 5.16(d) and 5.17(d) Edinburgh (55.9°N)
- tables 5.16(e) and 5.17(e) Glasgow (55.9°N)
- tables 5.16(f) and 5.17(f) Leeds (53.8°N)
- tables 5.16(g) and 5.17(g) London (51.5°N)
- tables 5.16(h) and 5.17(h) Manchester (53.3°N)
- tables 5.16(i) and 5.17(i) Newcastle (55.0°N)
- tables 5.16(j) and 5.17(j) Norwich (52.7°N)
- tables 5.16(k) and 5.17(k) Nottingham (53.0°N)
- tables 5.16(l) and 5.17(l) Plymouth (50.3°N)
- tables 5.16(m) and 5.17(m) Southampton (50.8°N)
- tables 5.16(n) and 5.17(n) Swindon (51.8°N).

The CIBSE tables also assume:

- constant internal temperature with cooling plant operating 10 hours per day (7.30–17.30)
- sunny spell of 4–5 days' duration
- *fast response (lightweight)* buildings; see Table 11.6 for characteristics.

For *slow response (heavyweight)* buildings, correction factors included at the end of tables 5.16(a – n) in CIBSE Guide A should be used.

Using these factors the cooling load due to glazing can be determined from:

$$Q_{sg} = C_f C_c Q_{st} A_g \tag{11.40}$$

Where:
Q_{sg} is the cooling load due to solar gain
C_f is the correction factor for building response
C_c is the correction factor for air temperature control
Q_{st} is the cooling load given in tables
A_g is the glazing area

The type of glazing greatly influences the solar gain, which is related to the G-value (total transmission) of the glazing, and so a simplification to Eq. 11.40 is to determine the correction factor for the glazing (G-value = shading coefficient × 0.87) and then use Figure 11.7 to determine the relevant correction factor to apply to the solar cooling load figures quoted in tables 5.16(a–n) for the relevant building response.

11.6.6   *Cooling load due to air infiltration (ventilation)*

The sensible cooling load due to infiltration/ventilation, Q_I, is calculated from Eq.11.22.

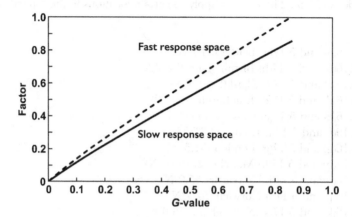

Figure 11.7 Cooling load: correction factor for unshaded glazing

Source: Adapted from CIBSE Guide A (2015),which also includes correction graphs for glazing with external, mid-pane and internal blinds.

The latent heat load due to infiltration/ventilation, Q_{VL}, can be calculated from:

$$Q_{VL} = \rho \frac{N}{3600} V(g_{ao.t} - g_{ai})h_{fg} \tag{11.41}$$

Where:
p is the density of air
N is the air changes per hour
V is the room volume

$g_{ao}.\tau$ and g_{ai} are the moisture content of external and internal air, respectively, at the time of load calculation.

h_{fg} = latent heat of evaporation of water.

If the values of ρ and h_{fg} are assumed to remain approximately constant with changes in ambient and room conditions, Eq.11.41 becomes:

$$Q_{VL} \approx 0.8 \, NV(g_{ao.\tau} - g_{ai}) \, (W) \tag{11.42}$$

Where: room volume is in m³ and moisture content in grams of water vapour per kilogram of dry air (g_w/kg_{da}).

These calculations for the cooling load are based on controlling the internal operative temperature. For control of the air temperature, the previous equations can be used after substituting the following:

- The operative temperature, θ_c, is replaced by the indoor air temperature, θ_{ai}
- F_{cu} and F_{cy} are replaced by F_{au} and F_{ay}, room conductance and admittance factors, respectively, for air temperature (°C) and are given by:

$$F_{au} = \frac{4.5\Sigma A}{4.5\Sigma A + \Sigma(AU)} \tag{11.43}$$

$$F_{ay} = \frac{4.5\Sigma A}{4.5\Sigma A + \Sigma(AY)} \quad (11.44)$$

11.7 Building heating load calculations

If used to calculate heating loads, the admittance procedure described in the previous section for cooling load calculations can cause an under-sizing of heating systems and a delay in achieving the building design internal operative temperature (or internal air temperature).

To avoid under-sizing of heating systems, the CIBSE Simple Model is recommended for heating load calculations.

11.7.1 CIBSE Simple Model for heating load calculations

According to this model, the heating load has the same value as the total of heat losses through the building, Q_{t}, which is the sum of the heat losses through the building fabric and losses due to ventilation/infiltration. It is calculated as follows:

$$Q_{t} = [F_{1cu} \Sigma AU + F_{2cu} C_v] (\theta_c - \theta_{ao}) \quad (11.45)$$

Where:
Q_{t} is the total building heat loss (W)
θ_c is the operative temperature (°C)
θ_{ao} is the outside air temperature (°C)
F_{1cu} and F_{2cu} are factors related to the heat emitter sources given by:

$$F_{1cu} = \frac{3(C_v + 6\Sigma A)}{\Sigma AU + 18\Sigma A + 1.5R(3C_v - \Sigma AU)} \quad (11.46)$$

$$F_{2cu} = \frac{\Sigma AU + 18\Sigma A}{\Sigma AU + 18\Sigma A + 1.5R(3C_v - \Sigma AU)} \quad (11.47)$$

R is the radiant fraction of the heat emitter source (typical values are given in Table 11.7).

Table 11.7 Typical values for the radiant and convected fractions of the heat source

Emitter type	Heat emission (%)	
	Convective	*Radiative (R)*
Forced warm-air heaters	1.0	0
Natural convectors, convector radiators	0.9	0.1
Multi-column radiators	0.8	0.2
Double/treble-panel radiators, double-column radiators	0.7	0.3
Single-column radiators, floor warming systems, block storage heaters	0.5	0.5
Vertical and ceiling panel heaters	0.33	0.67
High-temperature radiant systems	0.1	0.9

Source: Adapted from CIBSE Guide A 2015

As the F_{1cu} and F_{2cu} values are close to 1, it is an acceptable engineering approximation to estimate the total building heat losses through the fabric and ventilation/infiltration from:

$$Q_t = \sum AU(\theta_c - \theta_{ao}) + C_v(\theta_c - \theta_{ao}) \tag{11.48}$$

It should be noted that when the spaces surrounding the considered space are at different temperatures, heat transfer (losses or gains) will occur. Equation 11.48 can still be used, but with a modified U-value, U', as follows:

$$U' = \frac{U(\theta_c - \theta_c')}{(\theta_c - \theta_{ao})} \tag{11.49}$$

where θ_c' is the operative temperature of the opposite side of the partition.

Note: It is common for building services engineers to neglect solar and internal heat gains when calculating heating loads. The reason for this is to allow the selection of heating systems capable of providing the required heating loads under extreme conditions.

11.8   Summary

In this chapter, ways of calculating building heat gains and losses and cooling and heating loads arising from these gains have been explained. The concept of building admittance has been introduced, and its use to determine the thermal behaviour of buildings has been highlighted. The CIBSE admittance procedure outlined in the chapter is detailed in CIBSE Guide A (2015), which also provides the data tables required for its application in cooling and heating load calculations. The CIBSE Simple Model for the calculation of heating loads has also been introduced.

12 Building electric power load assessment

12.1 Introduction

This chapter introduces simple methods of load assessment for electric power systems in buildings. The need for load assessment as part of the design development process has been outlined in Chapter 2. It is also a necessary step in the selection of an energy strategy to minimise carbon emissions, as outlined in Chapter 3. A brief overview provides the context for some of the key issues for power system infrastructure design related to selection of equipment locations in buildings. The specific requirements for load assessment are outlined, together with a review of the nature of load patterns and profiles for electricity in buildings. The main methods for load assessment are described, including guidance on applying diversity factors. In practice, early estimation is usually based on load per unit area or unit of accommodation types. Typical load densities are provided for a range of systems and types of functional spaces in buildings. In commercial or public buildings, the electric power load is usually dominated by the main items of the mechanical plant in the heating, ventilation and air conditioning (HVAC) systems. Simple methods of early-stage load assessment are described for the main items of the mechanical plant, together with methods for some of the more problematic electrical loads, such as process loads. An introduction is provided for the more complex data processing loads and those supported by uninterruptible power supply (UPS) systems. The relevance of day, night, summer and winter load variations is outlined. A sample tabulation method is provided to illustrate a simple initial load assessment. Most electrical systems have a considerable proportion of non-linear load that gives rise to harmonics. This chapter concludes with a brief discussion on the impact of harmonics on equipment sizing and capacities.

Thermal load assessment has been covered in Chapter 11. Thermal loads are directly related to external conditions, the thermal performance of the building envelope, internal gains and occupancy levels, together with any processes within the building. Electric power loads are more directly related to floor area and are also a function of the HVAC systems that address the thermal loads and occupancy levels and processes. So an electric power load assessment cannot be undertaken without suitable information on the HVAC systems proposals. Electric power loads can be best understood on a system-by-system basis. It should be emphasised that load assessment is about power load flow at a point in time and should be clearly distinguished from energy demand, which is about accumulated energy consumption over a period of time. However, designers should recognise that an awareness of load profiles will be necessary for various aspects of decision-making for energy strategies and equipment selection and sizing.

It should be noted that the approach shown here is based on certain assumptions and simplifications, as outlined in Section 12.12. As this section relates to load assessment, for the purposes of simplicity, the schematic diagrams have largely been kept as indicative concepts only (so, for example, they do not indicate circuit breakers/protective devices).

12.2   Basic elements of a power system infrastructure

Load assessments should be related to the relevant parts of a power system infrastructure, as well as to the overall building or site and the incoming power requirements. Similarly, the planning of a power system needs to take account of the locations and magnitudes of the main loads, so power system planning and load assessment are inextricably linked, as outlined in Chapter 2, Section 2.5.1. It is useful, therefore, to start by reviewing those aspects of infrastructure that are most relevant to a load assessment.

Nearly all major power distribution networks operate with alternating current (AC) and consist of a multitude of interconnected items of generation, transmission and distribution equipment. In the UK this is known as the 'grid' or 'National Grid'. The AC frequency is 50 Hz (with close tolerances). The grid provides a reliable and stable supply of electricity. The UK transmission system comprises numerous dispersed power stations feeding an extensive transmission system, which in turn feeds local distribution to consumers. Transformation of AC power is the basis of modern interconnected grid systems and extensive local distribution. Distribution in cities and towns is typically at 33,000 or 11,000 V. This is usually defined as being within the 'medium voltage' (MV) band, although by tradition, building services engineers will often refer to this as 'high voltage' (HV) and sometimes use both terms interchangeably. The term HV is used here. The generation, transmission and distribution systems are planned to meet the relevant identified load (and to some extent the anticipated future load at each location), with equipment capacities selected accordingly. Inevitably different parts of the distribution infrastructure will have different levels of spare capacity as the load pattern changes over time. In some locations the additional load required for a proposed new development might require reinforcement of the local network to a greater or lesser extent. It is often the case that the developer of the site will incur some or all of the associated costs for the extent of reinforcement works that are required.

Figure 12.1 shows a fictitious 11,000-V ring main, in indicative format, as might be used in a part of a town or city to feed a variety of consumers – buildings or other facilities – or could be used for a single site with a significant load. This chapter relates to the load assessment for individual buildings or developments, but it is important to understand that the same principles apply further to each 'upstream' part of the system, i.e. towards the transmission network and the sources of supply. In the case of a ring main as shown, the individual maximum demand estimates for each building will be relevant, but should be considered in relation to the specific load patterns and when the anticipated maximum demands occur for each building. It should be evident that there are considerable differences in the daily (diurnal) and seasonal load patterns for these building types. For example, the office building will have its highest load during the daytime, with a considerable reduction in the early morning and evening, while the hotel is likely to have its highest load during the early morning and evening, with a much reduced load during the day. The college building is likely to have a minimal load during the summer. The design of the 11,000-V infrastructure will relate to the anticipated maximum demand and load pattern for the ring main and should include a suitable allowance for future spare capacity and expansion.

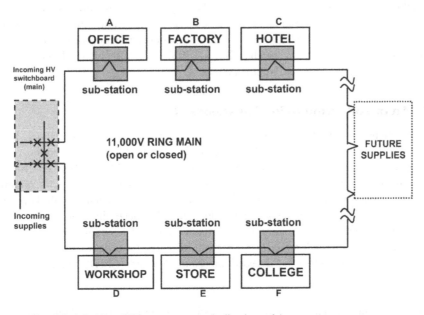

Figure 12.1 Simplified fictitious HV power system (indicative only)

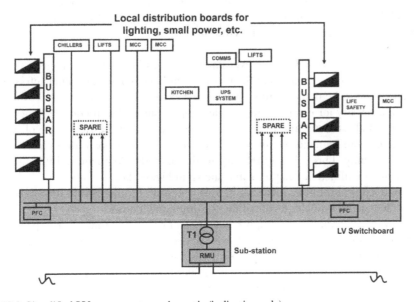

Figure 12.2 Simplified LV power system schematic (indicative only)

Figure 12.2 shows a greatly simplified low-voltage (LV) power system schematic, in indicative format, as might be used in an office building. The system has been shown with a sub-station, with a transformer fed from a ring main unit (RMU) on the HV side, an LV switchboard, rising main busbars and various distribution boards, motor control centres (MCCs) and other equipment. Load assessment is required at all parts of a system in order to size all items of equipment, but is of particular importance in relation to the capacity of the LV switchboard, transformers and the capacity of the incoming supply. The actual ratings of equipment will, of course, be part of the design development process covering all the necessary regulations and criteria (such as BS7671 (BS7671 2018)), of which the load assessment is only one aspect, albeit an important aspect.

12.3   An introduction to load assessment

Load assessment is required for a number of reasons. Its primary reason is so that decisions can be made for the power system design, the capacities for plant and equipment and the incoming supply provision based on the maximum predicted load (BS7671 2018; CIBSE 2004b). Assessment of the load profile is required for design decisions affecting the strategy for minimising carbon emissions, such as assessing the viability of co-generation (together with thermal load profiles) and the relative potential contribution for renewable electricity generation.

A key requirement is to estimate the maximum demand of a system, or individual parts of a system. The estimated load figure for the whole building is usually called the 'estimated maximum demand' (EMD) or the 'assessed maximum demand' (AMD). This assessment will be used to:

- determine the capacity of the components in a power system (and hence the physical sizes, so it will be of relevance to space planning)
- allocate a suitable capacity of incoming electricity supply and determine the associated availability of service capacity and the energy costs from the utility supplier (and hence the contract charges for the client/customer/consumer).

The design decisions based on load assessment must consider the likely usage of the installation (CIBSE 2008a). As these parameters are fundamental to the initial infrastructure design, they need to be made at an early stage, when very little information is available. It is therefore necessary for engineers to be able to apply suitable techniques so that they can make basic initial estimates and exercise suitable judgement at the early stage of a project and then update and amend the assessment as the design develops with a greater level of detail and certainty. This is highly relevant to the brief development and iterative design development, as outlined in Chapter 2 and shown in Figure 2.6.

For a load assessment to be meaningful, it is necessary to understand the nature of the individual loads and how these contribute to summated loads for 'upstream' parts of the system, towards the source of supply. An important factor is that electric power loads are dynamic. For example, lighting loads will vary with occupancy and usage patterns and the changing daylight level; small power loads will vary with occupancy patterns and activities during the day; passenger lift and escalator loads will vary with inter-floor traffic patterns; and HVAC equipment loads will vary in relation to a mixture of external temperature and humidity, incident solar irradiation, occupancy patterns and internal gains. At any point in a system, it is necessary to establish the 'worst-case scenario' for power demand. This is rarely just a simple summation of the maximum load for each individual circuit. Instead, it usually involves a more nuanced approach reflecting the likely dynamic load scenario – both initially and in the future – so that assessments properly represent anticipated load patterns. The potential impacts of incorrect load assessment need to be fully understood, because they illustrate the importance of obtaining a realistic estimate for maximum demand at an early stage.

If the demand estimate is oversized, several issues could potentially arise:

- unnecessary capital cost and a waste of resources (including embodied energy/carbon)
- potentially oversized equipment, which could result in operating inefficiencies (and hence additional energy cost and carbon emissions)
- waste of space, and hence cost
- increased maintenance cost
- increased service charge and/or penalty.

If the demand estimate is undersized, several issues could potentially arise:

- equipment capacities could be insufficient and the design would not meet relevant codes
- safety issues, such as overheating failures, faults or fire
- inadequate space for plant (in sub-stations, switchrooms, risers, etc.)
- the supplier's system cannot satisfy demand, with business implications (for example, potential nuisance tripping of supplies).

Furthermore, design errors would be compounded if the erroneous load assessment data in W/m^2 is used for calculating internal gains, as this could affect HVAC equipment sizing, and hence efficiencies. If the anticipated load profile is incorrect, it could have a misleading influence on decisions made for the building/site energy strategy.

It should be recognised that considered engineering judgement is necessary, ideally based on experience from previous projects of a similar type. However, it should also be recognised that load estimation is not an exact science. Historically, engineers have tended to adopt a cautious approach. There is much experience of predicted loads not materialising in practice, perhaps as a result of an overcautious approach and a lack of insight and dialogue with the client regarding the likely operational pattern. Alongside the need to understand loads in existing buildings and avoid being overcautious, it is necessary to make suitable provision for future flexibility and for potential changes in building usage, although this aspect is beyond the scope of this book.

The need to address climate adaptation in design, particularly in relation to HVAC equipment, is also an issue in electric power load assessment, but is beyond the scope of this book.

12.4 Load patterns and profiles

Load estimation would be relatively easy if all electric power loads were of fixed magnitude, purely resistive and switched on and off for known periods. But in reality, most of the loads in buildings are more complex and dynamic:

- variable in magnitude (to some extent)
- partly reactive
- often non-linear, sometimes with significant harmonic content
- switched on and off for known or unknown periods
- sometimes transient.

Unfortunately, to make the assessment more complicated, the largest loads in buildings can often be the most difficult to estimate reliably at the early stage, thereby requiring a

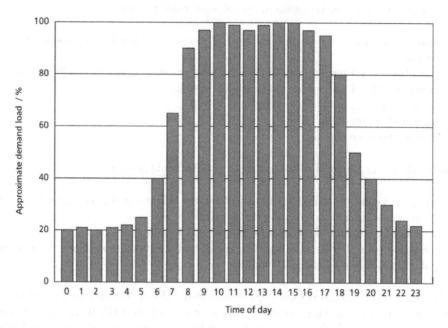

Figure 12.3 Typical load profile for an office building

Source: Reproduced from CIBSE Guide K (2004b) with the permission of the Chartered Institution of Building Services Engineers

particular focus in initial estimates. It should be noted that load estimation deals with steady-state conditions, so transient conditions can largely be ignored. Two particular aspects of load variation should be appreciated – diurnal (i.e. through the course of the day) and seasonal. These aspects are, of course, relevant to thermal systems as well.

Figure 12.3 shows a typical load profile for an office building, with approximate load variation through a 24-hour period. It can be seen that the peak period lasts for about ten hours. For approximately eight hours during the night the load reduces to a minimum of about 20% of the peak; this represents the 24-hour 'base load' (i.e. the minimum load that is continuous for 24 hours). A full assessment would include further 24-hour profiles to show the seasonal variations. In the case of a site development with a number of buildings, while each building's power infrastructure would, of course, be required to satisfy its own load profile, the infrastructure for the whole site would have to satisfy the accumulated profile.

A useful way to understand load build-up for an individual building is to look at the typical breakdown of the EMD load in terms of systems. It should be noted that this is different from a breakdown of the annual energy consumption. Figure 12.4 shows the diversified loads for a typical office building on a business park, with the mechanical services dominant at 44%, followed by lighting and small power as the other significant system loads. Figure 12.5 shows a comparative breakdown for a typical healthcare building. The generally less intensive HVAC systems, and the significant loads for catering and medical equipment, result in a smaller proportion for the mechanical services, while lighting represents a higher proportion. It should be noted that these diagrams represent relative proportions of total loads in the sample buildings, rather than the absolute loads.

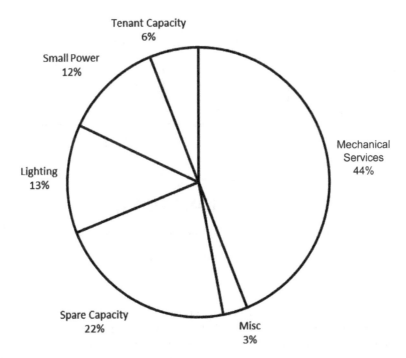

Figure 12.4 Maximum demand breakdown example: business park office building

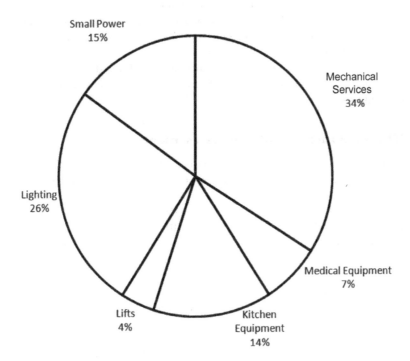

Figure 12.5 Maximum demand breakdown example: healthcare building

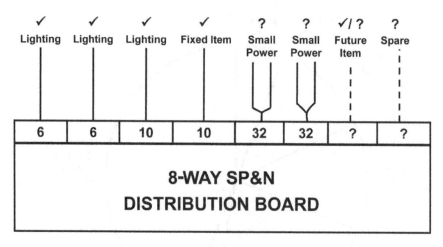

Figure 12.6 Load assessment issues for a simple distribution board

As we attempt to estimate loads, it is worth noting that even at the completion stage for a project, it is difficult to determine the whole load with any degree of certainty. The load pattern in practice may not reflect the anticipated usage in the client brief at the outset of the project. The load pattern will only emerge with full occupation and usage (in accordance with the design brief), hence the importance of post occupancy feedback and monitoring to inform the design. If an example of a simple distribution board is taken, as shown in Figure 12.6, it can be seen that while the loads might be known for lighting circuits and fixed items of equipment, the load for small power ring circuits (feeding 'unregulated' loads), future and spare items cannot be known with any certainty.

It is clear that some circuit loads are easy to estimate, while others are not. But to size equipment, it is necessary to make an objective assessment of the unknown factors, using engineering judgement.

12.5   The main methods of load assessment

Various methods are available to assess loads, either separately or in combination. Some of these are only applicable when the design has been well developed. Examples are:

- 'rules of thumb', or other guidance using typical load densities for types of buildings, types of areas within buildings or types of systems (CIBSE 2004b; DOH 2007; Hawkins 2011)
- applying diversification to estimated design loads or connected loads
- average figures for similar buildings or systems based on monitored in-use records
- client guidance
- manufacturers' information
- information from the mechanical systems designer
- information from specialists.

Most assessments will require a mixture of these methods at different stages of the design. Because initial assessment is required at an early stage, it usually involves applying 'rules of

Table 12.1 Typical load densities and unit loads by building type

Type of building	Typical load density
Office building (air conditioned)	80 W/m^2 to 100 W/m^2
Hotel (air conditioned)	4000 W/room (*)
School	32 W/m^2
Warehouse (standard)	22 W/m^2
University	25 W/m^2 to 35 W/m^2
Residential accommodation	0.5 kW to 1.8 kW/dwelling (*)

Note: * These figures could be diversified for large multiples.

thumb' and diversity. As the Building Regulations in England and Wales provide certain limiting design criteria (HMG 2013, 2016), these can also be useful for assessing loads.

Table 12.1 shows some useful 'rules of thumb' load density figures for whole buildings or groups of buildings. These are typical average figures for use at the earliest stage of a project, when there is limited information on the facilities to be included. In each case the figures assume that there is no electric heating. The figure for residential accommodation relates to average loads for typical sizes of standard quality accommodation. Higher figures would, of course, apply to larger properties;properties with a large number of residents; luxury houses and apartments; and properties with air conditioning, large audio-visual systems or other power-consuming features. For residential loads, the nature of accommodation has become more varied, particularly with the spread in occupancy numbers, reflecting societal changes. This has made estimation of loads per unit of accommodation difficult. A report by Imperial College London addressed this by using a demand diversification analysis, based on data from smart meters in more than 3,000 households during 2013. The large dataset allowed patterns of consumption to be correlated with demographic, i.e. household income and occupancy. This showed a significant range of diversified peak demand for different demographic circumstances, ranging from 0.54 kW for single-person occupancy with adverse economic status to 1.78kW for occupancy by three or more people with affluent economic status. The study suggested that while estimates of diversified peaks for winter household demand in the UK has traditionally been 1.5 kW to 2 kW, a more suitable figure would be 1 kW (Konstantelos *et al.* 2014).

All load assessment figures should be reviewed and updated as the brief and design develop with more certainty and the specific equipment is determined.

12.6   Diversification and diversity factors

The numerous items of electrical equipment in a power system will each have their own pattern of operation, but when the overall load is examined, the pattern will be seen to be diverse, with different items reaching their peak loads at different times. This is known as diversification or diversity of operation and is applied in load assessments where the overall EMD at any point in the system is a lower figure than the sum of the individual maximum connected loads.

The key factors that give rise to diversity are:

* multiplicity of similar items which are unlikely to be used concurrently
* disposition of spaces on a site or in a building, which are used at different times

- usage pattern, operation or process, including control features to minimise energy consumption
- time of day (diurnal variations)
- seasonal influence.

Several multiplying (or dividing) factors are used in load assessment calculations to reflect the dynamic and non-coincident nature of multiple loads on a system, which means that aggregate loads do not equal the sum of individual loads. The definitions of these factors can be confusing and contradictory, and they are used in different ways by different sectors in the engineering industry. The tradition in building services electrical engineering is to use 'load diversification' to seek the diversified load by applying a 'diversity factor' (DF):

$$\text{Diversified load} = \text{Maximum system load} \times \text{DF} \tag{12.1}$$

The diversified load is the actual maximum demand, whereas the maximum system load is the summated maxima of all the individual loads (sometimes known as the 'total connected load'). DF applied in this way will be used in this section. It is always less than 1 (but might, very rarely, be equal to 1). It is sometimes also known as 'coincidence factor' or 'demand factor'. To avoid confusion, it should be noted that it is actually the reciprocal of the officially defined 'diversity factor', which is also known as the 'simultaneity factor'.

The application of diversity can be understood by reference to the hypothetical LV distribution system shown in Figure 12.7, which has been drawn to illustrate the point. This shows ten items of distribution equipment, or load centres, with the main switchboard identified as item 10. It is necessary to assess the load at each item of distribution equipment, as well as for the main incoming supply. If we take the distribution board labelled 1, then each circuit might have design currents I_A, I_B, etc., as shown. While it would depend upon the nature of

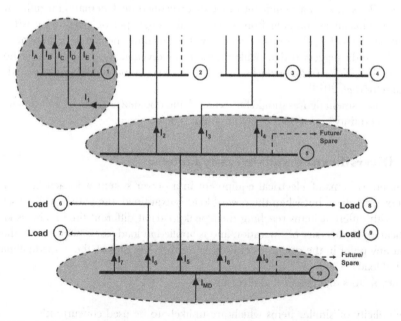

Figure 12.7 Design loads for distribution centres (relates to diversified summation)

the loads, the maximum demand for the distribution board, I_1, is likely to be less than the sum of the individual circuit design currents, as they would not necessarily all occur at the same time. Therefore, the design load, I_1 would be:

$$I_1 = (Ia + Ib + Ic + Id + Ie)\,(DF_1\,SA_1) \qquad (12.2)$$

Where:
I_1 = design current for distribution board 1
Ia, Ib, Ic, Id and Ie = design currents for final circuits a, b, c, d and e, respectively
DF_1 = diversity factor for distribution board 1
SA_1 = spare allowance factor for distribution board 1

There would be some diversity, which would depend on the nature of the circuits, perhaps 0.7, for example. It would be necessary to allocate some spare capacity for the spare circuit(s) allocated and some future load growth, perhaps 20%. So, the design load for distribution board 1 would become:

$$I_1 = (Ia + Ib + Ic + Id + Ie)(0.7)(1.2) \qquad (12.3)$$
$$I_1 = (Ia + Ib + Ic + Id + Ie)(0.84) \qquad (12.4)$$

Distribution Boards 2, 3 and 4 might be similar to distribution board 1 and might have similar loads. However, when we assess the load at switchboard 5, this might not be the sum of the design currents for I_1, I_2, I_3 and I_4 (plus the future/spare provision), as these design currents might not all have their maxima at the same time. Thus, there will be some further diversity, and the design current would be:

$$I_5 = (I_1 + I_2 + I_3 + I_4)(DF_5\,SA_5) \qquad (12.5)$$

In this case it might be considered that the most appropriate DF is 0.6 and the future and/or spare allowance should be 15%; thus, the design current for switchboard 5 would be:

$$I_5 = (I_1 + I_2 + I_3 + I_4)(0.6)(1.15) \qquad (12.6)$$
$$I_5 = (I_1 + I_2 + I_3 + I_4)(0.69) \qquad (12.7)$$

The same approach will apply in assessing the total load at switchboard 10, the design current of which represents the building's maximum demand I_{MD}, and so on to the external supply infrastructure. There is no exact science to the application of DFs, which relates more to engineering judgement based on an understanding of likely patterns of operation.

In this example design currents have been used for simplicity in relation to the schematic diagram, but design kVA figures would normally be used instead.

12.7   Load assessment by system

Lighting and small power loads tend to be evenly distributed within buildings and are fairly simple to assess using a load density approach. However, the key loads that need to be assessed require more careful consideration. Guidance on these load assessments is included in the following sections.

Catering equipment loads will be dependent on the specific brief, proportion of electric to gas equipment and the electric equipment selection. These can be of considerable magnitude and are usually associated with high ventilation air flow rates, so the total load related to catering facilities can be considerable. These loads are a special case and are not considered here.

12.8   Lighting

Assessing the load for a general illumination lighting system is relatively easy compared with other systems. However, it is important to recognise that a lighting load:

- varies with time due to the occupancy pattern
- varies with control usage, i.e. through the use of dimming, time switching, daylight-linking or other lighting management features, which is an important energy efficiency feature, as described in Section 10.4.

Reliable application factors were in use for many years for conventional office buildings with lighting by luminaires with T5 fluorescent tubes, where load density was often taken as 9–11 W/m², and diversity as 90% (0.9). While these figures might still be valid where fluorescent lighting is still used, the rapid rise in development and usage of light-emitting diode (LED) lighting has resulted in reduced load densities, typically 6–8 W/m². However, this might require adjustment, as there are various approaches to the spacing, layout and selection of LED luminaires to achieve the desired lighting outcome. Lighting load densities are useful for initial assessment, but it is important to only use them where appropriate. In particular, task lighting should be included where applicable. This is often supplied from separate distribution boards and would therefore be allocated as a small power load to the appropriate part of the distribution system. For conventional fluorescent lighting applications, broad guidance on the target range of power densities is given in Table 12.2, which shows average installed power densities for different lamps and illuminance levels. These figures relate to efficient lamps and luminaires in good-quality installations with high surface reflectances and a high degree of installation maintenance (CIBSE 2002). As noted in Chapter 10, fluorescent lighting is largely being replaced by LED lighting, so these figures relate more to legacy installations.

It is important when applying lighting W/m² figures to be aware of the key factors that will increase the power density compared with a conventional arrangement. These include increased mounting height (and/or room geometries that affect the 'room index'), increased

Table 12.2 Typical average lighting power densities for different lamps and illuminance levels (prior to widespread usage of LED light sources)

		Lamp type (pre-LED)		
		Fluorescent triphosphor	*Compact fluorescent*	*Metal halide*
Task	300	7 W/m²	8 W/m²	11 W/m²
Illuminance	500	11 W/m²	14 W/m²	18 W/m²
(Lux)	750	17 W/m²	21 W/m²	27 W/m²

Source: Derived from CIBSE (2002): table 2.5

illuminance levels, less satisfactory maintenance, and a design requirement for a high level of colour rendering. Each of these factors will increase the power density figure, and a combination of these factors could result in a considerable increase in W/m².

The load for external lighting should only be included in the overall maximum demand assessment, where its usage will coincide with periods of worst-case demand. For buildings such as sports stadia and entertainment centres that can have their peak activities in the evening, this is likely to be concurrent with their maximum demand period.

12.9 Small power

Small power systems mainly consist of socket outlets and therefore have a variable load, although they can also contain fixed items of equipment with a more predictable load. There is no precise method of estimating the load for circuits with socket outlets. The nature of the circuits is to allow flexibility of usage to suit user requirements (i.e. 'unregulated' loads), so the load will vary accordingly. However, the load will nearly always be only a relatively small proportion of the circuit capacity. The best approach is to understand the likely usage through discussion with the client; observation of usage in similar locations; or, preferably, data logging of metered power usage in similar areas.

Small power usage will vary depending on the type of functional space. In some spaces there might be a considerable number of sockets, but their usage might be primarily for cleaning purposes outside the main occupancy period. In such cases, the load contribution might not be appropriate in a maximum demand assessment. Examples would be sports halls, foyer or circulation areas and classrooms. Careful consideration should be given to those areas where the sockets provide power to continuous functional facilities as part of the business process. This would include offices, but also spaces such as laboratories. For offices it is necessary to know the types of desk-based equipment, such as PCs and monitors, and ancillary equipment such as printers and photocopiers. It is also necessary to know the density of occupation and the intensity of equipment usage. Many offices have under-floor power distribution via busbars to floor boxes, or directly to desks, with multiple outlets, providing high flexibility and the potential to connect a dense coverage of equipment. Some sense of the occupancy density and intensity of usage will assist in the assessment, which could be based on a unit area or a Watts per person basis.

As with most other electrical equipment, the actual average power consumed by PCs, monitors and other office equipment is, on average, often well below the nameplate rating. The load density will be influenced by the population density, which can vary widely. It might be the case that the anticipated nameplate equipment loads in a localised area might total to 30–45 W/m², and in the past some designers used these figures. There is little reliable information about small power load densities that reflects equipment and usage patterns in contemporary offices. A study for the British Council for Offices (BCO) in 2014 found that loads could be up to 40% lower than traditional assumptions. It found that for high occupation densities (8 sq.m per person), power consumption rarely exceeds 19 W/sq.m (BCO 2014). A study in 2005 (based on surveys at 30 air-conditioned offices between 2000 and 2002) found that small power loads averaged 17.5 W/sq.m, with a range of 6–34 W/sq.m of treated floor area (Dunn and Knight 2005). There is a clear indication that load density has been reducing due to improvements in the energy efficiency of office IT equipment and due to a move towards more flexible and mobile working; and that load density is related to occupancy density. However, there is also a recent tendency for some offices to have two monitors per desk, which might counteract the reduction to some extent.

For standard occupancy densities, it would seem that about 15 W/m^2 would be a suitable estimate at the early design stages and should then be revised as more information becomes available during later stages, with a strong likelihood that it would be reduced. A figure of 10–15 W/m^2 might be sufficient for non-intensive general office areas, perhaps rising to 15–20 W/m^2 for more densely populated spaces. However, at the early design stage, a figure of 15 W/m^2 might be suitable when applied to the whole floor area, including circulation areas. The exceptions would be for highly intensive use, such as dealers' rooms, where loads of 350–550 W per desk, or 50–75 W/m^2, can be applicable in certain localised areas, but might require diversification.

It is important to understand that load assessment figures may be quite different from the capacity that needs to be allocated. An example is where the installed systems and their power infrastructure require capacity ratings to satisfy industry-sector body standards, such as recommendations of the BCO.

As stated in Section 12.8, task lighting loads should be allocated where fed from distribution boards for small power. Task lighting often has a high diversity of usage, so an assumption would need to be made for a suitable diversity factor.

12.10   Mechanical plant

For most modern commercial or public buildings with HVAC systems, the mechanical plant will usually represent a dominant proportion of the maximum electric power demand. It is therefore essential that the engineering systems are designed in a coordinated manner, with close liaison and clear communication between the mechanical and electrical systems designers. The largest loads are normally chillers (and their ancillary equipment), fans and pumps, all of which are motor loads, and hence are inductive loads. The overall figure for an air conditioning system will depend on many factors, including the cooling load, occupancy levels, ventilation rates and type of systems selected. Typical power loads for HVAC systems can often be in the range of 50–70 W/m^2, but can also be outside this range for certain types of systems.

12.10.1   Cooling systems

Chillers are regularly the largest items of load in the electrical distribution systems for air-conditioned buildings. It is most important for the electrical designer to understand the details of the cooling equipment, and all the other components in the cooling system, so that the correct total electrical load figures can be applied. The mechanical designer will usually describe chillers in terms of maximum cooling power, i.e. kW or MW of cooling capability, sometimes written as kWc or MWc. Electrical designers must not confuse this with electric power, sometimes written as kWe or MWe. The relationship between the two is described by the coefficient of performance (COP) where:

$$\text{COP} = \frac{\text{Cooling power output (at evaporator)}}{\text{Electrical power input (to compressor motor)}} = \frac{\text{kWc (thermal)}}{\text{kWe (electrical)}} \quad (12.8)$$

A higher COP represents a better performance or energy efficiency of the cooling production process. The COP will vary with the type of cooling equipment. It might also vary with the operating conditions and the proportion of load, so part-load and seasonal aspects need to be addressed.

Typical COP ranges for chillers are:

Air-cooled heat rejection: COP of 2.8–3.2
Water-cooled heat rejection: COP of 3.2–3.8

For example, for a 1200-kWc water-cooled chiller with a COP of 3.6:

$$\text{Electrical power} = 1200 \text{ kWc}/3.6 = 333 \text{ kWe} \tag{12.9}$$

The chiller plant will usually be associated with ancillary equipment for heat rejection, such as condenser water pumps and cooling towers or condenser fans, depending on the specific cooling and heat rejection arrangement. Chillers operate with capacity control to satisfy system demand at part-load, and this should be addressed in the load calculations. Seasonal part-load operation may have a lower COP, so this should be considered, where appropriate.

Once the power input has been established for each chiller, it is necessary to understand how the overall cooling system works, so that the concurrent loads can be determined. Figure 12.8 shows an arrangement of a central plant in a chilled water system concept schematic, indicative for this purpose. The questions that should be addressed for the worst-case concurrent operation are:

- How many chillers operate together, and are they sharing the load equally?
- At what proportion of load are they operating?
- Do they have integral pumps?
- What is the arrangement for any separate heat rejection in terms of condenser fans, cooling towers, pumps and so on?
- How many primary pumps and secondary pumps run together; do they have variable-speed drives; and what is their operating point (to determine the power absorbed)?

The electrical load pattern can only be understood by fully understanding the dynamic operation of the mechanical system to meet the cooling demand and determining the worst-case concurrent electrical load and the time(s) at which it occurs.

12.10.2   *Heating systems*

For conventional gas-fired 'wet' heating systems, the only items requiring electric power are the boilers (for the burners and ancillaries) and the circulating water pumps. Power requirements for boilers will be a function of the equipment rating and characteristics. Power requirements for heating pumps in multiple boiler systems will require the same considerations as outlined earlier for the primary and secondary pumps in multiple chiller systems.

If electricity is used directly as the fuel for all or part of the heating for a building, the electric power load assessment will be related to the relevant thermal heating load assessment. For heating by heat pumps, see Section 12.10.3.

Other items of electric heating equipment can be included in HVAC systems, which often have significant loads. These include frost coils (located in the air intake to air handling units to prevent freezing of filters) and heater batteries (which might be present in close-control air handling units, for example, in computer rooms). It is important not to overlook these in the load assessment.

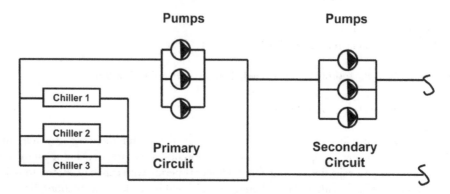

Figure 12.8 Example chilled water system (simplified and indicative only)

12.10.3   Heat pumps

Where heat pumps are used for heating, similar considerations apply to electrical power input as described earlier for chillers, but using the relevant COP for heating. This figure will vary with the external temperature. The worst-case figure for total winter peak summation will usually relate to the external temperature at the time when the rest of the load is at a peak, which may not necessarily be the worst-case external temperature. For reverse-cycle heat pumps, the load assessment should use the relevant COP figures when in cooling mode and heating mode, and these should be added to the worst-case summer and winter EMD figures accordingly.

12.10.4   Fan coil units and other distributed equipment

HVAC equipment can often be in the form of distributed items that are either stand-alone or part of a system that includes a central plant. Examples are fan coil units (FCUs) and fan-assisted terminal boxes for variable air volume (VAV) systems. In all cases the electric power loads should be sought from the mechanical designer, together with an indication of the operating mode, and hence the load figure that will be concurrent with other worst-case figures. It should be noted that VAV terminal boxes can have heating coils that are electric or use the LTHW heating system. With decarbonisation of the grid, it is likely that heating coils will be increasingly selected for electric operation. This can have a major impact on the load.

As outlined in Chapter 10, electronically commutated/direct current (EC/DC)FCUs have different power characteristics from conventional FCUs.

12.10.5   Fans and pumps in HVAC systems

The absorbed motor power associated with fans and pumps will depend on the specific parameters of the air and hydronic systems, respectively, and the relevant operating points on the fan or pump characteristic curves. The mechanical designer will advise on the power ratings, and it is important to understand how the systems work, so that the individual operational loads and the concurrent loads can be determined accordingly. The electrical power input to the fan or pump will consider the serial efficiencies of the fan or pump, the motor

Table 12.3 Typical pump/fan plant efficiencies

Item	Efficiency ratio
Fan or pump (small)	0.6–0.7
Fan or pump (large)	0.7–0.75
Motor efficiency	0.9–0.95
Variable-speed (inverter) drive	0.96–0.98
Belt drive	0.95

and the drive arrangement. Table 12.3 shows the range of typical plant efficiencies for fans and pumps.

For example, for a large pump operating at 50 kW, fed from a variable-speed drive and with efficiencies at the upper end of the range, the overall electric power input, P, would need to allow for the pump efficiency ratio (0.75), the motor efficiency ratio (0.95) and the VSD efficiency ratio (0.98)). So the input power P would be:

$$P = 50/(0.75).(0.95).(0.98) = 71.6 \text{ kW} \qquad (12.10)$$

For new non-domestic buildings, an estimate of power loads for fans in relation to area or ventilation flow rate can be made at an early stage (where the Building Regulations for England and Wales will apply). This approach to estimation takes account of the total power for all the supply and extract fans for a building. The compliance guide for Building Regulations Part L2A (HMG 2013) sets a limit for the specific fan power (SFP), which therefore provides a useful figure for use in load assessment, as it relates to guidance criteria in order to achieve compliance. Section 3.8.2 gives the definition of SFP and shows the limiting figures in terms of Watts per litre per second.

Therefore, from an awareness of the building population and the design ventilation rate in litres per second per person, the maximum total fan power for the building can be estimated. For example, if an office building has a design population of 1,500 people and a maximum ventilation air flow rate of 14 litres per second per person is assumed, the total air flow rate would be 21,000 litres per second. For a central balanced mechanical ventilation system with heating and cooling, the SFP is 1.6 W per l/s. Therefore, the maximum total fan power for supply and extract is 33,600 W or 33.6 kW.

This approach could be used as an installed load in assessments and could be subject to diversification.

12.10.6 Humidification

Humidification can be provided to an air conditioning system in a number of ways. One of these is to inject steam into an air stream from an electrode boiler. This is a purely resistive heating load, which can be a significant load in relation to the other loads in a building. It is primarily a winter load. If an air-conditioned building does not have humidification included in the original design, it is important to be aware of any future possibility of humidification being provided and how this requirement has been defined at the briefing stage. This is so that the required future spare capacity can be included in the load assessment, if appropriate.

12.11   Data processing loads

It is likely that the areas of buildings with the highest power load densities will be rooms used for data processing or similar facilities. This can cover a range of functions, such as computer rooms, communications centres, dealers' areas and tele-hosting facilities. This is a major growth area, and the trend is likely to continue. It is associated with fast-changing technology, such as blade servers, and novel approaches to cooling to meet the high cooling density requirement in an energy-efficient way. There is often considerable difference in opinion as to the most appropriate load density figures to use for these facilities, and it is essential to assess each case separately and seek project-specific information from the client. Among many key issues to decide upon are the proportion of the area to which the load densities apply and the most appropriate allowances to include for spare capacity and load growth. Table 12.4 shows typical load density values (probably on the high side) for technical equipment.

Clients' estimates of load density at the briefing stage are often considered to be on the high side. In some cases their figures are derived from the 'nameplate rating' of their existing or proposed equipment and therefore represent an overcautious approach. The stated ratings on nameplates will usually be the worst-case maximum load for the equipment. This is often considerably higher than the average power consumption during normal usage. It is always necessary to seek clear guidance at the commencement of the project, so that a figure can be established as part of the brief development process. For these functional spaces, it will be necessary to determine the resilience (i.e. the redundancy or availability) criteria during the brief development, as this will have a fundamental impact on the power infrastructure design, the efficiencies and hence the resultant load (see Section 2.4.3).

Data processing loads vary widely depending upon the application and the technology. There is a relationship between the power density of the data processing equipment and the type of cooling system that is most appropriate to satisfy the cooling load density. As the cooling load will, itself, give rise to a power load, it is essential that the power and cooling arrangements are considered together, so that a suitable overall power load can be estimated for both operating together.

Power densities of up to 1500 W/m^2 are not uncommon and would usually be associated with conventional cooling systems. Some clients set a brief for power densities up to 3,000 W/m^2 or higher. At load densities approaching these levels, it is likely that non-standard cooling systems will be required, necessitating care when estimating the corresponding power loads for cooling. Clients might also set a brief in relation to anticipated server cabinet loads, together with a notional layout. In such cases, careful consideration is required, as the figures might be based on nameplate ratings rather than the actual running load. Loads are often defined as 'rack' loads, which in data centres can typically range from about 5 kW per rack up to 20 kW per rack (but can be well above this figure). However, it is

Table 12.4 Typical load densities for data processing equipment in communications equipment areas (does not include associated cooling equipment)

Communications equipment area	Typical load density (W/m^2)
Dealer floor (in the area allocated as dealer space)	400–500
Main equipment room (MER)	600–1,200
Secondary equipment room (SER)	350–500
Data centre (in the space allocated to equipment)	500–2,000 (and occasionally up to 3,000 or even higher). A useful average figure is 1,500

difficult to provide conventional cooling solutions for rack loads above 6 kW. The key design decision will be to determine realistic operational loads for the equipment and the diversity that can be applied. This will then represent the heat gain to the space and determine the cooling load, and thus the likely cooling solution, and hence the power required for cooling in the worst-case load scenario.

It is also necessary to take account of miscellaneous small items of equipment that are fed using 'Power over IP'. This will represent a load on the relevant circuits supplying the cabinets, as opposed to small power circuits.

It should be recognised that, unlike most other loads, data processing loads are often continuous, running for 24 hours a day, every day of the year. This will be relevant to the diurnal load profile and seasonal load pattern. See also Section 12.13 on harmonics.

12.12   Loads supported by UPS systems

Assessment of loads associated with UPS systems is beyond the scope of this book, but it is necessary to have some awareness, as they can be among the largest loads in a building. Data processing facilities, as described earlier, and other business-critical facilities, usually receive their power supply via an online UPS system. This will have an impact on the load as seen by the incoming power supply.

It might be assumed that the grid will always supply a perfect and continuous sinusoidal AC waveform. However, in reality, the power supply from the grid is not a perfect sinusoidal waveform, but is often distorted by spikes and harmonics. Furthermore, it can be subject to supply failures (brown-outs or black-outs) and transients. The degradation of waveform quality has been exacerbated in recent years by the increase in non-linear loads, which create harmonics. The harmonic currents feed back into the supply and cause distortion of the voltage waveform.

An online UPS system performs two main functions: a) continual conditioning of the power supply (to improve the quality of waveform) and b) maintaining the power supply continuity during power failures by providing a short-term standby supply for the duration of the 'autonomy' period.

The increasing importance of maintaining continued operation of electrical equipment to business operations has led to a major expansion in the use of UPS systems for what have become known as 'business-critical systems'. There are various types of UPS systems, all of which contain fairly complex items of equipment, including power electronic devices to rectify and invert AC mains. Most large UPS units take the existing AC mains and convert/regenerate a new AC supply for the critical load. To develop the design, it is normal to assess the specific operating mode in some detail with UPS equipment suppliers. Supporting a critical or sensitive system via an online UPS system will affect the load seen by the supply. If there is a test load, such as a load bank, the arrangement and timing for usage needs to be understood so that this can be considered. However, in most cases, connection of test loads would not be concurrent with other worst-case loads.

Most systems will be designed to achieve a certain level of resilience and will have a redundancy in modules so that the load can be maintained if one or more modules fail. Therefore, to assess the load related to a UPS system, it is necessary to know the key design parameters, including the resilience criteria, as outlined in Chapter 2. It is usually the case that there will be multiple parallel UPS modules in a UPS system – and possibly more than one system – to meet the criteria. In a simplistic sense, the factors that will influence the maximum demand load and that need to be considered include:

- the design load and the allowance for growth in the brief
- the efficiencies of the UPS modules for the part-load condition, i.e. the losses in the modules
- the worst-case battery-charging condition concurrent with the design load (particularly after partial discharge and re-charging)
- any test load that might be run concurrently with the main UPS load
- the impact of harmonics on the supply side load, considering any front-end harmonic filtration.

UPS module efficiencies would be typically up to about 96%, but will vary with loads, as described in Chapter 10. The selection of the numbers and ratings of modules will have an impact on the overall load, and hence the carbon impact. The maximum demand assessment for the system must, of course, take account of the worst-case simultaneous load for the system in relation to the range of operating scenarios. Such loads always merit detailed assessment and discussion with the manufacturers of the specialist equipment.

12.13 Parasitic loads

For certain types of electric power equipment, there will be an additional load related to ancillary equipment that is required to facilitate the operation of the main equipment. These can be considered 'parasitic' loads and, if fed from the equipment under consideration, they will reduce the net power available to meet the intended primary load. One example would be a UPS system that requires ventilation and/or cooling equipment to ensure the room temperatures in which the UPS modules and batteries operate are maintained below prescribed maximum levels, and possibly other loads for control and monitoring facilities. Another example would be ancillary equipment for a combined heat and power (CHP) plant or other generating equipment. Parasitic loads must be included, as appropriate, within assessments for worst-case concurrent loads.

12.14 Lifts and escalators

The vertical transportation equipment within buildings – lifts and escalators – can often represent a considerable proportion of the installed load. However, while due account will need to be taken of the loads for sizing the local power infrastructure that serves them, the value of load that should be carried forward to the total building maximum demand requires engineering judgement. For passenger lifts, the motors do not work continuously and do not work at constant load, so load assessment is not straightforward.

Figure 12.9 shows an indicative diagrammatic representation of two LV switchboards that illustrate the comparison between the typical cumulative load for a group of passenger lifts and for a group of escalators. In both cases the total connected load is 80kW. For the passenger lifts the average power will be small compared with the connected load, although there is no definitive guidance. For the purposes of notional assessment, a suitable diversity figure might be 0.25. So the total average load, Pav, would be:

$$\text{Pav} = [(4).(15) + (2).(10)].(0.25) = 20 \text{ kW} \tag{12.11}$$

As this demand only occurs at peak periods for lift passenger traffic, which do not usually coincide with peaks in other loads, this load is often omitted from the maximum demand

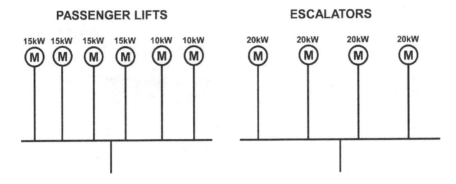

Figure 12.9 LV switchboards supplying lifts and escalators (diagrammatic)

figure for the building. Escalators have a different operational pattern. They tend to run continuously and, in many cases, the load will not vary significantly throughout the operational period of the day. For the escalator motors shown, a diversity of 0.9 might be applied, so the total load Pav would be:

$$\text{Pav} = (4).(20).(0.9) = 72 \text{ kW} \tag{12.12}$$

The load figure for escalators would usually coincide with the peaks for other loads, should it be included in the EMD for the building.

12.15 Chargers for electric cars

The transition to electric vehicles has already created a need for charging facilities in buildings. The extent to which charging will be required in any particular type of building is not yet clear, as there is likely to be a mix of facilities at homes, at workplaces and in public spaces. Some local authorities in the UK are requiring charging facilities in buildings as part of planning applications. Chargers can typically range in power rating from about 3 kW to about 75 kW for fast-speed chargers.

12.16 Load assessment tabulation method

Because an initial load assessment is largely about allocating loads to parts of buildings, or to particular electrical systems, and then obtaining a diversified summated figure, the simplest way to undertake a load assessment is using a tabulation or spreadsheet technique. However, prior to creating a load table, it is worthwhile producing a set of simplified 'load curves' for each system, showing the proportion of the load over a typical 24-hour period (for summer and winter). A simpler alternative would be to create a load table showing the significant time periods when each of the system loads occurs. Diagrams of this form can be useful to estimate the time at which the worst-case concurrent load occurs. Figure 12.10 gives an example for an office building. This shows assumed periods when each system load might be at its normal level, periods at which loads might be at part of their normal level, and periods when the extent to which the load would be at its normal level is uncertain and could only really be determined from a client's guidance on the intended functional operation. For the

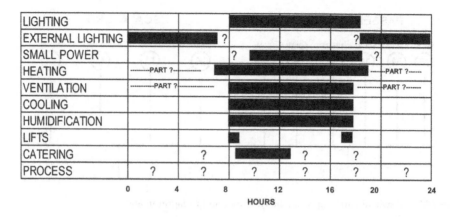

Figure 12.10 Typical significant time periods for electrical loads in a conventional office building

HVAC systems, the level of electrical load would, of course, be related to the thermal or ventilation demand, and hence to the external ambient conditions to some extent. In the example shown, the peak load is likely to be early to mid-afternoon when the peak cooling load occurs and would exclude external lighting, lifts and possibly most of the catering load. Separate diagrams should be created for each building type and for summer and winter to identify the concurrent loads that should be included when estimating the summer and winter peaks. In certain buildings the maximum demand might occur in Spring or Autumn, rather than Summer or Winter.

Table 12.5 demonstrates an example format of an initial load assessment tabulation for an office building. A table in this simple format could be used to obtain initial estimates of summer and winter peak loads at an early design stage – perhaps when the concept for the building form has commenced and the client has advised in basic terms the type of accommodation required. The example figures in the table are for a fictitious commercial high-quality office building in the UK with these basic requirements:

•   Total office area of 12,000 sq.m (gross internal area, GIA)
•   1,000 occupants, with a typical space allocation of 10 sq.m per person
•   Catering facility operating throughout main office hours
•   Separate computer centre, approx. 200 sq.m (GIA), operating continuously.

At the initial stage, when little information is available, the assessment can only be based on applying load densities where they are applicable, or making 'educated guesses', on the understanding that all figures will be updated as the design progresses, the brief is clarified and more information becomes available. For the fictitious example estimate shown, individual loads have been allocated by system, as follows:

   Lighting: It could be assumed that this would be primarily LED luminaires with dimming control. A suitable load density to use throughout this stage would be 8 W/ sq.m, which could reduce as the design progresses and specific layouts and arrangements are identified.

External lighting: A judgement would need to be made based on the extent of external areas to be covered. It could be assumed that this would be primarily LED luminaires. At this stage, an allocation of 5 kW would seem to be reasonable.

Small power: Using 15 W/sq.m as a load density for all areas would be a useful figure to use initially (as outlined in Section 12.9), in anticipation of potential reduction as the design progresses and specific arrangements are identified.

Lifts: Without any information on the numbers and capacities of lifts at an early design stage, a notional undiversified figure of 80 kW has been allocated; but as the load does not really contribute to EMD, it should be omitted from worst-case summations.

Catering: At this stage, a notional figure of 50 kW has been allocated. As the design progresses with specialist input and the proportions of gas and electric appliances is determined, the figure can be amended.

Heating (pumps): At this stage, before the HVAC strategy is determined, a notional figure of 50 kW has been allocated for pumps for a traditional gas boiler system. (If a heat pump is being provided for heating, the relevant load requirement should be determined from the heating load and the COP for heating.)

Cooling (chillers/heat rejection/pumps): At this stage a load density of 45 W/sq.m could be used for all areas, and might be expected to reduce as the design develops.

Ventilation fans: With a design population of 1,000, assuming 14 litres per second per person, and using the Part L2A SFP limit of 1.8 W per litre per second per person, the total power for all fans, P, would be $P = 1,000 \times 14 \times 1.8 = 25,200$ W.

Humidification: At this stage, before the HVAC strategy is determined, a notional figure of 80 kW has been allocated. This is a winter load.

Computer centre: Without any further information, a total all-inclusive load density of 2,000 W/sq.m could be assumed; at this stage this could be considered to include for a separate air conditioning system and the additional load related to UPS/harmonics. A modest amount of diversification could be assumed at this early design stage.

Miscellaneous other loads: There will be a load associated with hot water services (HWS) and with hand-drying in toilets. The magnitude of the load will depend on specific arrangements that might not be decided until a later stage of the project, but could be minimal if there is a central, non-electric HWS system and no electric hand dryers. There might also be other specialist facilities and equipment, the need for which might only become apparent as the brief is developed. It is useful to include an allocation for miscellaneous loads within the additional allocation for 'spare' and 'future capacity', i.e. a load contingency factor.

Suitable diversification can be applied for each system, except the computer centre. Power factors should be close to unity for contemporary lighting, small power and fans (assuming they have variable-speed drives), but it would be useful to use 0.95 at this early stage.

Where appropriate, load densities are used and multiplied by the gross internal area (GIA) – because this is an initial estimate – to give an undiversified load. (At later design stages, the net internal area should be used; see Section 12.17). A diversity factor is applied to give a diversified load in kW, and an average power factor is applied to give a maximum kVA figure for the system. Consideration can then be given to any adjustments required for the winter and summer peak demands. In this case, the tabulation indicates that the EMD will occur in summer, the dominant influence being the cooling system.

A suitable table should be produced for each load centre and should be updated as the design progresses and more information becomes available.

Table 12.5 Example load assessment tabulation format (early stage – very basic)

Ref	Service	GIA (m²)	Load density (W/m²)	Undiversified load (W)	Diversity	Diversified load (kW)	Average PF	Maximum kVA	Winter Peak (kVA)	Summer Peak (kVA)	Notes
1	Lighting	12,000	8	96,000	0.8	76.8	0.95	81	81	60	
2	Ext. lighting	–	–	5,000	1.0	5	0.95	5	5	–	
3	Small power	12,000	12	144,000	0.7	100.8	0.95	106	106	106	
4	Lifts	–	–	80,000	0.3	24	0.8	30	–	–	Ignore for EMD
5	Catering	–	–	50,000	0.6	30	0.90	33	33	33	Assume no summer heating
6	Heating (pumps)	–	–	50,000	1.0	50	0.8	63	63	–	Assume minor winter cooling
7	Cooling (chillers, pumps, etc.)	12,000	45	540,000	0.9	486	0.9	540	50	540	
8	Ventilation fans	–	–	25,000	1.0	25	0.95	26	26	26	Winter load only to include UPS, harmonics, etc.
9	Humidification	–	–	80,000	0.7	56	1.0	56	56	–	
10	Computer centre (all inclusive)	200	2,000	400,000	0.8	320	0.9	355	355	355	
								Sub-totals	775	1120	
					Add for miscellaneous and spare/future capacity (20%)					224	
								E.M.D.		1344	kVA (summer peak)

(apply further diversification, if considered appropriate)

Note 1: This example assessment format assumes a conventional heating system with gas boilers.

Note 2: This example assessment format assumes that there are no charging facilities for cars.

12.17   Brief note on assumptions and simplifications

The simplified approach outlined here assumes balanced, linear loads. This is relevant when calculating design current from the total kVA load. Modern building loads often have significant harmonic content, which makes demand assessment more complicated. It is worthwhile noting the following points:

- It is helpful to use the GIA when applying load densities in initial load assessments, as this will tend to provide a design margin. More detailed assessments at later design stages can use separate load densities for the net internal area (NIA) and for circulation and other areas. As more detail becomes available, specific load densities should be used for each individual treated area.
- Always be clear as to whether you are dealing with kW or kVA values, single-phase or three-phase.
- kW loads can always be added numerically because they are in phase. Strictly, kVA loads can only be added numerically if the power factor is the same. Like the currents, they are phasors (rotating vectors) and can only be added vectorially. However, as most power factors in buildings are similar lagging values, designers tend to simply add them numerically. For the level of detail required in an initial assessment, this is usually sufficient, but it is important to recognise the limitations.
- If power factor correction is installed, take account of this at the appropriate location in the system.
- AC parameters are always expressed as 'root mean square' (RMS) values.

12.18   Brief note on the impact of harmonics

The impact of harmonics is mentioned here in the briefest sense only, to provide some awareness, as this can be an important factor in the load assessment for buildings.

The simplified approach to assessing power loads outlined here tends to assume linear parameters, i.e. a pure sinewave for the supply voltage and, therefore, a pure sinewave for the current. For modern electrical systems in buildings, this is far from being the case. The voltage supply waveform (and hence the current) sometimes contains imperfections, but of more relevance, many loads are non-linear and create harmonics, including:

- fluorescent lighting (with high-frequency control gear)
- electronic equipment that incorporates switched-mode power supplies to create the required DC supply
- variable-speed drives (variable frequency inverters) supplying motors for fans, pumps, lifts, etc.
- UPS systems
- LED lighting control gear.

With the presence of harmonics, the current waveform will be distorted and the RMS value (equivalent heating effect) will be higher than for a linear load of the same power. As a consequence, all system components which carry harmonic currents (cables, busbars, switchgear, transformers) would have to be selected with suitable design ratings. Therefore, it is important to determine the true RMS load and size equipment to suit. This is also relevant to service capacity and regulations for limiting harmonics at the point of common coupling.

12.19   Summary

In this chapter, the key factors related to load assessment for power systems in buildings have been considered. These include:

- Load assessment must be correctly related to the items of equipment in the power system, and an awareness of the loads at different locations will influence the power system design, so these are interrelated aspects of design development.
- Load assessment is not an exact science and requires engineering judgement based on careful consideration of the load pattern for all items.
- The design decisions based on load assessment will determine the key design parameters of a system.
- Overestimating or underestimating the load can have considerable detrimental impact.
- At the early design stage, the main method of load assessment is the application of load densities for building or system types.
- It is essential to have a clear understanding of the mechanical systems (HVAC) loads, as these will have a dominant influence on the overall load in many types of buildings.
- It is necessary to understand the factors affecting diversification and apply diversity where appropriate.
- The load pattern should be understood in relation to diurnal and seasonal variations.
- Load assessment is required for each item of distribution equipment, as well as for the overall load.
- Load assessments should be undertaken in a clear, tabulated manner and updated as the design develops.
- It should be recognised that many buildings have UPS systems that present complexities in load assessment, and the presence of harmonics arising from UPS systems and other non-linear equipment will have an impact on the load and equipment sizing.

It should be noted that this chapter has covered the load assessment for the main power system under normal operating conditions with a grid-derived electricity supply. There are separate considerations for estimating the load for emergency or standby supply (such as life-safety or business-critical loads), with the power provided by generator(s), for which reference should be made to other publications.

13 Space planning and design integration for services

13.1 Introduction

The active engineering systems will comprise a myriad of items of plant, central equipment, distribution elements and terminal components. Although the terminal components will be located within, or immediately adjacent to, the spaces being treated, the major items of plant and distribution elements need to be provided with dedicated spaces within the building. The nature of the mechanical and electrical equipment components used in active engineering systems in buildings requires that the spaces allocated are markedly different from spaces used for occupancy. Such equipment can be potentially hazardous due to the energy systems used and, due to the need to restrict access to competent operational personnel, will, in nearly all cases, require being located in discrete and dedicated enclosures or spaces. The division of spaces within the building can therefore in the simplest sense be considered to comprise primary spaces for the occupational needs of the building, spaces for access and circulation to the primary spaces and secondary spaces for active engineering plant and equipment. The plant spaces can be subdivided into spaces for the building owner's M&E equipment and spaces for the energy or services providers' equipment, for which the determining criteria will be their own standard space requirements and access regulations. Where the building is split into landlord and tenant areas, there might also be a requirement for subdividing plant spaces.

The assessment, selection and allocation of suitable spaces for the M&E services comprise one of the most important design activities for the building services engineer and are collectively usually known as 'space planning'. A not inconsiderable proportion of the building volume is usually dedicated to M&E equipment, and the 'lead designer' for these spaces is, in effect, the building services engineer, rather than the architect. The importance of making appropriate design decisions for space planning cannot be overestimated, not only to allow for the initial design and installation but also for the future usage of the building. As discussed in Chapter 3, the space allocation will have a key determining influence on the operation and maintenance. As such, it will influence the longevity of equipment, functional and energy performance and the overall effectiveness of the active systems and is, therefore, a key consideration for an energy-efficient building.

As with many aspects of interdisciplinary design, there is a requirement for a carefully balanced, astute approach that considers the many other design imperatives and demands on the allocation of spaces. The process of space planning has traditionally been a matter requiring extensive negotiation, primarily between the building services engineer and the architect but also – often to a lesser extent – with the civil and structural engineer. But in a wider sense it involves all parties, as the outcome is central to realising the whole-life performance, so the interdisciplinary influences shown in Figure 2.4 for the building more generally will also apply for space planning for services. For example, the building services engineer will be trying to achieve satisfying outcomes for functional performance, sustainability, energy, ease of installation and operation and maintenance and will want to do this in an economical way. The architect will want to allocate such space in a way that fits with the wider spatial rationale for all spaces and allows realisation of aesthetic aspirations and satisfying planning constraints. The civil/structural engineer will want to maintain the integrity of the structure arising from the disposition of masses and the openings made in parts of structural elements. The client and cost consultant will be protecting the 'net' prime space proportion and will be keen to assess the cost impacts of allocated spaces and structural development in relation to the cost plan and business case. The client and his FM team will want an enduring solution that can be easily operated and maintained to provide a continuous high level of performance. The contractors will want ease of installation and 'buildability'. As with the building design more generally, all parties will actually want to achieve all of these outcomes to create a building that can perform well throughout its life.

It is most important that the building services engineer makes the case for their space allocation needs with clarity, and the requirement should be properly outlined and recorded. To do this, the building services engineer will normally be required to demonstrate that the options proposed have been the outcome of a logical assessment process. The starting point is to have a clear strategy from which the planning can commence and for the proposed allocation to be demonstrable through drawn information and presented as a logical and auditable design exercise. The designer's duties for health and safety, under the construction design and management (CDM) regulations, have been outlined in Chapter 2 and are relevant to all aspects of space planning for services.

Because this book is about design for energy efficiency, the focus here is on space planning for the main energy-using systems – the heating, ventilation and air conditioning (HVAC) and electrical systems. The space planning will also have to include an allocation for other services, such as sanitary systems; hot and cold water systems; fire engineering and other life-safety systems; and a variety of communication, control and alarm systems. The space planning for lift services, covering lift shafts and lift motor rooms (where required), forms a fundamental element of the space planning for the structural cores of the building and is beyond the scope of this book. The space planning for drainage systems will relate to the locations of toilets and/or bathrooms and will also influence the design of cores. They are a special consideration in space planning due to the need to seek vertical continuity, and as such they will have a major influence on the arrangement of the cores. Aspects of design integration are included throughout the chapter. A brief introduction is provided to builder's work for services, which is another important requirement for achieving good whole-life performance.

13.2   Space planning strategy

So that spaces for services can be successfully integrated into the building design in a holistic manner, it is essential to develop an appropriate strategy at the initial design stage. The key issues for space planning strategy can be considered to be:

- building services design objectives
- architectural form and structural design
- suitability of locations for all main plants

- internal and external access routes
- operational aspects, including health and safety
- fire strategy and means of escape
- economics
- planning issues, acoustics and vibration considerations
- resilience.

Each of these aspects is described in the following sections.

13.2.1   *Design objectives*

The design objective, from the building services perspective, is to allocate spaces for all equipment in suitable locations to facilitate ease of initial installation, subsequent routine operation and maintenance and future upgrading or replacement. There should also be a suitable allowance for anticipated expansion or spare space capability. The nature and extent of such allowances will vary between projects and should be identified at the briefing stage. The requirement is to ensure that throughout the process of design development, spaces are allocated for all relevant items of equipment. As outlined in Chapter 2, the system schematics should be maintained as the primary representation and statement of intent for the active system designs and will be developed in line with the load assessments to provide notional equipment capacities. The series of engineering system schematics then form an auditable or control checklist for the spaces that will be required.

The proportion of the building allocated to plant spaces can vary widely, as different buildings can have wide variations in thermal and electrical load densities, ventilation air requirements, complexities of plant arrangements and so on. For buildings with simple engineering systems – such as residential accommodations, warehouses and naturally ventilated offices – it might be in the region of 2–4% of the gross floor area. For buildings with a more intense services provision – such as fully air-conditioned commercial offices – it might be in the region of 4–10% of the gross floor area. For buildings where the functional activity requires services at a more complex or industrial level – such as research laboratories or data centres – the proportion could be 15–20% of the gross floor area or more. In some buildings it will be more appropriate to relate space allocation to overall volume rather than gross floor area. It must be emphasised, however, that the space planning exercise is about much more than simply estimating a total proportion of gross floor area, although that is an important metric for cost-effectiveness. It is, more specifically, about allocating *suitable* spaces in *suitable* locations for the types of systems and building operations involved. And while total internal area is important, it should be seen more as a three-dimensional design issue, relating spatial coordination to ergonomics, to identify the most appropriate mix of spaces. In reality, because different building forms will provide differing amounts of roof space for services, the internal area required is not necessarily a particularly representative measure of the total space requirement.

The notional capacities of equipment – which can be used, in conjunction with manufacturers' literature, to determine generic equipment sizes – provide the base information from which the space planning process commences. To achieve these objectives, spaces will be required both for major plant and for distribution components. The requirement tends to fall, therefore, into these distinct categories: a) plant rooms, b) vertical distribution (risers) and c) horizontal distribution. An important feature is, of course, the continuity and connectivity of spaces to allow for the connections between the main equipment, main distribution (both horizontal and vertical) and more localised distribution to terminal equipment. Horizontal distribution tends to fall into two distinct categories: a) distribution for primary elements, which is best located outside of the occupied zone (primary occupied spaces), and b) distribution for final branches and circuits, which is usually located within the floor or ceiling zones of occupied spaces. Where practical, services spaces should be designed to allow for zoning divisions, and with some degree of flexibility and expansion, so the client's objectives in that regard can be an important part of the design brief.

13.2.2   *Architectural form and structural design*

The building form will have a considerable influence on the planning of spaces for services. The shape itself will influence the arrangement of plant spaces and distribution routes. A tall, narrow building will lend itself to a very different space planning strategy compared to a low, wide building. In a tall building, the space planning will be dominated by vertical distribution, with plant areas likely to be restricted to the basement, ground floor and roof area. As the other floor plates will have core space that limits the prime functional space at each level, it is preferable not to make any further reduction. But as the roof area can be severely limited compared to the overall floor area, as the height increases, there is often a requirement for plant spaces to be allocated on intermediate floors as well. There is a limitation in external access, which is, in practice, only directly available at ground and basement levels. The best access to outside air is at roof level. For a low, wide building, the space planning will be dominated by horizontal distribution. There will be more options for plant areas with external access and large areas of roof space for access to outside air.

   The architectural aspirations for the elevations will also influence the locations available for plant rooms, as will the visual appearance and planning criteria related to the roof area. Most plant rooms will require louvres in external walls to a greater or lesser extent, depending on the specific plant requirements. Louvres would require careful visual integration into the elevation or facade treatment and can be a limiting factor in achieving a satisfying architectural solution. It is often the case that an architect would propose that a continuous louvred 'band' be provided, for example, for the whole of a floor level in a tower, or for one facade, rather than a more *ad hoc* arrangement of discrete louvres to match distributed plant requirements.

   The development of the structural design will also have a primary influence. The building is likely to have one or more primary cores containing the primary vertical structure, or 'spine(s)'. These are areas of the building where the main continuous vertical elements are co-located, such as the lift shafts, lift lobbies and staircases, together with the main services risers and other shafts. The development of the cores is an essential early-stage design activity involving the structural engineer, architect and building services engineer. The allocation of space for risers is a key determining factor, alongside allocation of lift shafts and lobbies, together with means of escape and other features – such as any risers and sprinkler services – arising from the fire strategy. The cores form an important part of the overall structural frame. Toilet areas are usually located within the core areas, and hence there is a requirement for riser space for sanitary services pipework and a need to maintain vertical continuity. The structural frame will have columns and beams that could be of reinforced concrete or steelwork. The floor slabs will have a cross-sectional profile that might be flat or might have a deeper shaped profile to provide structural strength. There might be reinforced concrete downstand beams or solid or perforated steel beams. All of these factors will influence the options available for plant spaces, services risers and horizontal distribution zones. An essential design integration activity for the building services engineers is to understand the limitations in the structural design advised by the structural engineer. This relates to the need to achieve the specific structural integrity, strength and rigidity that might be required by the structural design intention, while seeking planned holes or slots within slabs and beams (if required) to allow vertical and horizontal distribution of engineering systems.

The structural engineer will require details of the typical mass and footprint area of all equipment, together with information on any vibration that is likely to arise. The disposition of M&E plant and equipment will be a major consideration in the structural design, particularly the heaviest items, such as water tanks, fuel tanks, chillers, boilers and transformers. From a structural point of view, these items would preferably be located at basement or ground level to avoid any additional structural impact (and cost) from locating them higher up in the building. However, that may not be the most practical location from the services strategy perspective.

13.2.3   *Suitability of locations*

For most buildings, only certain areas lend themselves to being appropriate as plant areas due to either economics, ease of access or proximity to outside air. In most buildings, prime space is allocated for the primary occupancy functions of the building. Services spaces tend to be located in lower-grade or 'back-of-house' areas, rather than displacing the primary occupation. Figure 13.1 shows the typical locations available for major plants, and in many cases it will be a balance between basement/ground floor and roof areas, with only minimal intrusion into other areas. It should be emphasised that the selected locations must provide the necessary height as well as floor area. Many individual plant items, and the preferred 'stacked' plant arrangements, will require significantly more than the normal floor-to-ceiling heights for occupied spaces.

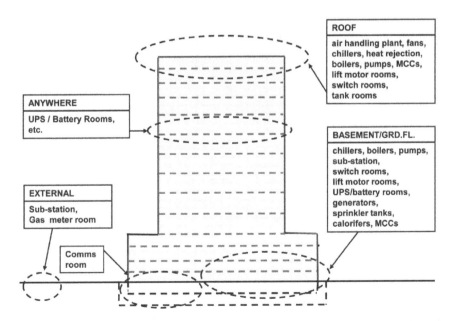

Figure 13.1 Typical locations for major plant items

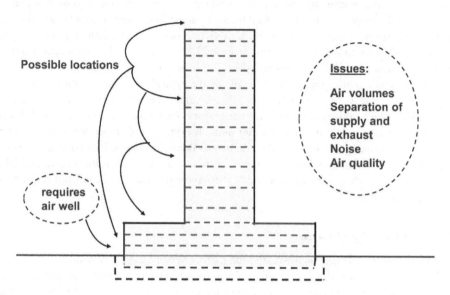

Figure 13.2 Potential locations for air supply and exhaust equipment

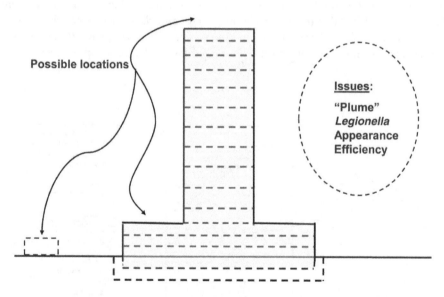

Figure 13.3 Potential locations for heat rejection equipment

The suitability of different locations will depend on the type of equipment being considered. Figures 13.2–13.4 cover some of the key issues, most of which are simple, common sense design factors.

A number of possible locations could, in theory, be considered for air supply and exhaust equipment, as shown in Figure 13.2. For central systems with significant air volumes, the

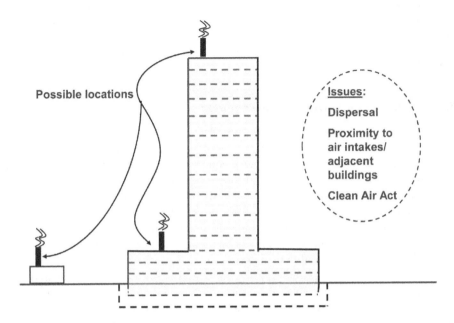

Possible locations

Issues:

Dispersal

Proximity to air intakes/ adjacent buildings

Clean Air Act

Figure 13.4 Potential locations for fume exhaust discharge equipment

most practical location is likely to be in a clear roof space. However, the extent of roof space is limited on tall, slim buildings and can result in extensive distribution ductwork. A proportion of the plant could be located elsewhere, such as in the basement, but this is often inappropriate and impractical. This is due to the difficulty in obtaining access to the required quantity and quality of outside air. Where a basement solution is the only option, it can be made to work, but it would require an air well, or a means of drawing in/discharging ventilation air from higher up in the building. Intrusion into prime space would, however, cause limitations in space planning for occupied areas and require louvres on the facade. It would also involve acoustic and access issues.

Heat rejection equipment for cooling systems requires a clear roof area that will provide suitable air movement for efficient operation and be acceptable in terms of appearance. The uppermost roof area is the ideal location. For wet cooling towers there are significant issues related to the 'plume' or cloud of condensation arising and avoiding any possibility of *Legionella*.

There are good safety reasons to locate boilers and generators in basement areas, as this limits the extent of potentially hazardous fuel distribution (gas or oil) within the building. The disadvantage is the routing of flues to a suitable exhaust location, which is usually the roof. This would require some space at each floor level and would have to adhere to fire regulations and consider noise breakout aspects. Similarly, there are good reasons to locate chillers at roof level to achieve effective heat rejection and minimise the presence of refrigerants within the building. The disadvantage can be noise and vibration close to the prime top-storey areas.

Fume exhaust discharge equipment – for example, flues from boilers and generators but also exhaust ventilation from catering areas and laboratory gases – must be located so that dispersal satisfies the requirements of the relevant environmental regulations. The fumes must not impinge upon air intakes or adjacent buildings. As indicated in Figure 13.4, only the uppermost roof area is likely to be satisfactory, but this can cause problems with long flues from lower areas due to excessive pressure drop, and there would also be cost considerations and an increase in the space required in risers at each level, which would reduce the prime occupied space.

13.2.4   Internal and external access routes

All plant areas require access for the initial installation to be undertaken and for subsequent operation and maintenance throughout the life of the building. The access provision is for competent personnel and for replacement of equipment. The access for people will require doors from suitable circulation routes and should relate to the escape routes in accordance with the fire strategy. As operational personnel will be bringing tools and equipment, these access routes should ideally be within or from 'back-of-house' areas. The access for equipment will require sufficient spaces for moving all items of plant into their final locations during the construction stage. Such access might be through de-mountable panels or similar and so might be of a temporary rather than permanent nature. The access requirements during the operational stage must be sufficient for the largest individual components that will be replaced, either as a routine replacement procedure or as a consequence of failure. These considerations will often involve aspects such as road access, roof access, goods lifts and craneage. This may therefore merit a strategy in itself, and plant replacement strategies are regularly produced as a part of operational and maintenance strategies, which are sometimes prepared at the design stage.

13.2.5   Operational engineering

In addition to the necessary access and egress provision outlined in the previous section, the arrangement and disposition of equipment within the space must be such that the required operational activities can be undertaken safely and effectively. This will require clear access to all items where activities of a regular or routine nature take place. The access space should be sufficient for the tasks undertaken and is a three-dimensional ergonomic design consideration covering the personnel and equipment activities. The nature of activities for plant replacement is sometimes set out in a plant replacement strategy document at the design stage.

For electrical equipment, there is a specific need to provide adequate space in switchrooms and similar areas for the installation and replacement of individual items. There should also be sufficient accessibility for the required range of activities, including operation, maintenance and testing, along with inspection and repair. The arrangements should satisfy the requirements for operation or maintenance gangways, or similar, in relevant codes and regulations, such as BS7671 (IET 2018).

13.2.6   Fire strategy and life-safety systems

Most buildings will have a fire strategy that defines the features incorporated to resist the spread of fire and smoke and allow safe escape of occupants. This normally results in specific features relevant to the space planning, such as:

- fire compartmentation for spaces, comprising walls, floors, doors and cavity barriers in ceiling voids

- certain building compartments will have a specified fire rating, typically one-hour, two-hour and sometimes four-hour rating
- defined 'means of escape' (MOE) routes.

The planning of MEP services needs to consider the fire strategy. For example, it may not be possible to locate certain types of plant rooms next to designated MOE corridors. Special features are required where penetrating fire-rated components in order to maintain integrity, such as fire/smoke dampers in ventilation ductwork and fire-rated packing for electrical trunking, cable trays and similar.

The fire strategy might require specific engineering systems to effect detection and alarms related to a fire condition and fire protection for the occupants and structure in the form of smoke control systems (to keep the escape routes clear of smoke), smoke clearance systems, sprinkler systems and/or fire suppression systems. There might be standby generation and fuel storage associated with these systems. These life-safety systems will also require space allocation, which can be considerable. These systems are not part of the normal energy-consuming systems and are not covered specifically in this chapter.

13.2.7 Economics

An essential aspect of building design is the need to provide sensible and economic usage of the space. The primary purpose of the building is to house the occupancy function for which it is intended. Any space that is not allocated for the primary function detracts from the economic viability, so a balance needs to be struck. It is usual to relate the usable and non-usable spaces in terms of 'net-to-gross' ratios, particularly for office buildings. The net area is, in simple terms, the gross area minus the non-usable space, although more formal definitions are in use in the industry. For commercial buildings, it is likely that clients and developers would aim for net-to-gross ratios in the order of 75% or above to be economically viable. However, this will depend on the type of building and wider aspects of the function, financial assessment and business case.

13.2.8 Planning issues, acoustics and vibration

The wider planning issues relevant to building services engineering have been described in Chapter 2. For the most part, these relate to the impact of plant and plant rooms, either from a visual aspect or from aspects such as the acoustics and vibration impact during daytime and night-time operation, or fumes and pollution. The planning of spaces for plant must address the specific issues advised by the planning authorities, and this can be a primary consideration that can have a fundamental impact on the options available.

13.2.9 Resilience

The overall mechanical, electrical and public health (MEP) systems infrastructure should be resilient to the extent that is necessary to satisfy the design objectives. For certain projects, exacting criteria for resilience will form an important part of the briefing criteria. The selection of spaces for plant will be influenced by the need to house equipment to meet resilience criteria, and the numbers and locations of plant spaces might themselves be arranged to provide resilience in relation to incidents or failures in different parts of a building or site.

13.2.10 Selecting plant room and riser locations

A key factor in the iterative development of spaces for equipment is to ensure that the schematic diagrams for all of the engineering systems are maintained as the primary reference for proposals, with the development of plant spaces following the schematic proposals. This will ensure that suitable spaces are allocated for all components of the building services systems and are continually correlated with the systems proposals as they undergo iterative development.

In a generic sense, the typical space requirements for mechanical and electrical systems are quite similar, as shown in Figures 13.5 and 13.6. In both cases there are likely to be plant

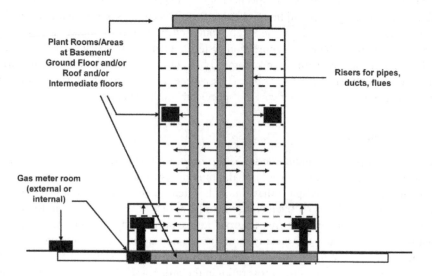

Figure 13.5 Mechanical systems: typical generic spatial requirements

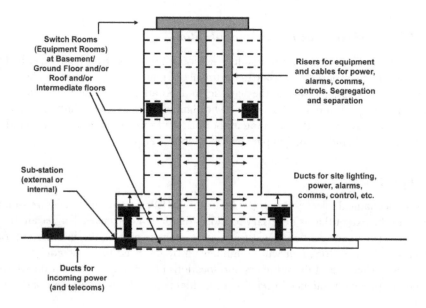

Figure 13.6 Electrical systems: typical generic spatial requirements

areas at ground and/or basement levels and at the roof level, together with other plant areas whose disposition would depend on the building form and function. In each case there would be a number of risers connecting the plant areas. The major difference is one of scale, with mechanical plant areas and risers usually considerably larger than those for electrical services, particularly due to the requirements of air handling units and ductwork distribution. However, the growth in the range of electrical systems has resulted in a general increase in the space required for electrical risers in modern buildings and more space for on-floor electrical cupboards for local distribution.

13.3   Space criteria for mechanical and electrical equipment

It is necessary to understand the physical considerations that apply to each item of the plant, so that appropriate space can be allocated for each item individually and for all items as an interrelated and coordinated arrangement. Figure 13.7 shows the main physical considerations that would apply to a generic item of plant. It is essential to develop a design as a three-dimensional exercise. It should be recognised that the height required for equipment, particularly air handling plant rooms, often presents considerable challenges. The mass is of major interest to the structural engineer (or civil engineer for substructure areas), as allowance will be required within the structural design. Suitable access is required for initial installation and for future operation and maintenance. At some stage the plant will need to be replaced, either in full or in part, for which suitable space is required. Depending on the item of equipment, it might be necessary to provide restricted access for reasons of health and safety. It might also be necessary for security reasons, particularly where the item might be of a business-critical nature, and this can also be important in multi-tenanted buildings.

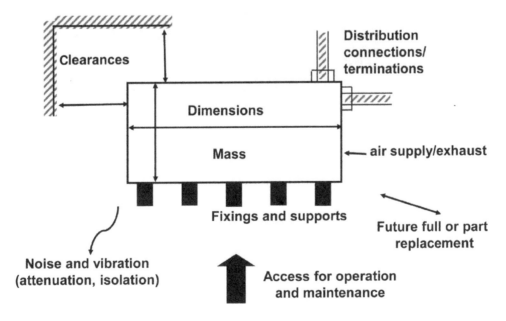

Figure 13.7 Generic physical considerations for location of equipment

All items of equipment and their distribution connections will require some form of supports or fixings, and these are shown generically in Figure 13.8. If an item of equipment causes significant noise and vibration, it will require some form of anti-vibration mountings (AVMs) and flexible connections to the rigid distribution elements.

Many issues related to architectural appearance and planning will restrict the potential locations for plant areas. Some of the key issues are shown in Figure 13.9. While the roof level is an important location for the plant, there are likely to be restrictions in terms of appearance – often requiring some form of louvred aesthetic screen – and overall height.

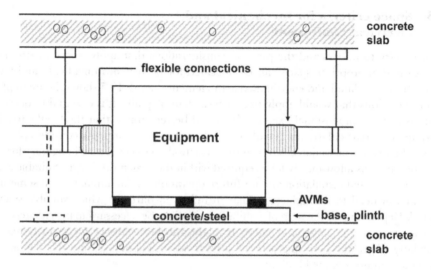

Figure 13.8 Generic supports, fixings and isolation

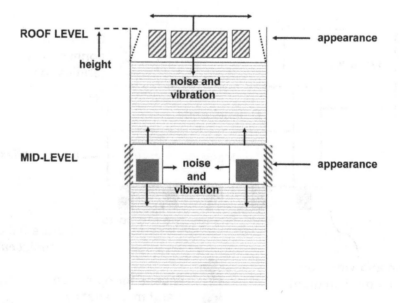

Figure 13.9 Main restrictions: architectural/planning

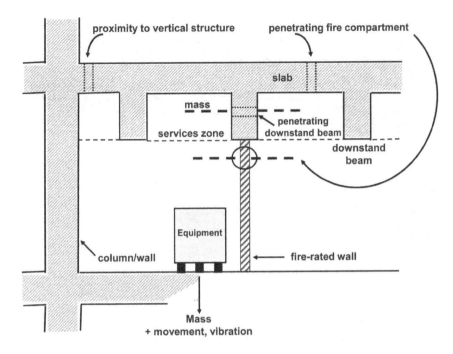

Figure 13.10 Main restrictions: structural

Noise breakout to adjacent premises, and noise and vibration impact on the upper floor, are also major issues. Similar considerations apply to mid-level plant areas.

There are also many restrictions in relation to the structural design. Some of the key aspects are shown in Figure 13.10. The mass of equipment has to be accounted for, together with any movement and vibration during operation. Similarly, any proposal for distribution components to penetrate the structural slab will require approval, particularly if it is in close proximity to a vertical structure – which is usually the case with risers. Risers are special cases and require particular attention and negotiation with the structural engineer and the architect. The sizes required for openings in the slab, and the specific structural solution, will require careful coordination to achieve an acceptable solution. An essential feature will be horizontal openings in vertical components (structural or architectural) at the riser for on-floor distribution, such as pipework, ductwork, trunking and cable trays. Any openings required in downstand beams will require approval and careful coordination, as will any penetrations of fire compartments.

Certain items of renewables equipment will, by their nature, require special considerations in space planning. Wind turbines require clear space so that the wind pattern provides effective energy yield and also allows sufficient clearance for health and safety and the maintenance activities. They are likely to require structural bases suitably integrated to avoid vibration problems. Photovoltaic panels require a location that provides good energy yield and avoids shading. The integration of wind generation and/or photovoltaics into a roof plant area therefore adds considerable challenges to the planning. Figure 10.17 shows a visualisation of photovoltaic arrays with air handling units and other equipment to provide a coordinated solution and good energy yield. Ground-source heat pumps require detailed

Figure 13.11 External services located below plant

Source: Courtesy of Hoare Lea

design integration with the civil engineering proposals for below-ground areas for the pipe-work systems together with space allocation for equipment.

In some cases, to provide a coordinated solution for equipment and distribution, the equipment is raised on a support platform to provide a distribution zone underneath. Figure 13.11 illustrates an example where pipework and valves are located in a zone below raised equipment and arranged such that no services rest on the roof finishes.

The photographs in the sections that follow show examples of equipment in plant areas during the installation stage, so they are not necessarily in a completed state.

13.4   Space planning for plant rooms

A useful checklist of generic factors to consider when assessing space requirements for all types of plant rooms is to:

- prepare notional layouts in block format to include for every item of the plant
- refer to manufacturers' technical literature for equipment shapes and typical dimensions
- seek a layout that is economic in relation to usage of space, including access
- compare initial services space allocation with other projects and 'rules of thumb'
- consider space for access, operation and maintenance, as well as plant replacement (either in full or in part)
- consider health and safety implications and CDM regulations in relation to the whole life of the facility
- consider zoning of systems and how this could influence plant arrangements
- consider means of escape provision and adherence to the building fire strategy; for plant rooms and roof areas, provide an alternative means of escape
- consider space for future additional plant, or extension of plant, such as switchboards.

A further general objective is to achieve a design arrangement whereby failure of a component of one system or service does not jeopardise the operation of other systems. A specific example is the typical arrangement of concentrated services (often mechanical, water and sometimes drainage) located close to electrical plant rooms. The design arrangement should protect the electrical equipment operation from failure of a water pipe to reduce the risk of electrical system failure (which can have major consequences). This can be a particular issue in basement plant areas, where there is often a concentration of services equipment in close proximity. Design features could include providing a floor with a bund in electrical plant areas, or extensive floor drainage in adjacent areas, together with plinths or platforms to raise electrical equipment above floors. For projects requiring high levels of resilience, it is likely that the risk of failures of one service impacting on other services would be formally evaluated using design risk assessments. Sensible design decisions at the outset can avoid the need for complicated risk reduction at a later design stage, when change can be difficult to achieve.

In addition to these generic factors, other specific factors will be relevant depending on the type of plant room, as outlined next.

13.4.1   Heating plant rooms

Specific considerations include:

- location of natural air inlets, or mechanical ventilation
- location of flues and related planning approval
- select a room separate from refrigeration and air handling plant
- floor drainage
- noise and vibration control.

Figure 13.12 shows the coordinated layout in a heating room with boilers on the left and pump sets in the centre. Ductwork is located at a high level with sufficient clear space below. The layout has been arranged with due regard to the locations of the circular structural columns, and clear access routes have been provided between items of the plant.

Figure 13.12 Typical heating plant room

Source: Courtesy of Hoare Lea

Figure 13.13 Pump sets
Source: Courtesy of Hoare Lea

Figure 13.13 shows an arrangement of twin pump sets. Each pump set is located on an inertia base, which itself is mounted on a concrete plinth to provide isolation from vibrations. Clear access space has been provided between each pump set.

13.4.2   Chiller plant rooms

Specific considerations include:

*   select a room separate from the heating and air handling plant
*   floor drainage
*   noise and vibration control.

13.4.3   Air handling plant rooms

Specific considerations include:

*   location of outdoor air and exhaust louvres
*   proximity of ductwork risers
*   floor drainage
*   clear height of plant room (if internal)
*   noise and vibration control, including the size of attenuation equipment.

Figures 13.14 and 13.15 show air handling units in roof-level plant areas during the installation stage. The units are mounted on raised structural platforms, providing access for operation and maintenance.

Figure 13.14 Air handling unit within screened plant area
Source: Courtesy of Hoare Lea

Figure 13.15 Air handling units
Source: Courtesy of Hoare Lea

13.4.4   *Electrical switch rooms and substations*

Specific considerations include:

- avoid locations that could be subject to flooding
- ownership, metering and access
- ducts for cabling, plant replacement strategy for transformers
- proximity to other services and electromagnetic compatibility (EMC) issues.

13.4.5   *Generator plant rooms*

Plant room considerations for the generating plant are complex and beyond the scope of this book. However, it is worth noting here some specific considerations, as in some buildings, these rooms can be a significant size. Acoustics issues are a major consideration.

- plant installation and removal: skid-mounted sets or engine and alternator
- ventilation equipment and air movement; louvres and attenuation
- height to allow for exhaust silencers and lifting beam for maintenance
- separate plant space for fuel storage and pumping
- ducts for cabling and pipework.

13.4.6   *LV switch rooms*

Specific considerations for low-voltage (LV) switchrooms are:

- switchboard access: front only *or* front and rear?
- switchgear maintenance and circuit breaker withdrawal
- switchboard extensibility: one end only *or* both ends?
- space for power factor correction and/or harmonic filtering equipment
- cable bending radii: initial installation and future (as this affects height)
- ducts for cabling
- wall-mounted equipment: distribution boards, control panels, etc.

13.4.7   *UPS rooms*

It is worth noting the key considerations, as these rooms can be of significant size in some buildings:

- equipment installation and removal for switchgear, uninterruptible power supply (UPS) modules, control panels, etc.
- operation and maintenance without disrupting business-critical systems
- separate room for battery racks
- ventilation equipment and air movement; louvres and attenuation
- ducts for cabling, etc.
- extensibility and expansion without disruption to business-critical systems.

13.5   Space planning for risers

There are two main considerations when planning spaces for risers: achieving vertical continuity (and retaining shape) throughout their height and the suitability of horizontal spaces interconnecting with risers to facilitate distribution within plant areas and the ceiling and floor voids of occupied spaces. Vertical continuity is required so that the sizes and shapes of distribution components can remain the same throughout their length and can be installed easily, without the need for any off-sets or changes of direction. Straight lengths of distribution components provide an economic solution, reduce time for installation and help minimise distribution energy losses.

The starting point for planning risers is to identify all of the risers required from the early-stage schematic diagrams of the engineering systems and make a preliminary assessment of their sizes. It might be necessary to make reasonable assumptions, as the full details will not be known until later in the design process. It is sensible to allow spare space in risers for future extensions or additions, perhaps allocating 10–15% of extra space. Where appropriate, reference should be made to any criteria for future expansion within the brief. In larger buildings, it is likely that there would be a number of mechanical risers, electrical risers, drainage risers and communications risers. These would ideally be separate. In smaller buildings, there might be a need to compromise by combining appropriate services in common risers (such as electrical and communications services), provided that certain separation criteria could be assured. Figure 13.25 shows a visualisation of a core of a building containing numerous services risers (see Section 13.7).

The iterative design exercises to explore riser options might result in corresponding developments of the system schematics, so that a satisfactory overall solution can be found for vertical distribution components and the spaces in which they are located.

13.5.1 *Mechanical risers*

The risers for the HVAC systems will, for most air-conditioned or mechanically ventilated buildings with central air handling plant, be the largest risers. This is primarily due to the need to contain the supply and extract ductwork, which is usually much larger in cross-sectional areas than pipework or cabling services. Similarly, the ductwork connections on each floor will normally be larger than pipework or cabling services, with limited scope for changes in direction. The risers for ductwork will therefore normally require the most space and require the most attention during the initial space-planning exercises. It should be noted that mechanical services risers at each floor level usually only contain ductwork and pipework, together with branch connections to serve the relevant floor. Unlike electrical risers, they do not normally contain wall-mounted items of equipment requiring access. The thickness of insulation for ductwork and pipework must be considered in assessing the total space required.

The key considerations that apply to the allocation of mechanical risers are the area, zone or tenancy to be served; the practical lengths of ductwork or pipework branches from riser mains; and the practicality of providing service connections from the riser to the ceiling and floor voids.

13.5.2 *Electrical risers*

Some of the same considerations apply to the allocation of electrical risers as for mechanical services. In the case of electrical risers, the design decision on the number of risers will need to consider the area, zone or tenancy to be served; the practical lengths of final circuit cabling from distribution boards; the practicality of equipment layout within the riser; and the practicality of providing services connections from the riser into the ceiling and floor voids. It might also be necessary to provide a degree of separation between different types of electrical service and independence – for example to separate high-voltage (HV) services from LV services – and to provide independent routes to business-critical or life-safety equipment.

An important difference between electrical and mechanical services risers is that while mechanical risers tend to house distribution elements only – such as ductwork and pipework – electrical risers also house a variety of items of equipment. Key factors to consider in the design of electrical risers are:

- initial and future rising services: trays, ladders, busbars and trunking
- initial and future equipment: distribution boards, control panels, contactors, harmonic filters, etc.
- access for operation, maintenance and other activities in accordance with relevant codes
- separation and segregation of services; for example, separating power from control, monitoring and communications (due to electromagnetic compatibility [EMC] issues), as outlined later.

Figure 13.16 shows a typical layout of equipment in an electrical riser in an office building. This would normally house the distribution boards for lighting and small power, but could also include local equipment related to fire alarm, public address, security, building management systems/controls, metering, outstations and other systems. It might also contain harmonic filters. Risers are also likely to include 'through services', for example, LV cables passing through the floor level to feed chillers or other equipment at roof level.

Figure 13.17 shows an example of an electrical riser on the ground floor during the installation stage. At the stage shown, this only houses the LV cables that form the incoming supply rising to the LV switchroom at roof level.

Figure 13.18 shows an example of an electrical riser in an office building. At the installation stage shown, this contains distribution boards and tap-offs from a rising main busbar, with space for additional equipment.

Separate spaces should be provided for communications rooms and risers.

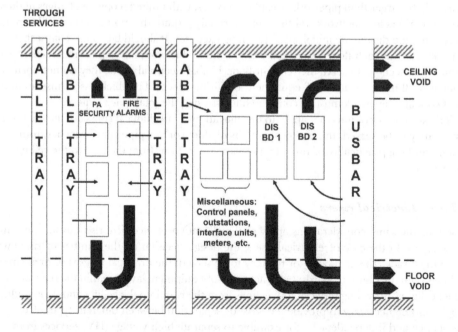

Figure 13.16 Electrical risers: typical equipment layout

Figure 13.17 Incoming electrical riser

Source: Courtesy of Hoare Lea

Figure 13.18 Electrical riser on office floor level

Source: Courtesy of Hoare Lea

Note on electromagnetic compatibility

It is necessary for electrical systems within buildings to be designed such that they achieve EMC. To achieve this, in simple terms, the following conditions must be satisfied:

- an item of equipment should not be adversely affected by its electromagnetic environment
- an item of equipment should not adversely affect the operation of other equipment.

Typical sources of electromagnetic interference (EMI) and radio frequency interference (RFI) are:

- power equipment, such as generators, transformers, cables and busbars
- lightning and transients
- transmission equipment, such as radar and radio.

Equipment that operates at low levels of power and voltage (such as electronic devices) is the most susceptible to interference, which could cause malfunctioning.

The best way to avoid interference is to locate the susceptible equipment so that it is separated by a suitable distance from the major power components. If separation is not possible, some form of metallic shielding is likely to be necessary, as illustrated in Figure 13.19. To reduce the likelihood of EMC problems, the cable installation arrangement should be close to symmetrical, and power cables should be separated from digital cables (CIBSE 2004b). EMC issues are complex and are beyond the scope of this book, but are mentioned briefly here as useful information in relation to the planning of spaces for electrical equipment.

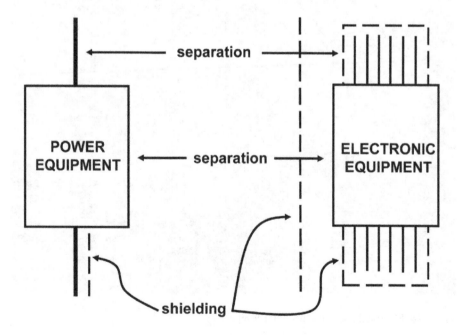

Figure 13.19 Separation and shielding for EMC

13.6   Planning horizontal distribution

The horizontal distribution elements that need to be housed tend to be of two types: a) primary infrastructure, connecting main items of equipment to risers or to other plants, and b) secondary distribution from risers that form the circuits or branches that feed terminal equipment. In some cases the horizontal space will house a mixture of both of these types of distribution components.

At the basement level or other parts of the building with plant rooms, it is often the case that the slab-to-slab heights will be well above those on occupied floors to provide the required heights in plant rooms. In such cases, primary distribution can often be located at a high level in plant areas and adjacent circulation routes.

In buildings or site developments with extensive plan areas, it is sometimes the case that the most appropriate design solution for the primary infrastructure horizontal distribution is to create a services tunnel. Figure 13.20 shows a tunnel with the pipework located on one side and cabling on the other side. This allows space for future services and access for operation and maintenance, with minimum disruption to the rest of the building. A services tunnel of this type would be a significant feature in the civil engineering design and requires consideration from the concept stage.

For the second type of distribution in occupied areas, there is usually a need to allocate spaces within ceiling voids and floor voids for horizontal distribution to terminal equipment. The specific arrangements at high and low levels will depend to a considerable extent on the architectural proposals for the finishes and the nature of the space. In some cases, a design solution could be to not have a ceiling void, and the architect might consider that the exposed services and structure are appropriate for the type of space. Such a solution might be sought for environmental reasons to benefit from the exposed thermal mass of the slab, but would usually require careful consideration from an acoustics perspective. Many modern buildings, however, are likely to require floor voids and ceiling voids with raised

Figure 13.20 Services tunnel

Source: Courtesy of John Pietrzyba

floors and suspended ceilings, respectively. The design iteration to agree to heights of ceiling voids and floor voids is an important part of the design process. There is an imperative to maintain sufficient clear floor-to-ceiling heights within the occupied zone, while minimising any further height that will add to the cost of the building. From a building services perspective, the objectives are the same as for other plant areas. Key considerations are the level of access provided by de-mountable ceiling panels or floor tiles and the three-dimensional arrangement of components within the void to minimise cross-overs that will add to the height requirement.

In most modern buildings, the floor void is a major services zone and requires similar considerations to ceiling zones in terms of design, particularly in relation to the clear depth available. For most office buildings, the raised floor void will contain power and data cabling at a minimum. Depending on the selected HVAC solution, the floor void could also contain pipework and ductwork. In certain types of buildings with intensive services requirements, such as data centres, the floor void will be the primary zone for services, and assessment of the depth will require detailed design attention.

The following photographs show examples of ceiling voids and floor voids with partially completed services installations at the construction stage.

Figures 13.21 and 13.22 give an indication of the depths required for ceiling voids and the arrangement of services components to coordinate with the structure and ceiling grid and coordinated services arrangements.

Figure 13.23 shows the floor void for a modern office building at the construction stage. This illustrates the depth of the services zone within the floor void and the ductwork, pipework and cable trays that had been installed at that stage.

Figure 13.21 Coordinated services in ceiling void prior to installation of chilled ceiling panels

Source: Courtesy of Hoare Lea

Figure 13.22 Services in ceiling void

Source: Courtesy of Hoare Lea

Figure 13.23 Services in floor void

Source: Courtesy of Hoare Lea

13.7   Integrated and coordinated solutions

It is necessary to achieve satisfactory design solutions for the most concentrated and complicated plant areas – usually roof-level plant arrangements and risers – embracing many of the concepts outlined earlier. This requires a high level of spatial coordination and integration with the work of other disciplines. The use of 3D design visualisations can help all parties understand the proposals and implications as the design iterates. These can allow exploration of alternatives so that all parties can fully appreciate the issues involved and seek a mutually acceptable resolution. Figure 13.24 shows a visualisation of a roof plant area with a complicated arrangement of equipment. Figure 13.25 shows a visualisation of a

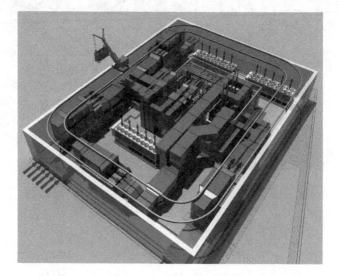

Figure 13.24 Visualisation of roof plant area

Source: Courtesy of Hoare Lea

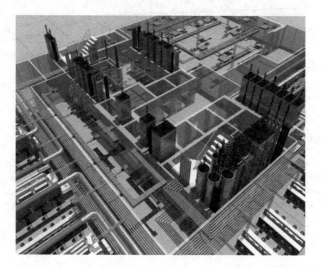

Figure 13.25 Visualisation of core area risers and lift shafts

Source: Courtesy of Hoare Lea

core and adjacent areas of a building, including numerous services risers, lift shafts and circulation areas. Figure 10.17 shows a visualisation of the coordinated layout of photovoltaic arrays with air handling units and other equipment in a roof plant area.

13.8 Implications of adaptation to climate change on space planning

The need to design for adaptation to climate change is likely to have an impact on the space allocation for equipment. In particular, if cooling loads increase due to higher ambient temperatures, the required capacities of cooling equipment (and associated air handling equipment) are likely to increase, and hence the space required. The higher ambient temperatures may also require different arrangements and physical sizes for heat rejection equipment, could affect air handling unit sizes and associated ductwork and pipework and affect power supplies for cooling, air handling and pumping equipment. With a warming climate, the required capacities for heating equipment might reduce. It is possible that the most appropriate provision would be to include a degree of flexibility and modularisation in the design to allow incremental changes to be incorporated. Overall, design for adaptation is likely to increase space requirements. This should be considered during the brief development in order to seek the client's view on the extent to which this should be addressed. Another possible implication of climate change is an increase in flood risk for some locations. In such cases it would be inappropriate to locate certain items of equipment – particularly electrical equipment – in basement areas, or at ground floor level, unless special flood protection arrangements were provided. At present there is little guidance on how these issues should be addressed, but this is likely to be an aspect of design that receives increased attention in the future.

13.9 Builder's work

In addition to the principal plant spaces (plant rooms and risers) there will be a need for a wide range of minor building work to facilitate the installation of the engineering systems. This must also be planned as an integral part of the building design and is generally known as 'Builder's Work in Connection' (i.e. in connection with the MEP services). This includes:

- bases, plinths, frames and supporting structures for equipment
- ducts and trenches for distribution components, such as pipes or cables
- pre-formed holes in concrete and other masonry construction for distribution components, such as ducts, pipes or cables
- pre-formed holes in steelwork for distribution components, such as ducts, pipes or cables
- components that are cast into the concrete structure for fixings and supports
- fire cladding and packing to maintain fire integrity where distribution components penetrate fire compartments.

All significant builder's work requirements should be advised to the architect and structural/civil engineer at an early stage so that they can be incorporated within the architectural and structural designs as they are developed.

Figure 13.26 shows the builder's work that has been provided for a roof plant area to allow the installation of large air handling units. A series of supports have been provided for the

Figure 13.26 Builder's work bases for large air handling units

Source: Courtesy of Hoare Lea

equipment. Located around the edge of the area allocated for the equipment is the framework for the architectural louvred screens.

It should be understood that some of the same considerations for builder's work also apply to the external areas around buildings. The aspects of external integration that usually require the most attention are the routes for major services distribution, such as drainage, power distribution, district heating pipework (if relevant) and gas and water pipework. All ducts and other external services need to be integrated and coordinated with the other external elements, such as roads, paths, paving, hard surfaces (such as car parks), planting and other landscape features.

13.10 Summary

Space planning is an essential part of the building services design process to allocate dedicated spaces for active engineering systems components. When planning space allocations, a strategy should be established at the outset to cover aspects such as the design objectives, building form, suitability of locations, operation, maintenance, fire strategy and economics. Suitable locations should be identified that are appropriate for the types of equipment in plant rooms and for risers for vertical distribution and zones for horizontal distribution. Generic space criteria for equipment will determine the size of space required. The main considerations for space planning of plant rooms have been described in general form, together with the special considerations for different types of plant rooms. The planning and sizing of risers is a key aspect of the overall space planning process. Risers have generic requirements, together with special considerations that apply to mechanical and electrical risers. The requirements for planning horizontal distribution have been outlined in brief and form an important aspect of the planning of spaces and height allocations. The planning of distribution elements should address factors that influence energy efficiency by minimising distribution losses, namely reducing lengths and changes in direction. Where roof-mounted wind power or photovoltaics are proposed, their space planning should address optimisation of energy yield and health and safety aspects and can represent a challenge in providing a satisfactory integrated solution with other equipment. The need to design for adaptation to climate change should be addressed in the space planning strategy. The identification of builder's work for services is an important design activity.

References

ABB (2010) *Power Factor Correction and Harmonic Filtering*, presentation slideshow.

Andrews, J. and Jelley, N. (2017) *Energy Science: Principles, Technologies, and Impacts*, 3rd edition, Oxford: Oxford University Press.

ASHRAE (2016a) Air-to-air energy recovery equipment, (Ch. 26), in *Handbook:HVAC Systems and Equipment*, Atlanta, GA: American Society of Heating Refrigerating and Air-Conditioning Engineers.

ASHRAE (2016b) Panel heating and cooling, (Ch. 6), in *Handbook: HVAC Systems and Equipment*, SI edition, Atlanta, GA: American Society of Heating, Refrigerating and Air-Conditioning Engineers.

ASHRAE (2016c) Thermal storage, (Ch. 51), in *Handbook: HVAC Systems and Equipment*, Atlanta, GA: American Society of Heating Refrigerating and Air-Conditioning Engineers.

ASHRAE (2017a) Sorbents and desiccants, (Ch. 32.5), *Handbook: Fundamentals*, SI edition, Atlanta, GA: American Society of Heating, Refrigerating and Air-Conditioning Engineers.

ASHRAE (2017b) Non-residential cooling and heating load calculations, (Ch. 18), in *Handbook of Fundamentals*, Atlanta, GA: American Society of Heating Refrigerating and Air Conditioning Engineers.

ASHRAE (2019a) *62.1 Ventilation for Acceptable Indoor Air Quality*, Atlanta, GA: American Society of Heating Refrigerating and Air Conditioning Engineers.

ASHRAE (2019b) Evaporative air cooling, (Ch. 53), in *Handbook: HVAC Applications*, SI edition, Atlanta, GA: American Society of Heating, Refrigerating and Air-Conditioning Engineers.

Barley, S. (2010) Escape to the city, *New Scientist*, 6 November, pp. 32–5.

Barney, G. (2006) *Towards Low Carbon Lifts*, CIBSE National Conference, London: CIBSE.

Bateson, A. (2008) Job satisfaction, *Building Services Journal*, 30(1), January.

Bateson, A. (2009) *Introduction to Energy Strategy Reports*, CIBSE Energy Strategy Seminar, London: CIBSE.

BBP (Better Buildings Partnership) (undated) *Measuring and Reporting Metering*, online at www.better buildingspartnership.co.uk (accessed January 2020).

BCO (2014) *Desk Power Load Monitoring*, London: British Council for Offices, online at www.bco.org.uk/ Research/Publications/Desk-Power-Load-Monitoring.aspx

BCO (2019) *Specification for Offices*, London: British Council for Offices.

BEIS (2019a) *UK Energy in Brief 2019*, London: BEIS, online at www.gov.uk/government/statistics/ uk-energy-in-brief-2019

BEIS (2019b) *Department of Business, Energy and Industrial Strategy, Greenhouse Gas Reporting: Conversion Factors 2019*, online at www.gov.uk/government/publications/greenhouse-gas-reporting-conversion-factors-2019 (accessed 24 January 2020).

Bellas, I. and Tassou, S.A. (2005) Present and future applications of ice slurries, *International Journal of Refrigeration*, 28(1): 115–21.

Birkeland, J. (2002) *Design for Sustainability: A Sourcebook of Integrated Ecological Solutions*, London: Earthscan, pp. 14, 64.

Blackwell, H. (2010) Presentation slideshow, Hoare Lea.

Bleicher, D. (2019) *Green Building and Sustainability*, 2nd edition, TG15/2019, Bracknell: BSRIA.

Bordass, B. and Leaman, A. (2011) Test of time, *Building Services Journal*, March 2012.

Boyle, G. (Ed.) (2004) *Renewable Energy: Power for a Sustainable Future*, Oxford, UK: Oxford University Press, pp. 10–11, 86, 96, 245–86.

BRE (1994) Natural ventilation in non-domestic buildings, *BRE Digest 399*.

BREEAM (2019) *The Building Research Establishment (BRE), Environmental Assessment Method for New Office Building Designs*, online at www.breeam.com/ (accessed 24 January 2020).

BS 8233 (2014) *Guidance on Sound Insulation and Noise Reduction for Buildings*, London: British Standards Institution.

BS EN5925 (1991) *Code of Practice for Ventilation Principles and Designing for Natural Ventilation*, London: British Standards Institution.

BS EN13786 (1999) *Thermal Performance of Building Components: Dynamic Thermal Characteristics: Calculation Method*, London: British Standards Institution.

BS EN ISO 7730 (2005) *Ergonomics of the Thermal Environment*, London: British Standards Institution.

BSRIA (2010) *Soft Landings Framework*, Bracknell, UK: Building Services Research and Information Association.

BSRIA (2012) *Topic Guide TG3/2012: Embodied Carbon*, Bracknell, UK: Building Services Research and Information Association.

BSRIA (undated) *Circular Economy*, online at www.bsria.com/uk/information-training/information-centre/circular-economy/ (accessed December 2019).

Carbon Trust (2010) *Introducing Combined Heat and Power*, London: Carbon Trust, online at ww.carbontrust.com/media/19529/ctv044_introducing_combined_heat_and_power.pdf (accessed 21 January 2013).

Carbon Trust (2012) *Metering: Technology Overview CTV061*, London: Carbon Trust.

Carbon Trust (2018) *Carbon Footprinting Guide*, online at www.carbontrust.com/resources/guides/carbon-footprinting-and-reporting/

Carbon Trust (undated a) *How to Implement External Lighting*, Carbon Trust.

Carbon Trust (undated b) *How to Implement LED Lighting*, Carbon Trust.

Christopherson, R.W. (1997) *Geosystems: An Introduction to Physical Geography*, 3rd edition, Upper Saddle River, NJ: Prentice Hall, pp. 90–110, 306–12.

CIBSE (2000a) *AM13 Mixed Mode Ventilation*, London: CIBSE.

CIBSE (2000b) *TM25 Understanding Building Integrated Photovoltaics*, London: CIBSE.

CIBSE (2002) *Code for Lighting*, London: CIBSE, pp. 39–42, 70–90.

CIBSE (2004a) *Guide F: Energy Efficiency in Buildings*, London: CIBSE, pp. 3–7, 2.1–5, 3.1–2, 3.4–7, 4.1–7, 5.11–14, 6.1–6, 17.1–4, A2, A3.

CIBSE (2004b) *Guide K., Electricity in Buildings*, London: CIBSE, pp. 3.1–5, 8–20, Ch. 11.

CIBSE (2005a) *AM10:Natural Ventilation in Non-Domestic Buildings*, London: CIBSE.

CIBSE (2005b) *KS4 Understanding Controls*, London: CIBSE.

CIBSE (2006) *TM22 Energy Assessment and Reporting Methodology*, London: CIBSE.

CIBSE (2007) *Guide L: Sustainability*, London: CIBSE, pp. 2–8, 16–19, 30–3, 38.

CIBSE (2008a) *Guide M: Maintenance Engineering and Management*, London: CIBSE, pp. 2.1–2, 8.5, 11.5–6, Section 6.

CIBSE (2008b) *TM46 Energy Benchmarks*, London: CIBSE.

CIBSE (2009a) *Knowledge Services KS14: Energy Efficient Heating*, London: CIBSE.

CIBSE (2009b) *TM 39 Building Energy Metering*, London: CIBSE.

CIBSE (2009c) *Guide H: Building Control Systems*, London: CIBSE.

CIBSE (2015) *Guide A: Environmental Design*, London: CIBSE.

CIBSE (2016) *CIBSE Guide B2, Ventilation and Ductwork*, London: The Chartered Institution of Building Services Engineers, pp. 2–69.

CIBSE (2018) *DE9: Application of Soft Landings and Government Soft Landings in Building Services Engineering*, London: CIBSE.

CLG (2010) *Approved Document F: Ventilation (2010 Edition)*, online at www.gov.uk/government/publications/ventilation-approved-document-f (accessed 27 October 2019).

Coley, D. (2008) *Energy and Climate Change: Creating a Sustainable Future*, Chichester, UK: John Wiley & Sons, pp. 31, 36, 83, 69, 77–113, 118–25, 439, 456–8.

CONTAM (2019) *Multizone Air flow and Contaminant Transport Analysis Software*, online at www.nist.gov/el/energy-and-environment-division-73200/nist-multizone-modeling/software-tools/contam (accessed 15 November 2019).

Cook, M. and Short, A. (2005) Natural ventilation and low energy cooling of large, nondomestic buildings, *International Journal of Ventilation*, 3(4), March.

Court, D. (2011) Presentation slideshow, Hoare Lea.

DECC (2012) *Energy Consumption in the UK, Service Sector Data Tables*, updated, online at www.decc.gov.uk/en/content/CMS/statistics/publications/ecuk.aspx (accessed 31 July 2012).

Department of Health (DOH) (2007) *Electrical Services Supply and Distribution: Part A: Design, HTM 06-01*, London: TSO.

Dunn, G.and Knight, I. (2005) Small power equipment loads in UK office environments, *Energy and Buildings*, 37(1):87–91.

EN 16798-1 (2019) Energy performance of buildings: Ventilation for buildings part 1: Indoor environmental input parameters for design and assessment of energy performance of buildings addressing indoor air quality, thermal environment, lighting and acoustics: Module M1-6, BSI.

EN 16798-3 Part 3 (2017) Energy performance of buildings: Ventilation for buildings part 3: For non-residential buildings: Part 3: For non-residential buildings: Performance requirements for ventilation and room-conditioning systems (Modules M5-1, M5-4).

EN 16798-3 Part 7 (2017) Energy performance of buildings: Ventilation for buildings part 7: Calculation methods for the determination of air flow rates in buildings including infiltration (Modules M5-5).

European Commission. *Climate-Friendly Alternatives to HFCs*, online at https://ec.europa.eu/clima/policies/f-gas/alternatives_en (accessed 25 January 2020).

Foley, J. (2010) Boundaries for a healthy planet, *Scientific American*, 302(4): 38–41.

Foster, R.E. (1996) Evaporative air conditioning fundamentals: Environmental and economic benefits worldwide, Proceedings of the International Institute of Refrigeration Conference, IIR Commissions, B1, B2, E1, Aarhus, Denmark, pp. 1.6–1.10.

GLA (Greater London Authority) (2016) *The London Plan, Policy 5.2 Minimising Carbon Dioxide Emissions*, online at www.london.gov.uk/what-we-do/planning/london-plan/current-london-plan/

Goudie, A. (2000) *The Human Impact on the Natural Environment*, 5th edition, Oxford: Blackwell Publishers, pp. 326–37, 348–55.

Grosso, M. and Raimondo, L. (2008) Horizontal air-to-earth heat exchangers in northern Italy: Testing, design and monitoring, *International Journal of Ventilation*, 7(1): 1–10.

Hammond, G. and Jones, C. (2011) *Embodied Carbon: The Inventory of Carbon and Energy (ICE)*, edited by Lowrie, F. and Tse, P., University of Bath with BSRIA, ICAT. Bracknell: BSRIA.

Hamworthy Heating, online at www.hamworthy-heating.com/Knowledge/Articles/ErP-compliant-boilers (accessed 17 November 2019).

Hawkins, G. (2011) *BSRIA BG9 Rules of Thumb: Guidelines for Building Services*, 5th edition, London: BSRIA.

Heiselberg, P. (Ed.) (2002) Principles of hybrid ventilation, *IEA Energy Conservation in Buildings and Community Systems Programme*, Annex 35, Aalborg University, Denmark.

HMG (2013) *Non Domestic Building Services Compliance Guide*, online at www.gov.uk/government/publications/conservation-of-fuel-and-power-approved-document-l (accessed 24 June 2019).

HMG (2016) *Building Regulations Approved Document Part L2A*, Conservation of Fuel and Power in New Buildings Other Than Dwellings (2013 edition with 2016 Amendments), online at www.gov.uk/government/publications/

Hollmuller, P. (2011) Air/ground heat exchangers for heating and cooling: Dimensioning guidelines, AIVC Conference, Brussels, 10–12 October.

Hopfe, C.J. and McLeod, R.S. (Eds.) (2015) *The Passivhaus Designer's Manual: A Technical Guide to Low and Zero Energy Buildings*, Abingdon: Routledge.

Horsley, K. (2010) Guidance document, Hoare Lea.

HSE (2015a) *Managing Health and Safety in Construction: Construction (Design and Management) Regulations 2015, Guidance on Regulations*, London: Health and Safety Executive.

HSE (2015b) *Designers: Roles and Responsibilities*, Health and Safety Executive, online at www.hse.gov.uk/construction/cdm/2015/designers.htm

HSE (2015c) *Principal Designers: Roles and Responsibilities*, Health and Safety Executive, online at www.hse.gov.uk/construction/cdm/2015/designers.htm

Hunn, B.D. and Peterson, J.L. (1991) Cost-effectiveness of indirect evaporative cooling for commercial buildings, *ASHRAE Transactions: Research*:434–47.

HVCA (1988) *DW/142 Specification of Sheet Metal Ductwork*, Heating and Ventilating Contractors Association, London, on-line at www.scribd.com/doc/190658737/DW-142-Specification-for-Sheet-Metal-Ductwork-Addendum-A (accessed 16 May 2020).

IEC (2014) *60003430-1 Rotating Electrical Machines Part 30-1: Efficiency Classes of Line Operated AC Motors (IE Code)*, International Electrotechnical Commission, online at www.iec.ch/search/?q=60034

IET (2007) Solar power factfile, *IET*, online at www.theiet.org/factfiles/energy/solar-power-page.cfm (accessed 23 January 2013).

IET (2008) Combined heat and power (CHP) factfile, *IET*, online at www.theiet.org/factfiles/energy/combined-heat-page.cfm (accessed 21 January 2013).

IET (2009) Government policy submission S832: Smart metering for electricity and gas, *IET*, August, online at www.theiet.org/publicaffairs/submissions/s832.cfm

IET (2011) *Smart Grids: The Wider Picture*, Stevenage: The Institution of Engineering and Technology, online at www.theiet.org/factfiles

IET (2013) *Energy Principles*, Stevenage: The Institution of Engineering and Technology, online at www.theiet.org/factfiles

IET (2015) *Code of Practice for Grid Connected Solar Photovoltaic Systems*, Stevenage: The Institution of Engineering and Technology.

IET (2016) *Guide to Metering Systems*, London: The Institution of Engineering and Technology.

IET (2018) *BS 7671: IET Requirements for Electrical Installations, IET Wiring Regulations*, 18th edition, Stevenage, UK: The Institution of Engineering and Technology.

IET (undated) *Engineering Priorities for Delivering Net-Zero*, online at theiet.org/impact-society (accessed December 2019).

IIR (International Institute of Refrigeration) (2005) *Handbook on Ice Slurries*, edited by Kauffeld, M., Kawaji, M. and Egolf, P.W., Paris: IIR.

IPCC (Intergovernmental Panel on Climate Change) (2007) *4th Assessment Report*, online at www.ipcc.ch/publications_and_data/ar4/syr/en/contents.html (accessed 17 December 2012).

IPCC (2018) *IPCC Special Report Global Warming of 1.5 Degrees C*, online at www.ipcc.ch/sr15/

Iten, M., Liu, S. and Shukla, A. (2016) A review on the air-PCM-TES application for free cooling and heating in the buildings, *Renewable and Sustainable Energy Reviews*, 61: 175–86.

IWBI (2016) *The Well Building Standard v1. International Well Building Institute*, New York, online at www.wellcertified.com (accessed 27 October 2019).

King, D. (2010a) *Engineering a Low Carbon Built Environment*, London: Royal Academy of Engineering.

King, D. (2010b) Urban fallacy, *CIBSE Journal*, June, London: CIBSE.

Kolokotroni, M. (1998) *Night Ventilation for Cooling Office Buildings*, IP04/98, BRE.

Kolokotroni, M. (2001) Night ventilation for cooling office buildings: Parametric analyses of conceptual energy impacts, *ASHRAE Trans*, 107(1): 479–90.

Konstantelos, I., Sun, M. and Strbac, G. (2014) *Quantifying Demand Diversity of Households, Report for the 'Low Carbon London' LCNF Project*, Imperial College London, online at www.researchgate.net

Kukadia, V. and Upton, S. (2019) *Ensuring Good Indoor Air Quality in Buildings*, Watford: BRE.

Levermore, G.J. (2002) The exponential limit to the cooling of buildings by natural ventilation, *Building Services Engineering Research and Technology*, 23(2): 99–125.

Liddament, M.W. (1993) *A Review of Ventilation Efficiency*, Brussels: AIVC TN 39.

Liddament, M.W. (1996) *A Guide to Energy Efficient Ventilation*, Brussels: AIVC.

Liddament, M.W. (2009) Editorial: Summer comfort and cooling: An INIVE workshop, *International Journal of Ventilation*, 8(1), June.

Makstutis, G. (2010) *Architecture*, London: Lawrence King, pp. 88–92, 104–60, 203–6.

Makstutis, G. (2018) *Design Process in Architecture*, London: Laurence King Publishing Ltd.

McCrae, A. (2013) *Renewable Energy*, Marlborough: The Crowood Press Ltd.

Met Office (2020) online at www.metoffice.gov.uk/about-us/press-office/news/weather-and-climate/2020/confirmation-that-2019-concludes-warmest-decade

Met Office (undated) online at www.metoffice.gov.uk/weather/climate-change/effects-of-climate-change

Moe, K. (2008) *Integrated Design in Contemporary Architecture*, New York: Princeton University Press.

Mundt, E. (Ed.) (2004) Ventilation effectiveness, in *REHVA Guidebook No. 2*, Brussels: REHVA.

NASA (2019) online at www. Jpl.nasa.gov/news/news.php?feature=7556 (accessed 11 December 2019).

National Statistics (2019) *National Energy Efficiency Data-Framework (NEED)*, London: Department for Business, Energy and Industrial Strategy, online at www.data.gov.uk

Passivhaus Trust (2019) *What Is Passivhaus?*, online at www.passivhaustrust.org.uk/what-is-passivhaus-php

Pearce, F. (2010) Earth's nine lives, *New Scientist*, 27 February, pp. 31–5.

Pearson, A. (1997) Liquid desiccant ventilators, *Building Services Journal*, May: 47–8.

Pearson, D. (2005) *In Search of Natural Architecture*, London: Gaia Books, p. 72.

Pegg, I. (2007) *Assessing the Role of Post-Occupancy Evaluation in the Design Environment: A Case Study Approach*, PhD thesis, Brunel University.

Perino, M. (Ed.) (2008) Annex 44 integrating environmentally responsive elements in buildings: Responsive building elements, *IEA Energy in Buildings & Communities, State-of-the-Art Review*, Vol. 2A, online at www.iea-ebc.org/projects/project?AnnexID=44 (accessed 13 November 2019).

Public Health England (2018) *Health Matters: Air Pollution*, online at www.gov.uk/government/publications/health-matters-air-pollution/health-matters-air-pollution

Quaschning, V. (2005) *Understanding Renewable Energy Systems*, London: Earthscan, pp. 9–10, 44, 141–201, 232.

RIBA (2013) *RIBA Plan of Work 2013 Overview*, London: RIBA, online at www.ribaplanofwork.com (accessed August 2019).

Roe, M. (2011) Presentation slideshow, Hoare Lea.

Romm, J. (2018) *Climate Change: What Everyone Needs to Know*, 2nd edition, New York: Oxford University Press.

Royal Society (2017) *Decarbonising UK Energy: Effective Technology and Policy Options for Achieving a Zero-Carbon Future*, London: The Royal Society (with the British Academy and the Royal Academy of Engineering).

Russett, S. (2010) Energy performance of vertical transportation systems, *Elevation Magazine*, Spring.

Santamouris, M. (2004) *Night Ventilation Strategies*, Brussels: AIVC VIP 4.

Santamouris, M. (2006) *Use of Earth to Air Heat Exchangers for Cooling*, Brussels: AIVC VIP 11.

Santamouris, M. and Asimakopoulos, D.N. (1996) *Passive Cooling of Buildings*, London: James & James.

Schild, P. (2004) *Displacement Ventilation*, Brussels: AIVC IP5.

Skistad, H. (Ed.) (2002) *Displacement Ventilation, Guidebook No. 1*, REHVA Guidebooks, Federation of European Heating and Air-Conditioning Associations, online at www.rehva.eu (accessed 13 November 2019).

SLL (2016) *Lighting Guide LG14: Control of Electric Lighting*, London: CIBSE.

Smith, M., Whitelegg, J. and Williams, N. (1998) *Greening the Built Environment*, London: Earthscan, pp. 2–19, 70.

Society of Light and Lighting (2004) *Code for Lighting*, London: Society of Light and Lighting.

Spon's Mechanical and Electrical Services Price Book (2018) edited by AECOM, 49th edition, London: Taylor and Francis, 837 p (accessed 24 January 2020).

SSL (2012) *Code for Lighting*, London: CIBSE.

Stasinopoulos, P., Smith, M.H., Hargroves, K. and Desha, C. (2009) *Whole Systems Design: An Integrated Approach to Sustainable Engineering*, London: Earthscan, pp. 3–8, 23–30, 75–6.

Sutcliffe, H. (1990) *A Guide to Air Change Efficiency*, Brussels: AIVC TN 28.

Tassou, S.A., Lewis, J.S., Ge, Y.T., Hadawey, A. and Chaer, I. (2010) A review of emerging technologies for food refrigeration applications, *Applied Thermal Engineering*, 30(4), March: 263–76.

Ticleanu, C. and Littlefair, P. (2019) *Quality Indoor Lighting for Comfort, Health, Well-Being and Productivity*, Watford: BRE.

UKGBC (2019) *Net Zero Carbon Buildings: A Framework Definition*, London: UK Green Building Council, online at www.ukgbc.org/wp-content/uploads/2019/04/Net-Zero-Carbon-Buildings-A-framework-definition.pdf

UN (1987) *Report of the World Commission on Environment and Development: Our Common Future*, p. 41, online at www.un-documents.net/our-common-future. pdf (accessed 4 January 2013).

UN (2015) *Transforming Our World: The 2030 Agenda for Sustainable Development*, A/RES/70/1, United Nations, online at https://sustainabledevelopment.un.org

UN (2019a) *Global Sustainable Development Report*, United Nations, online at https://www.un.org/publications

UN (2019b) *UN Environment Programme Emissions Gap Report 2019*, online at www.unep-wcmc.org/news/2019-emissions-gap-report

UN (2019c) *The Future Is Now: Science for Achieving Sustainable Development*, online at https://sustainable development.un.org/gsdr2019

Von Weizsäcker, E., Lovins, A.B. and Lovins, L.H. (1998) *Factor Four: Doubling Wealth, Halving Resource Use*, London: Earthscan, pp. 55–7.

Warwick, D.J., Cripps, A.J. and Kolokotroni, M. (2009) Integrating active thermal mass strategies with HVAC systems: Dynamic thermal modelling, *International Journal of Ventilation*, 7(4): 345–68.

Waskett, R. (2019) *The Biggest Daylight Development in a Decade*, Hoare Lea, online at https://hoarelea.com/2019/01/22/the-biggest-daylight-development-in-a-decade/

WHO (2010) Guidelines for indoor air quality: Selected pollutants (2010), The WHO European Centre for Environment and Health, Bonn Office, WHO Regional Office for Europe Coordinated the Development of These WHO Guidelines. ISBN 9789289002134.

Wilkins, C.K. and Hosni, M.H. (2000) Heat gain from office equipment, *ASHRAE Journal*, 42(6): 33.

Wines, J. (2000) *Green Architecture*, Los Angeles, CA: Taschen.

WMO (2019) online at https://eandt.org/content/articles/2019/11/greenhouse-gases-rising-faster-despite (accessed 29 November 2019).

WWEA (2019) *Wind Power Capacity Worldwide Reaches 597GW, 50.1GW Added in 2018*, World Wind Energy Association, online at https://wwindea.org/blog/2019/02/25

Zimmermann, M. and Anderson, J. (Eds.) (1998) Case studies of low energy cooling technologies, *IEA Energy in Buildings and Communities, Annex 28 Low Energy Cooling*, online at www.iea-ebc.org/projects/project?AnnexID=28 (accessed 13 November 2019).

Zimmermann, M. and Remund, S. (2001) IEA annex 28, low energy cooling systems: Ground coupled air systems, in Barnard, N. and Jaunzens, D. (Eds.), D*esign Tools for Low Energy Cooling: Technology Selection and Early Design Guidance*, UK: BRE, online at www.iea-ebc.org/projects/project?AnnexID=28 (accessed 13 November 2019).

Index